Fate and Effects of Uranium in Abandoned Uranium Mines

Martin Mkandawire

Fate and Effects of Uranium in Abandoned Uranium Mines

Springer

Martin Mkandawire
School of Science and Technology
Cape Breton University
Sydney, NS, Canada

ISBN 978-3-032-10446-5 ISBN 978-3-032-10447-2 (eBook)
https://doi.org/10.1007/978-3-032-10447-2

This Springer imprint is published by the registered company Springer Nature Switzerland AG
The registered company address is: Gewerbestrasse 11, 6330 Cham, Switzerland

In loving memory of my father, Nedson Mleza Chizotera Mkandawire, brother Gift Chizotera Mkandawire, and aunt Kellyness Masozi Chizotera Mkandawire

To my family and for my children, Chitsanzo and Chimango

Foreword

Uranium is one of those elements that never comes alone. It comes with a history of fascination, fear, politics, highly interesting physicochemical characteristics, and even art. If you have ever stood at a discharge point below a waste rock pile or a tailings pond, you will be familiar with its sometimes ordinary, sometimes unpredictable water chemistry. Abandoned uranium mining sites represent all these characteristics, and they usually do so for a long time. In this book, *Fate and Effects of Uranium in Abandoned Uranium Mines*, Martin Mkandawire addresses the persistence of uranium after mining stops: where it goes, what controls its mobility, and the consequences for water, sediments, and living systems.

Martin and I first met 24 years ago during the IMWA 2002 conference at Bergakademie Freiberg, Germany. He gave a presentation entitled "Uranium Attenuation from Tailing Waters by Floating Macrophyte *Lemna gibba* L.". At that time, we had no idea that our paths would cross so often in the future, or that two decades later he would write a book about uranium at abandoned mine sites. When I left my position as Industrial Research Chair at Cape Breton University in Nova Scotia, Canada, Martin succeeded me. Therefore, I am honoured to have been invited to write this foreword for him.

I have been studying uranium in mining-influenced water (MIW) for years, mostly in spaces where speciation, transport, and site hydraulics determine whether uranium remains in place or appears where it should not. The recurring lesson from those papers and book chapters was simple: "Uranium" is rarely the problem, but it is shorthand for something else. The real question lies in the system: redox boundaries that move, carbonates that buffer, colloids that carry, density stratification that prevents hydraulic transport, and the slow rearrangement of solids that appears static until it does not. This is why a book like this is as important to practitioners as it is to researchers. It treats uranium in MIW as a hydrogeochemical and ecological issue rather than a slogan.

In the preface, Mkandawire plainly lays out his motivation: older summaries have become outdated, methods have improved, and an updated synthesis is long overdue. He explicitly frames this volume as a revision and expansion of his 2005 publication. He also situates the topic firmly in the real world by explaining why society keeps returning to questions about uranium: new projects, often in struggling economies and legacy sites created before modern closure standards existed. If you work with mine water, you will already be aware that the "post-mining" phase is not a neat chapter break. In many places, it is the longest operational phase of the site, just without a budget that reflects its duration.

This book is broad in scope, but it is clear about its aims. It states from the outset that it is about the fate and effects of uranium in abandoned mines and that these outcomes are determined by the geological background, climatic changes, and interactions with biotic and abiotic components. It also acknowledges the practical truth that the dominant risks are often routed through water. While uranium may be misunderstood by the public, at the site scale it is often well understood by water via complexation, redox transformations, sorption, and transport. This helps to keep the discussion anchored. This book does not claim that one mechanism applies to all sites and repeatedly reminds the reader that site-specific investigations are indispensable.

A useful editorial decision, from my point of view, is the deliberate structure: each chapter is designed to be readable in isolation, even if this results in some repetition. In practice, technical books are rarely read from cover to cover but are more often used as a working reference during an assessment, literature review, remediation design, or supervision meeting with a student who needs a reliable starting point. You can open a chapter on topics such as underground mine flooding, pit lakes, tailings, in situ leaching, toxicology, radioecology, or natural attenuation and find enough context to proceed without having to refer to earlier sections for definitions.

The content focus is also sharpened by what is intentionally left out. Mkandawire notes that arsenic is not addressed in his book because the available literature has become too extensive. The consequence of this exclusion is that more space is devoted to uranium itself, including groundwater transport dynamics, nanoscience aspects, and biogeochemical behaviour in sediments. This matters. Too many site discussions devolve into a generic "multi-contaminant" fog where nothing is resolved. Focusing does not deny complexity—it makes it workable.

The author's background aligns well with the topic, covering solid-state chemistry, mine water management, remediation, and nanotechnology, as well as having a publication record spanning technical volumes and writing about writing. This diversity is evident in the manuscript. The author switches comfortably from fundamentals to mechanism and from mechanism to practical applications at a site. There is also a human element—his dedication to his family and the context of why the project had to be completed. While you don't need this information to calculate a distribution coefficient, it is essential for understanding why certain scientific problems remain unresolved for decades.

Finally, a personal note that remains professional: Mkandawire acknowledges his colleagues and explicitly mentions my role in encouraging this work and providing a foreword. I am glad he followed through. The uranium story in abandoned mining sites is not getting any simpler. Despite measuring more, resolving smaller scales, and modelling faster, we still encounter the same old field reality: water finds pathways, chemistry defines form, and time takes its course. This book provides readers with a solid, practical framework for considering uranium in MIW without assuming that the sites will cooperate.

Christian Wolkersdorfer
Department of Environmental, Water, and Earth Sciences,
Tshwane University of Technology
President and member of the Executive Council of the
International Mine Water Association, Member of the
Academy of Science of South Africa (ASSAf)

Preface

As research continues to uncover more exciting information about uranium behaviour and its interactions with biotic and abiotic systems in abandoned uranium mines, some of the data published a decade ago is outdated, while other findings are now obsolete. Technology has advanced significantly over the past few decades, and analytical procedures used to study uranium dynamics and biogeochemical behaviour in the waters of abandoned mines have also improved considerably. As a result, updating knowledge on uranium dynamics, biogeochemistry, and toxicity in abandoned uranium mines is now necessary. Additionally, many sectors of society remain very interested in understanding the potential dangers posed by uranium in abandoned mines, especially as some Western uranium mining companies have established operations in developing countries and have taken advantage of their need for foreign investment to stimulate their economies. Therefore, I have decided to update my previous compilation of research work, which I published as a book in 2005.

However, I was not inspired to begin this recompilation until my father asked me from his bed to do so, as one of his last wishes for me to fulfil. Unfortunately, he passed away shortly after I started organizing the material in 2014. His passing made it difficult for me to continue a project he wanted to see completed before his death because I lacked a clear purpose; he was a purpose in everything I did. However, as I healed and received several inquiries seeking information and advice on different aspects of uranium, I realized the importance of continuing to compile research findings and new, revealing information about uranium into an updated book. So, I am doing this now, partly out of respect and as a tribute to a wonderful man who was my hero and mentor, and who dedicated his whole life to shaping who I am, even if I have sometimes fallen short. I also recognize that there is limited accessible information about uranium, especially its environmental dangers, which can sometimes lead to justified but exaggerated fears or improper handling of uranium-related issues. Therefore, I hope to contribute to a better understanding of uranium in the environment, particularly concerning the massive, abandoned uranium mines, mostly a legacy of the Cold War arms race.

I have studied uranium, radionuclides, and metalloids in abandoned uranium mines for over 30 years. Throughout my research journey with uranium, I have met several esteemed colleagues over the past three decades. I have collaborated with some of them on various projects. I deeply respect their work and contributions to uranium research. Considering my father's final wishes, I realized that the expertise of some of these distinguished colleagues would significantly enrich and elevate the quality of this book. Therefore, I consulted a few colleagues about what they would like to see as a focus in this compilation. A consensus was to address issues related to: uranium mining and processing, especially the chemical dynamics and interactions with biological systems; how to manage risks and hazards arising from the chemical and decay behaviour of uranium and its daughter elements; and why there is an emphasis on depleted uranium and nuclear waste disposal, while significant dangers and risks at mining sites are often overlooked.

Most of the colleagues I consulted believe that uranium mining will increase again in the coming decades, despite recent disasters, such as Fukushima. While fossil fuels are on the verge of depletion, there is still no realistic, dependable renewable energy solution to feed a hungry world. There will be more abandoned mines because the new uranium mining boom is likely to occur in countries with struggling economies, which will find it challenging to address the associated problems. Therefore, lessons can be learned from the current global legacy of uranium mining, as shown by many abandoned mines, primarily the result of unregulated mining that thrived after World War II. After World War II, uranium mining production mainly served the nuclear arms race during the Cold War.

Known health risks associated with uranium mining were often ignored, even in developed countries, with governments failing to inform workers and nearby communities of potential dangers or to regulate mining to reduce contamination. Consequently, high incidences of specific ailments began to appear in surrounding communities, and workers often failed in court cases seeking compensation (e.g., radon-related illnesses were not recognized at the time). Despite efforts to clean up uranium sites worldwide, especially in developed nations, significant problems stemming from the legacy of uranium development remain in former mining and processing regions. Thousands of abandoned mines remain unremediated; we must ask ourselves: how can the new generation do it better?

As a result, this book has omitted all discussions of arsenic, which the previous edition included alongside uranium, because research data on both elements in abandoned uranium mines has become overwhelming since the last compilation. This has allowed me to focus more on uranium itself, including groundwater transport dynamics, the nanoscience of natural uranium, and the biogeochemical behaviour of uranium in tailings pond sediments.

As I mentioned, mentors and colleagues have inspired me to undertake this work. I was first introduced to research in abandoned uranium mines by my long-time mentor, a good friend and collaborator, Ernst Gert Dudel. Gert has been an inspiration, a guide, and an encyclopaedia on uranium interactions in aquatic ecosystems. Through him, I expanded my knowledge from my quantum chemistry training to learn principles of limnology, toxicology, and ecology related to abandoned uranium mines. Similarly, I cannot imagine how my career would have evolved without Wolfgang Pompe. Wolfgang has been more than just a mentor to me; he lifted me when I was down and lost hope, guided me when I was confused. He never stopped believing in me, and in him, I found a shield that protected my family and me and provided spiritual guidance. As a solid-state physicist, Wolfgang showed me that a career built with passion is possible. He inspired me to pursue a unique career combining solid-state chemistry, mine water remediation, and nanotechnology. Thank you sincerely for the support and guidance you have given me, shaping my career alongside Gert and Wolfgang.

In writing this book, I found the comfort I very much needed, as I diverted my thoughts from the long sickness and loss of my hero, Nedson Mleza, my father, who passed away, followed 5 months later by the death of my dear and loving aunt, Kellyness Masozi, and 8 months later by my beloved younger brother, Gift. It felt as if the world had closed in on me, but my Lord kept his hands on me, and I found solace in continuing my work.

Sydney, NS, Canada — Martin Mkandawire

Acknowledgement

I am deeply indebted to the many individuals whose support, insight, and encouragement made this book, *Fate and Effects of Uranium in Abandoned Uranium Mines*, possible. I extend my sincere gratitude to Dr. Mumuni Amadu for his careful review of the manuscript and for offering incisive and constructive criticism that strengthened the work in countless ways. My heartfelt thanks also go to my brother, Dr. Joseph Mkandawire (aka Dr. JCN), whose meticulous reading, language and grammar corrections, and thoughtful feedback significantly improved the clarity and coherence of the text.

I am grateful to my son, Chitsanzo Mkandawire, for reviewing several chapters and providing constructive feedback and criticism that sharpened the arguments and enriched the narrative. I also wish to acknowledge Dr. Murugan Ramasamy for his valuable review of selected chapters and for his constructive criticism, which helped refine key sections of the book. My appreciation extends to Dr. Marzieh Baneshi, whose skill and creativity in producing, drawing, and adapting many of the illustrations and graphics brought essential visual clarity to complex concepts. I further extend my thanks to my friend Mr. Jamie Tunnicliff for reviewing some chapters and offering thoughtful comments that strengthened the manuscript.

My students, far too many to name individually, have each contributed in meaningful ways, through research assistance, challenging questions, and fresh ideas that continually pushed me to think more deeply and broadly. I am profoundly thankful to my wife, Dr. Msaukiranji Mary Mkandawire, whose teaching and guidance in molecular biology have often made me appear more knowledgeable than I truly am. My daughter, Chimango Mkandawire, deserves special recognition for keeping me grounded and steady during some of the most challenging periods of this journey, as does my son Chitsanzo, whose support has been equally unwavering.

I am also grateful to my colleagues, too numerous to list, who have contributed, directly and indirectly, to shaping this work in ways both visible and subtle. I owe particular thanks to Prof. Dr. habil. Christian Wolkersdorfer for his encouragement, wise counsel, and for graciously writing the foreword to this book.

Above all, I acknowledge with humility that without my trust in the Lord, none of this would have been possible.

Competing Interests The author has no competing interests to declare that are relevant to the content of this manuscript.

About This Book

This book has been deliberately structured so that each chapter can function as a self-contained unit, particularly when accessed in electronic format. Chapters are therefore designed to be read independently without requiring prior familiarity with other parts of the volume. As a consequence of this stand-alone design, readers may encounter some repetition of key background concepts, definitions, or methodological explanations across chapters. This approach is intended to enhance clarity, accessibility, and ease of use for readers consulting individual chapters for specific topics.

Contents

About the Author

Martin Mkandawire is a Professor of solid-state chemistry and a former Industrial Research Chair in Mine Water Management and Remediation at Cape Breton University, Nova Scotia. He holds a *Doctor Habilitus* (with a *venia legendi*) and a *Doctor rerum naturalium* (magna cum laude) from Technische Universität Dresden. His research spans environmental remediation, mine water management, and nanotechnology. He has authored and edited the following books: *Application of Nanotechnology in Mining Processes*: *Beneficiation and Sustainability*; *Technologies for Sustainability in Energy and Environment*; *Ecowriting*: *Advice to ESL on Effective Scientific Writing in Environmental Science and Engineering*; and the poetry anthologies, *Moulting Serpents* and *Moulted Serpents*.

Chapter 1
Introduction

Abstract This chapter provides a comprehensive overview of uranium, tracing its discovery as a chemical element and examining its natural occurrence within diverse geological environments. It explains the fundamental principles of uranium radioactivity and outlines the major types and classifications that define its industrial relevance. This chapter also explores the wide-ranging uses of uranium, from energy production to specialized technological applications. Building on this scientific foundation, it analyses the market dynamics shaping global uranium economics, including key drivers, emerging trends, and the structure of supply chains and marketplaces. Finally, this chapter surveys the world's major uranium deposits, highlighting their geological characteristics and strategic importance. Together, these sections offer an integrated understanding of uranium as both a natural resource and a critical commodity in the global energy landscape.

Keywords Radioactivity · Radionuclides · Nuclear fuel · Uranium market · Mineral deposits · Energy resources · Uranium geology · Supply chain · Market trends

1.1 Scope of this Book

The scope of this book is the fate and effect of uranium in abandoned mines. The fate of a chemical encompasses its life cycle from production to the end. In nature, fate refers to the destiny of a pollutant after it is released into the environment. The fate of uranium in abandoned mines is determined by factors ranging from geological background to climatic conditions, as well as its interactions with biotic and abiotic systems in the site's media.

Uranium is a metallic element often grossly misunderstood by the general public. Mentioning uranium provokes fears of a highly radioactive and hazardous element, based on stories about the devastation caused by nuclear bombs. However, uranium is a naturally occurring radioactive element. It is the most massive element in the periodic table that exists naturally, after which all elements are artificial. It is

M. Mkandawire, *Fate and Effects of Uranium in Abandoned Uranium Mines*,
https://doi.org/10.1007/978-3-032-10447-2_1

found in tiny amounts in nature as minerals. However, some known geological formations contain uranium at high concentrations, namely the anomaly zones. In its natural state, in undisturbed environments, uranium ore does not affect the environment or surrounding people because rock and soil block its harmful effects. Thus, human activities, especially mining and processing, are responsible for most uranium migration around natural environments. However, natural processes also contribute; for instance, when uranium-bearing rocks weather, uranium becomes part of the soil and can be easily washed into rivers and lakes.

The legacy of uranium mining that fuelled the Cold War arms race between the West and the East's nuclear energy programs is thousands of abandoned uranium mines worldwide. Abandoned mines are sites where mining activities occurred but where proper mine closure and reclamation were not completed. Most uranium mining activities occurred before mine closure regulations were developed. These abandoned uranium mines are currently causing concern as sources of environmental pollution. Generally, uranium is found as a readily soluble dioxide cation in most environments. Environmental exposure can occur in air, water, and soil, but the most significant concerns arise from contamination of the water system. Uranium is chemotoxic and radiotoxic; however, these dangers are usually negligible compared to the toxicity of uranium decay products. Uranium ore concentration and associated radioactivity vary widely in mining sites and natural formations. During decay processes, the parent uranium and its decay products, and their subsequent decay products, release a series of new elements and radiation, including radium and radon, alpha and beta particles, and gamma radiation. Alpha particles cannot pass through human skin, while gamma radiation passes more efficiently. Thus, abandoned uranium mines are complex systems with varying pollution sources, toxicity issues, and geochemical interactions.

Therefore, this book endeavours to explain the likelihood of uranium migrating and posing a danger. This will be accomplished by understanding how uranium's complex geochemical interactions with its biotic and abiotic milieu, under specific physicochemical conditions, determine the extent of risk posed by abandoned uranium mines. This book also provides a general overview of uranium behaviour in abandoned mines while keeping in mind that the factors influencing the fates and effects of uranium are very site-specific as well as dependent on the hydrogeological background, the type of mining procedure (e.g., underground, open-cast, or in situ leaching methods), and the ore grade and type.

1.2 Uranium: History and Uses

1.2.1 The Element and Its Discovery

Uranium is the most massive element found in nature. It was discovered in the mineral pitchblende by the German chemist Martin Heinrich Klaproth in 1789, while analysing mineral samples from the Joachimsthal silver mines in the present-day Czech Republic. Klaproth named the discovery after the newly discovered planet Uranus (discovered in 1781). However, it was not until 1841 that Eugène-Melchoir Péligot, a French chemist, isolated pure uranium by heating uranium dioxide with potassium in a platinum crucible. Apart from its value to chemists, the only significant use for uranium throughout the 1800s was to colour glass and ceramics—uranium compounds were used to give vases and decorative glassware a yellow-green colour.

In the periodic table, uranium is represented by the symbol U and belongs to the actinide group (i.e., actinide has no group number). It has an atomic number of 92, meaning 92 protons and 92 electrons in its atomic structure, but its neutron number can range from 141 to 146. It is a soft, metallic silver-white metal in its pure form, but in its natural, unrefined state, uranium is chemically bound to oxygen as triuranium octoxide (U_3O_8), the most stable and least radioactive form of uranium oxide. Uranium is present in more than 160 known oxygen-containing minerals. It turns yellow in the air, and with prolonged exposure, it can become brown-black and develop an oxide layer. It has a high density of 19.16 g cm^{-3}, a low melting point of 1133 °C, and a high boiling point of 3930 °C. Uranium can have three temperature-dependent allotropic transformations, namely α-uranium at temperatures below 667 °*C*, β-uranium at temperatures between 667 °*C* and 775 °*C*, and γ-uranium at temperatures between 775 °*C* and 1133 °*C*. Allotropism is the ability of a chemical element to exist in different forms in the same physical state. In other terms, allotropes are different structural modifications in the solid state of an element. The molar volume of uranium is 12.49×10^{-6} m^3 mol^{-1}.

Generally, uranium is air-resistant at ordinary temperatures, but an oxide layer mostly coats it in its metallic form. As a result, it dissolves in most acids and slowly dissolves in hot water, generating hydrogen. Nevertheless, uranium is chemically very stable. It reacts with halides to form uranium halide complexes. Consequently, uranium can form compounds in oxidation states from 3+ to 6+. The reduction potentials are given in Fig. 1.1, while details of the chemical and physical properties of uranium are summarized in Table 1.1.

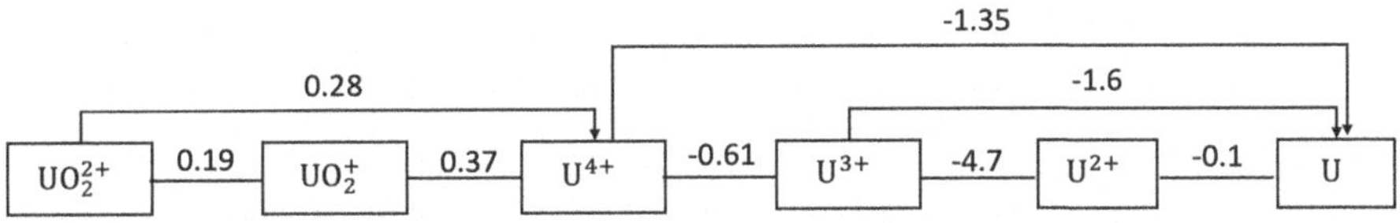

Fig. 1.1 Standard reduction potentials E_o of uranium. (Modified from Meinrath, 1999) [1]

Table 1.1 Some critical physicochemical properties of uranium

Description			Characteristics			
(a)	Element classification		Metal			
(b)	Colour	(i) Pure uranium	Silvery-white			
		(ii) In the air (when newly oxidized)	Yellow			
		(iii) Old oxide	Dark brown to black			
I	Phase at room temperature		Solid			
I(e)	Melting point		1135 °C, 1408 K			
(f)	Boiling point		4130 °C, 4403 K			
(g)	Several stable isotopes		0			
(h)	Hardness		6.0 months			
(i)	Reaction with air		Mild ⇒ U_3O_8			
(j)	Ionization energy		6.194 eV			
(k)	Electron shell configuration		1 s2			
			2 s2	2p6		
			3 s2	3p6	3d10	
			4 s2	4p6	4d10	4f14
			5 s2	5p6	5d10	5f3
			6 s2	6p6	6d1	
			7 s2			
(l)	Oxidation states		+6, +5, +4, +3			
(m)	Oxide(s)		UO, UO_2, UO_3, U_2O_5, U_3O_8			

In nature, uranium predominantly has valences of +4 or + 6, and it is never found in its elemental form. Some general characteristics of uranium in its different oxidation states include the following:

- U (III): The trivalent uranium is violet in the aqueous solution obtained from tetravalent uranium employing potent reducing agents. It is subject to rapid oxidation in the air.
- U (IV): Tetravalent uranium has low solubility in water; however, it has an emerald green colour in an aqueous solution when fully dissolved. It is very stable in reducing environments. Known tetravalent uranium compounds include the oxide UO_2 and the four halides UF_4, UCl_4, UBr_4, and UI_4. They are solids with high melting points above 500 $°C$. It predominantly occurs in primarily formed rocks.
- U (V): The pentavalent uranium has a pale lilac colour in an aqueous solution. Like trivalent uranium, it is easily oxidizable and rarely exists in nature.
- U (VI): The hexavalent uranium is yellow in an aqueous solution. In nature, it occurs exclusively as a uranyl cation UO_2^{2+} . This group preferably binds to oxygen-containing ligands (phosphate, sulfate, carbonate, and water as hydroxide). They usually form insoluble orthophosphates, orthosilicates, orthoarsenates, and orthovanadates upon interaction with phosphate (V), silicate, arsenate (V), and vanadate (V). Uranyl acetate and uranyl nitrate are examples of soluble uranium salts. Uranium dioxide is a black, crystalline powder. Also known com-

pounds are uranium trioxide UO_3, triuranoctoxide U_3O_8, and uranyl peroxide UO_4.

Naturally occurring uranium is a mixture of primarily three significant isotopes: 0.0054% of ^{234}U, 0.7204% of ^{235}U, and 99.2742% of ^{238}U, and it rarely varies more than 0.01% from these average compositions without artificial intervention. However, there are about 21 different uranium isotopes, of which a few are chemically engineered, summarized in Table 1.2.

1.2.2 Uranium Occurrence in Nature

Uranium, in nature, occurs over an extensive range of concentrations (Table 1.3). Uranium elements occur at an average concentration of 2.8 parts per million (ppm) in the Earth's crust, but their traces occur almost everywhere. Found in equal abundance to tin, it is more abundant than gold, silver, or mercury and slightly less than cobalt, lead, or molybdenum [2, 3].

Uranium typically occurs in its highest concentrations in the mineral uraninite (UO_2), also called pitchblende, because of its black colour. Some critical uranium mineral ores include brannerite ($(U,Ca)(Ti,Fe)_2O_6$), carnotite ($K_2(UO_2)_2VO_4 \bullet 1\text{-}3H_2O$), autunite ($Ca(UO_2)_2(PO_4)_2 \bullet 10H_2O$), heinrichite ($Ba(UO_2)_2(AsO_4)_2 \bullet 10H_2O$), zeunerite ($Cu(UO_2) \mid AsO_4]_2 \bullet 8\text{-}12\ H_2O$), coffinite ($U[(SiO_4),(OH)_4]$), and uranophane ($Ca[UO_2 \mid SiO_3OH]_2 \bullet 5H_2O$). Figure 1.2 has a picture of selected uranium minerals. Uranium is also found in phosphate rock ($Ca_3(PO4)_2$), lignite (brown coal), and monazite sand ($(Ce, La, Th, Nd, Y)PO_4$). Since there is little demand for uranium metal, uranium is usually sold in sodium diuranate ($Na_2U_2O_7 \bullet 6H_2O$), also known as yellowcake triuranium octoxide (U_3O_8). Vast amounts of uranium also occur in the oceans but have very low concentrations (see Table 1.4).

1.2.3 Uranium Radioactivity

Antoine Henri Becquerel, a French physicist, first observed uranium's radioactivity in 1896. However, the full significance of his discovery was realized when his student, Marie Curie, correctly interpreted the results and named the new phenomenon "radioactivity". Most of the first 82 elements in the periodic table have considerably stable isotopes. However, elements beyond 82 have unstable isotopes, which naturally decay into more stable isotopes or new daughter elements. As they do so, they release energy through subatomic particles or electromagnetic waves, collectively known as radiation. Thus, uranium isotopes continually change into other elements and emit particles after natural radioactive decay.

Table 1.2 Other known isotopes of uranium and their decay modes (Source: National Nuclear Data Center, USA)

Mass number	Half-life	Decay mode
217	16 milliseconds	Alpha decay
218	0.51 milliseconds	Alpha decay
218_m	0.56 milliseconds	Alpha decay
219	42 microseconds	Alpha decay
220	No data available	Alpha decay (suspected)
		Electron capture (suspected)
221	700 nanoseconds	No data available
222	1.0 microseconds	Alpha decay
223	18 microseconds	Alpha decay
		Electron capture
224	0.9 milliseconds	Alpha decay
225	95 milliseconds	Alpha decay
226	0.35 s	Alpha decay
227	1.1 min	Alpha decay
228	9.1 min	Alpha decay
		Electron capture
229	58 min	Electron capture
		Alpha decay
230	20.8 days	Alpha decay
		Spontaneous fission
		Emission of a neon-22 nucleus
231	4.2 days	Electron capture
		Alpha decay
232	68.9 years	Alpha decay
		Spontaneous fission
233	1.592×10^5 years	Alpha decay
		Emission of a neon-24 nucleus
		Spontaneous fission
		Emission of a magnesium-28 nucleus
234	2.455×10^5 years	Emission of a magnesium nucleus
		Emission of a neon nucleus
		Alpha decay
		Spontaneous fission
235	7.04×10^8 years	Alpha decay
		Spontaneous fission
		Emission of a magnesium-28 nucleus
		Emission of a neon nucleus
235_m	≈ 26 min	Isomeric transition
236	2.342×10^7 years	Alpha decay
		Spontaneous fission
237	6.75 days	Beta-minus decay

Table 1.2 (continued)

Mass number	Half-life	Decay mode
238	4.468×10^9 years	Spontaneous fission
		Alpha decay
239	23.45 min	Beta-minus decay
240	14.1 h	Beta-minus decay
241	No data available	Beta-minus decay (suspected)
242	16.8 min	Beta-minus decay
243	No data available	No data available

Table 1.3 Range of uranium concentrations in ore deposits and Earth

Grade	Concentration (ppm U)
Very high-grade ore, 20% U	200,000
High-grade ore, 2% U	20,000
Low-grade ore, 0.1% U	1000
Very low-grade ore, 0.01% U	100
Granite	4–5
Sedimentary rock	2
Earth's continental crust (average)	2.8
Seawater	0.003

Fig. 1.2 Some uranium ore minerals: the upper left is (**a**) Uraninite, *aka* pitchblende, from Niederschlema-Alberoda deposit, Saxony, Germany. (Courtesy of Geomartin); and right is (**b**) autunite, a secondary uranium mineral named after the town of Autun in France (Source: https://commons.wikimedia.org/w/index.php?curid=6199423); the bottom is (**c**) torbernite, a secondary uranium mineral (Source: https://commons.wikimedia.org/w/index.php?curid=3093268)

Table 1.4 Chemical constituents of the main uranium minerals

Mineral type	Name	Mineral formula
Primary minerals	Uraninite	$UO_2.x$
	Pitchblende	$UO_{2-x}(x = 0.2\text{-}0.6)$
	Coffinite	$U(SiO4)_{1-x}(OH)_{4x}$
	Brannerite	(U, CA, Y, CE)(Ti, Fe)$_2$O$_6$
	Davidite	(REE)(Y, U)(Ti, $Fe^{3+})_{20}O_{38}$
	Thucholite	Thorium- and uranium-bearing organic material
Secondary minerals	Autunite	$Ca(UO_2)_2(PO_4)_2{\cdot}8\text{-}12H_2O$
	Carnotite	$K_2(UO_2)_2(VO_4)_2{\cdot}1\text{-}3H_2O$
	Gummite	A mixture of uraninite and secondary uranium minerals of variable composition
	Selenite	$Mg(UO_2)_2(PO_4)_2{\cdot}10\ H_2O$
	Torbernite	$Cu(UO_2)_2(PO_4)_2{\cdot}12\ H_2O$
	Tyuyamunite	$Ca(UO_2)_2(VO_4)_2{\cdot}5\text{-}8\ H_2O$
	Uranocircite	$Ba(UO_2)_2(PO_4)_2{\cdot}8\text{-}10\ H_2O$
	Uranophane	$Ca(UO_2)_2(HSiO_4)_2{\cdot}5\ H_2O$
	Zeunerite	$Cu(UO_2)_2(AsO_4)_2{\cdot}8\text{-}10\ H_2O$

There are two forms of radiation, namely, non-ionizing and ionizing. Non-ionizing radiation has weak energy, like visible light, radio, and television signals, emanates from computer screens, and heats food in microwave ovens. In uranium, the subatomic particles emitted are alpha (α) particles, consisting of two protons and two neutrons, and beta (β) particles, electrons and neutrons, while the electromagnetic waves are gamma (γ) rays, similar to radio waves or light but typically emitted with much higher energies. Thus, uranium decay primarily produces ionizing radiation because the emitted α and β particles and γ rays have enough energy to remove electrons from atoms with which they interact, leaving the atoms charged or ionized. The decay series of the most important uranium isotopes, ^{238}U and ^{235}U, is summarized in Fig. 1.3.

The ^{238}U, the most abundant (99.27%) and stable uranium isotope in nature, has 146 neutrons and a half-life of 4.468 billion years. It decays into thorium-234 through alpha decay or spontaneous fission (refer to stability and decay mode of different uranium isotopes in Table 1.2). The second most abundant is ^{235}U, 143 neutrons, with a half-life of about 703.8 million years. It is the only known naturally occurring fissionable element, and its discovery has changed life on Earth. When a neutron strikes a fissionable element, its nucleus splits into smaller fragments (lighter nuclei) and releases energy. Therefore, fissionable material can produce enough free neutrons to sustain a nuclear chain reaction. Thus, ^{235}U is used as fuel in nuclear energy generation. However, the abundance of ^{235}U in naturally occurring uranium is too low to sustain a nuclear chain reaction.

When ^{238}U absorbs a neutron, ^{239}U forms have a half-life of about 23 min and decay into neptunium-239 (^{239}Np) through beta decay. ^{239}NP has a half-life of about 2.4 days and decays into plutonium −239 (^{239}Pu) through beta decay. Thus,

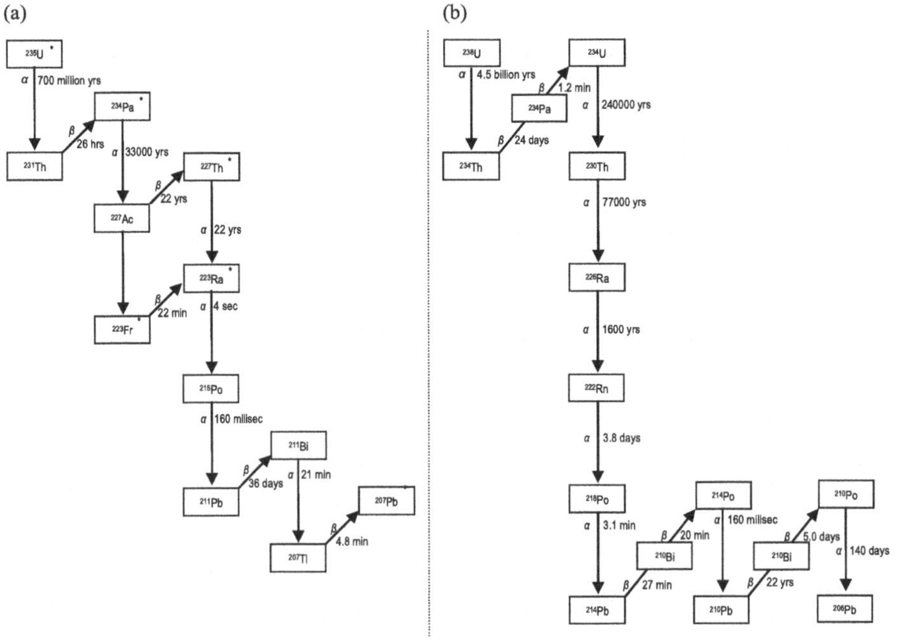

Fig. 1.3 The decay series of (**a**) ^{235}U and (**b**) ^{238}U: The α and β symbols indicate the alpha and beta decay, while the time indicates the half-life. The * symbol next to the isotope indicates the elements that emit γ particles, too [4–6]

exposure to neutrons can convert ^{238}U into the fissionable 239Pu, which can also be used as nuclear reactor fuel. ^{233}U is also a fissionable material used in nuclear reactors; however, it does not occur naturally. Therefore, to produce ^{233}U, atoms of thorium-232 (232th) are exposed to neutrons and 233th forms when 232th absorbs a neutron. It has a half-life of about 22 minutes and decays into protactinium-233 (^{233}Pa) through beta decay. ^{233}Pa has a half-life of about 27 days and decays into ^{233}U through beta decay. The conversion of ^{238}U into ^{239}Pu or 232th into ^{233}U is not a nuclear fusion because it does not involve using two smaller atoms to form a larger atom. Instead, fusion occurs in elements lighter than iron.

1.2.4 Types and Classifications of Uranium

The uranium found in ore, subsequently mined, is called natural uranium. Natural uranium contains about 99.28% 238 U, 0.71% 235 U, and about 0.0054% 234 U isotopes when mined. Since uranium is used primarily as fuel material in nuclear power, the content of ^{235}U in natural uranium is artificially increased because ^{235}U is the only isotope in nature that is fissionable by thermal neutrons. Increasing the percentage content of ^{235}U is called uranium enrichment and produces enriched uranium. In other words, enriching uranium increases the proportion of ^{235}U atoms that can be "split" by fission to sustain a nuclear reaction and release energy used to produce electricity. To produce enriched uranium, isotope separation removes a substantial portion of the ^{235}U for nuclear power, weapons, or other uses.

Enriched uranium for nuclear plants is about 5% ^{235}U, while weapons-grade enriched uranium is >90% ^{235}U. Enriching uranium may be completed through gaseous diffusion, gas centrifuge, or laser separation. The gaseous process uses molecular diffusion to separate a gas from a two-gas mixture. Thus, uranium is combined with fluorine to form uranium hexafluoride (UF_6) gas. The UF_6 gas is passed through a porous membrane barrier. The holes in the barriers are so small that there is barely enough room for the UF_6 gas molecules to pass through. The isotope enrichment occurs because the lighter UF_6 gas molecules (^{234}U and ^{235}U atoms) tend to diffuse faster through the barriers than the heavier UF_6 gas molecules containing ^{238}U. It takes hundreds of barriers before the UF_6 gas contains enough ^{235}U to be used in reactors. The gas centrifuge process involves many rotating cylinders in series and parallel configurations. UF_6 gas is introduced and rotated at high speed, creating a powerful centrifugal force so that the heavier gas molecules (containing ^{238}U) move towards the outside of the cylinder and the lighter gas molecules (containing ^{235}U) collect closer to the centre. Significantly more ^{235}U enrichment can be obtained from a single-unit gas centrifuge than from a single-unit gaseous diffusion stage. The laser separation process is based on photoexcitation principles (i.e., exciting molecules with laser light). The enrichment process involves three significant laser, optical, and separation module systems. Tunable lasers deliver a highly monochromatic light (i.e., light of a single colour). The light from these lasers can photoionize a specific isotopic species without affecting others. The affected species are then chemically changed, which separates the material.

After removing the enriched fraction, the remaining is the depleted uranium (DU), which contains only 0.2–0.4% ^{235}U. DU is refined from UF_6, the non-fissionable residue or byproduct of the uranium enrichment process. Natural uranium contains 99.2% by weight ^{238}U, while DU contains 99.8% by weight ^{238}U. Thus, DU is the ^{238}U isotope. A small proportion of other toxic metals, such as plutonium, may be present. DU is 60% less radioactive than natural uranium because ^{234}U and ^{235}U, which emit gamma rays and beta particles, have been removed.

Nevertheless, DU or ^{238}U remains a significant internal hazard because 238 U emits alpha particles at 4.2 and 4.15 MeV, which cause significant internal ionization and consequent cellular damage. Besides, its daughter products emit beta particles and gamma rays, which may cause additional radiological damage. Furthermore, since natural uranium has a low percentage of ^{235}U, the enrichment process produces large quantities of DU. For example, producing 1 kg of 5% enriched uranium requires 11.8 kg of natural uranium and leaves about 10.8 kg of depleted uranium with only 0.3% ^{235}U remaining. In addition, DU has a high density and is pyrophoric, two of the most significant physical properties that guide its use as kinetic energy penetrators, counterweights, and shielding or armour [7–12]. The uranium life cycle is summarized in Fig. 1.4 (Table 1.5).

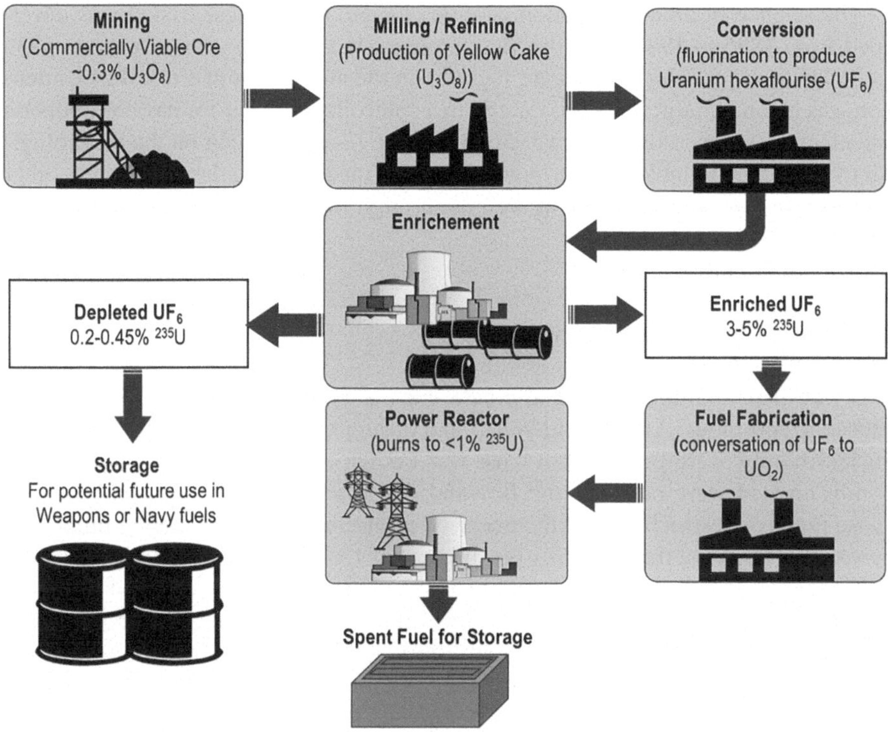

Fig. 1.4 The uranium fuel cycle involves many processing steps: the first step is the mining of uranium ore

Table 1.5 Distribution of main uranium isotopes in natural and depleted uranium

	Fractions of the isotopes	
Isotope	Natural uranium	Depleted uranium
^{234}U	0.006%	0.001%
^{235}U	0.72%	0.20%
^{238}U	99.28%	99.80%

1.2.5 Uses of Uranium

The most common assumed uses for uranium are in nuclear weapons and nuclear power plants because of the ability of the ^{235}U isotope to undergo nuclear fission naturally. Nuclear fission is when big atoms break into smaller atoms after being bombarded with neutrons. When neutrons hit the atom, they cause the nuclei to break apart, releasing large amounts of energy and an enormous explosion. This release of nuclear energy accounts for the power of nuclear weapons, such as atomic bombs. In addition, the energy released during fission is used to generate steam in steam boilers to produce electricity in reactors.

The science of atomic radiation, atomic change, and nuclear fission was developed from 1895 to 1945, with the most progress in the last 6 years. Between 1939 and 1945, most developments were focused on the atomic bomb. From 1945, attention was given to harnessing this energy in a controlled manner for naval propulsion and electricity generation. Since 1956, the prime focus has been on the technological evolution of reliable nuclear power plants. Consequently, there are many other uses of uranium as a chemical as well as through the production of radioisotopes, including the following:

1.2.5.1 Military Use of Uranium

It is true that uranium is used to create nuclear bombs and that the devastation caused is enormous. After World War II and during the Cold War, large numbers of nuclear bombs were produced in what was known as the arms race. The first two bombs created were named *Little Boy* and *Fat Man.* The *Little Boy* was made to cause massive destruction, like the one used in the Japanese city of Hiroshima. The destructive power of the *Little Boy* is estimated at 13–18 kilotonnes of TNT. The *Fat Man* bomb was designed to be far more advanced than the *Little Boy* and used plutonium as fuel. ^{239}Pu is synthesized via the following reaction using U and neutrons (n) via beta decay ($\beta-$) with neptunium (Np) as an intermediate. It is more fissile than ^{235}U. The *Fat Man* bomb explodes with approximately the force of 21 kilotonnes of TNT. However, the explosive power is not the most destructive part. Instead, the radioactivity and fallout that spread after detonation caused more long-term damage [13–15].

1.2.5.2 Fuel for Nuclear Power Generation

In the post-Cold War world, uranium is mainly used as fuel in nuclear power reactors for electricity generation, accounting for about 14% of global electricity. Over 31 countries use nuclear energy to generate electricity, and over 440 nuclear power reactors operate (excluding naval reactors) [2, 16]. Nuclear power plants boil water into steam, which turns turbines to produce electricity. The mechanism is precisely like electricity generation in plants that burn coal, oil, and natural gas, except that nuclear plants do not burn anything but obtain the heat needed to produce steam through a physical process called fission, which entails splitting uranium atoms in a nuclear reactor [2, 6, 17].

Nuclear fuel consists of small, hard ceramic pellets packaged into long, vertical tubes. Bundles of fuel are inserted into the reactor. Nuclear fuel consists of two types of uranium, ^{238}U and ^{235}U. ^{235}U is naturally fissable and can be directly used by nuclear reactors, while ^{238}U must first be converted to ^{239}Pu to become fissable. The vast majority of all nuclear power reactors require "enriched" uranium fuel, in which the proportion of the ^{235}U isotope has been raised from natural levels of 0.7% to about 3.5–5%. The enrichment process requires uranium to be in gaseous form.

On the way from the mine, it passes through a conversion plant that converts uranium oxide into uranium hexafluoride. For a reaction to occur, either uranium or plutonium is bombarded by neutrons, which initiates a chain reaction that releases energy and sends more neutrons to continue the reaction. The heat produced by fission turns the water into steam. The steam drives a turbine, which spins a generator to create electricity. The fission products remain in the ceramic fuel and undergo radioactive decay, releasing more heat. They are the primary waste products of the process [2, 16, 18].

There are two types of nuclear power plants: boiling-water and pressurized-water reactors. Pressurized-water reactors comprise more than 60% of nuclear power plants worldwide [2, 19]. It has water in its primary cooling/heat transfer circuit and generates steam in a secondary circuit. The boiling-water reactor generates steam in the primary circuit above the reactor core, though it remains under considerable pressure. Both types use water as a coolant and moderator to slow neutrons (see Fig. 1.5).

1.2.5.3 Non-radioactive Uses

Uranium is also widely used by the military for non-radioactive purposes, taking advantage of its physical properties, which make it very dense, hard, and flammable. For example, it is used as high-density penetrating ammo or artillery for military weapons, and these uranium bullets can penetrate armoured vehicles. In addition, depleted uranium, which is already used for nuclear power generation, is used. Thus, depleted uranium has lower radiation levels than natural uranium. However, it still possesses some radioactivity in its depleted uranium, which can be used in aircraft and medical equipment as a shield for radiation-generating devices.

1.2.5.4 Historical and Aesthetic Uses

Another use for uranium that is not common today but was widely used in the past is glass, dinnerware, and glow-in-the-dark material. Uranium compounds have been used to colour glass and ceramics for centuries. Uranium glass glows bright green under UV light and was a novelty item for party trays and platters. Fiesta dinner plates also used uranium but did not glow; they had very bright and vibrant colours. Scientific evidence shows that as early as A.D. 79, yellow glass found near Naples, Italy, was coloured by uranium oxide. Uranium trioxide (UO_3), an orange powder, was used in Fiestaware plates in the 1930s and 1940s. Other uranium compounds were used to make vaseline glass and glazes [20, 21]. Generally, ceramic glazes ranging from orange to bright red were used on items as varied as household crockery and architectural decorations. The uranium within these items is radioactive and should be treated with care. Uranium can still be found in small glassware such as beads or marbles, but the content must be minimized to reduce the radioactive hazard.

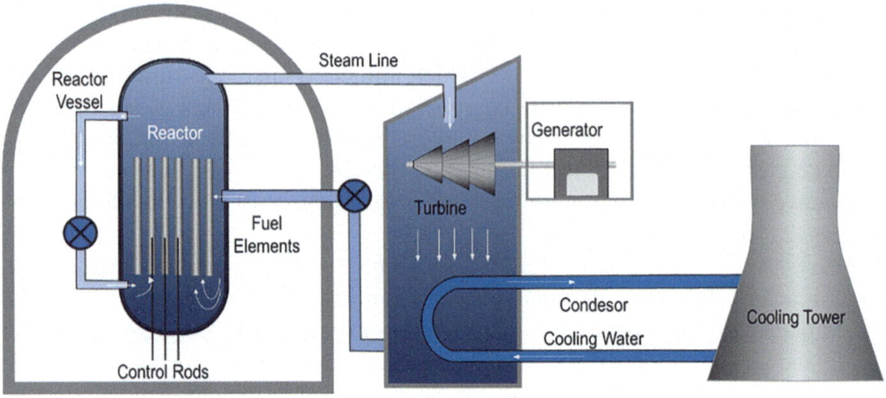

(b) Pressurized water reactors

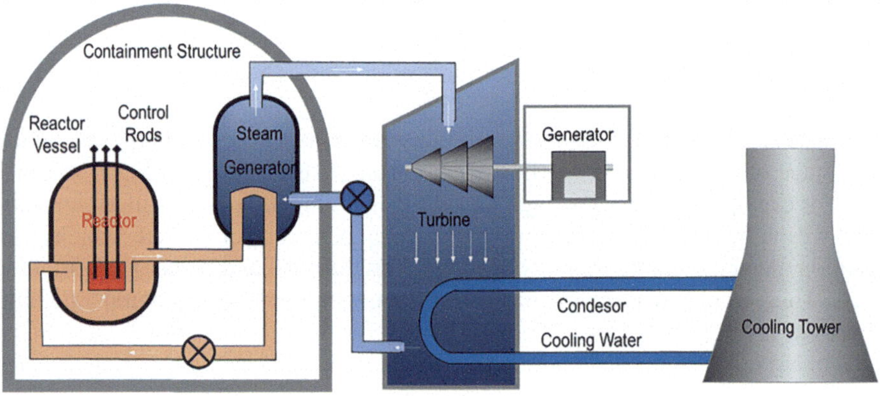

Fig. 1.5 Types of nuclear reactors

1.2.5.5 Electron Microscopy

In electron microscopy, uranium and lead are the most widely used heavy-metal stains, providing good contrast for ultra-thin sections of biological specimens. Uranyl acetate and lead citrate are the standard staining techniques for transmission electron microscopy (TEM). Uranyl acetate produces the highest electron density and image contrast and imparts a fine grain to the image due to the atomic weight of ^{238}U. The uranyl ions bind to proteins and lipids, including sialic acid carboxyl groups (e.g., glycoproteins and gangliosides), and to nucleic acid phosphate groups of DNA and RNA. Gangliosides are glycolipids concentrated on cell surfaces, while glycoproteins are abundant in all membranes and part of the glycocalyx. Thus, uranyl acetate produces good contrasting results in membranes, nucleic acids, and nucleic acids containing protein complexes, such as ribosomes. The only setback of uranyl acetate is its sensitivity to light, especially UV, which results in the formation of precipitates if exposed.

1.2.5.6 Radiometric Dating

Uranium is used in radiometric dating, often called radioactive dating, to determine the age of materials, such as rocks and other ancient objects. Other best-known radiometric techniques include radiocarbon and potassium-argon dating. Generally, radiometric dating compares the observed abundance of a naturally occurring radioactive isotope and its decay products using known decay rates. Thus, uranium-series dating techniques exploit the ability of radioactive uranium and thorium isotopes to decay into a series of unstable, radioactive "daughter" isotopes, which continue until a stable, non-radioactive lead isotope is formed. The "parent" isotopes have half-lives of several million years, enabling a dating range for the different uranium series of a few thousand years to 500,000 years.

There are two types of uranium-series dating techniques: the daughter-deficiency and the excess-daughter methods. In the daughter-deficiency method, the parent radioisotope is initially present without its daughter. The parent decays to the daughter through time until equilibrium is reached (i.e., equal amounts of each). The age of the specimen is determined by measuring the amount of the daughter product formed, provided that neither isotope has entered or exited the specimen after its initial formation. In the daughter-excess method, a greater amount of the daughter is initially deposited than of the parent. Over time, the excess daughter disappears as it is converted back into the parent. A sample or specimen's date is calculated by measuring the extent to which this has occurred.

The mathematical expression that relates radioactive decay to geologic time is:

$$D = D_0 + N(t)\left(e^{\lambda t} - 1\right) \tag{1.1}$$

where t is the age of the sample; D is the number of atoms of the daughter isotope in the sample; D_0 is the number of atoms of the daughter isotope in the original composition; N is the number of atoms of the parent isotope in the sample at time t (the present), given by $N(t) = N_0 e^{-\lambda t}$); and λ is the decay constant of the parent isotope, equal to the inverse of the radioactive half-life of the parent isotope times the natural logarithm of 2.

The half-life is delivered using a differential equation. Radioactive materials decay according to the law. The time of decay is proportional to the number of radioactive atoms. It can be expressed mathematically as:

$$t = -KN \tag{1.2}$$

where K is the decay constant, t is time, and N is the number of atoms. The minus sign indicates that each decay decreases the total number of atoms. Thus, at a given time t, the number of atoms $N(t)$ in the system is:

$$\frac{dN(t)}{dt} = -K \bullet N(t) \tag{1.3}$$

or

$$\frac{dN(t)}{dt} = -Kdt \tag{1.4}$$

Integrating both sides, it becomes:

$$lnN(t) = -Kt + C \tag{1.5}$$

C is the constant of integration that we can often ignore, but not here. When $t = 0$,

$$lnN(0) = C$$

Taking the exponential of both sides, we get

$$N(t) = N(0)e^{-Kt}$$

At t = one-half life, then,

$$\frac{N(t)}{N(0)} = \frac{1}{2} = e^{-Kt}$$

And:

$$\ln\left(\frac{1}{2}\right) = -\ln 2 = -Kt$$

so:

$$t = \frac{\ln 2}{K}$$

Some of the uranium-based dating techniques used currently include:

1. Uranium-uranium dating technique—compares two isotopes of uranium (U): ^{234}U and ^{238}U in a sample. It exploits the radioactive decay series of uranium to the stable isotope ^{206}Pb. Although analytically simpler, in practice $^{234}U/^{238}U$ requires knowledge of the ratio at the time the material under study was formed and is generally used only for samples older than the ca. 450,000-year upper limit of the 230th/^{238}U technique.
2. Uranium-thorium dating technique—compares two isotopes of ^{238}U and 230th. It provides accurate dating up to ca. 450,000 years.
3. The uranium-lead dating technique relies on two decay chains, the uranium series from ^{238}U to ^{206}Pb, with a half-life of 4.47 billion years, and the actinium series

^{235}U to ^{207}Pb, with a half-life of 710 million years. It is accurate up to the age of the solar system but problematic on samples younger than about two million years.
4. Protactinium-uranium dating technique—compares two isotopes of $^{231}Pa/^{235}U$. This daughter-deficiency method is used, for example, in dating carbonates. Living molluscs and corals take up dissolved compounds such as uranium isotopes but cannot take up protactinium, which is insoluble. So, protactinium—^{231} Pa—begins to accumulate via the decay of ^{235}U after the organism dies.

1.2.5.7 Other Uses

Uranium is mainly used as fuel in nuclear power reactors for electricity generation. Beyond providing about 14% of the world's electricity, there are many other uses of uranium through the production of radioisotopes, including:

1. **Medicine:** Radioisotopes are used for diagnosis and research. Radiodiagnostics can detect disease by injecting certain radioelements into the human body and observing their paths. Radiotherapy uses ionizing radiation to kill cancer cells, and nuclear medicine relies on radioactive drugs that target specific organs.
2. **Food-Processing Industry:** Radioisotopes are used to sterilize fresh products because irradiation kills parasites, pests, and bacteria.
3. **Industrial Sector:** Radioisotopes are used for industrial X-ray requirements. This technique is widely used to inspect safety and quality in metallurgy, automobiles, and aeronautics. Further, uranium oxide has also found limited application as an attachment to filaments in light bulbs. The compound reduces the speed at which an electric current enters the bulb. This reduces the likelihood of filament heating too fast and breaking.
4. **Space Industry:** When space probes are required to operate in places far away from the sun, the only available solution for the production of heat and electricity is the use of radioisotopes.
5. **Textile Industry:** Uranium compounds have been used as mordants in dyeing operations. A mordant is a material that helps a dye stick to the cloth.

1.3 Market and Economics of Uranium

1.3.1 Uranium Market Drivers

Uranium does not trade on an open market like other commodities. Instead, it is traded mostly through long-term contracts negotiated directly between buyers and sellers. However, like other commodities, its trade is demand-driven, primarily driven by a growing need for global energy security and low-carbon electricity generation. The nuclear power industry is currently driven by growing energy demand, driven by the increasing world population and industrial development (see Fig. 1.6).

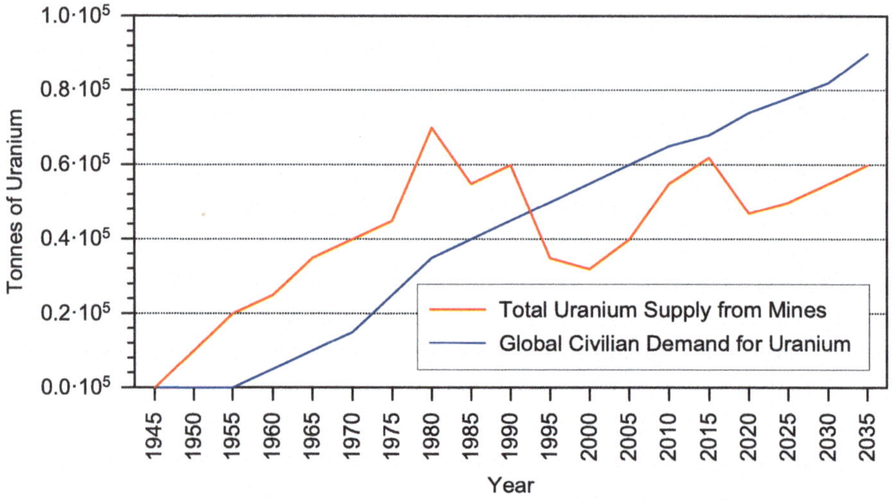

Fig. 1.6 The total uranium supply and demand for civil use from 1945 to 2035. (Data Sources: World Nuclear Association [2] and UDEPO [3])

Electricity consumption is forecast to increase by about 70% of current levels by 2035. Developing countries will use the most energy as their populations and living standards increase [2, 17, 22].

The demand for uranium is directly linked to the output of nuclear power plants. The global uranium consumption in 2012 was about 75 million kg, while mine production was about 68.9 million kg. The gap between production and demand is filled through secondary supply. Secondary supply is essentially inventoried by type. It includes inventories held by utilities and other fuel cycle companies, government inventories, used reactor fuel reprocessed, recycled materials from military nuclear programs, and uranium in depleted uranium stockpiles. Since 1985, world uranium production has been less than world utility uranium consumption. Several secondary sources have covered the resulting shortfall, which has steadily drawn down. For example, highly enriched uranium (HEU) derived from dismantling Russian nuclear weapons became a significant source of supply equivalent to a large mine until 2013. In addition, a limited amount of uranium from the US weapons program has been introduced into the market [23, 24].

After a decade of declining mine production, uranium output has generally risen since 1993 and now meets over 90% of power generation demand. Following the recovery of uranium prices around 2003, much activity was preparing to open new mines. The World Nuclear Association reference scenario estimated world uranium demand at about 65,220 t in 2016, and most of this will need to come directly from mines. However, due to the absence of Japanese consumption over the last couple of years and low prices, some of the stockpile buildup over 2013–2016 will come in as secondary supply in the next few years [17].

1.3.2 Market Trends

The uranium market dates back to the discovery of its radioactive properties in 1896 by Henri Becquerel. Marie Curie discovered radium in uranium ore in 1898. Radium was deemed a miracle cure for cancer and commanded prices as high as $2700 per gram until its market crumbled in the late 1930s. The demand for radium led to a rapid expansion in uranium ore mining in the early 1900s.

In 1939, Otto Hahn did the first nuclear fission reaction in Germany. Learning from Hahn's success, atomic scientists led by Enrico Fermi built the first nuclear reactor at the University of Chicago in 1942. An international team of nuclear scientists collaborated in the so-called Manhattan Project to develop an atomic bomb, which was tested at the Trinity test site in New Mexico in July 1945. A month later, these atomic bombs were used to destroy the Japanese cities of Hiroshima and Nagasaki. After the war ended, attention quickly turned to developing nuclear energy's peaceful uses, although military nuclear power symbolized the Cold War's arms race between the Western and Eastern blocs. The first practical use of nuclear power was in 1951 when an experimental nuclear reactor at a U.S. research centre in Idaho Falls lit four ordinary light bulbs. In 1954, the world's first commercial reactor produced power in Obninsk, Russia, followed by the world's first industrial-scale nuclear power station in 1956 at Calder Hall in Britain. The nuclear industries proliferated during the 1960s and 1970s [25–27].

The world's major source of uranium until the early 1950s was in the Belgian Congo. Later, to meet the fast-growing nuclear industry requirements, uranium mining was expanded in the USA, Canada, France, Australia, and Africa. Today, over two-thirds of the world's uranium production from mines is from Kazakhstan, Canada, and Australia. By 2015, Kazakhstan produced the largest share of uranium from mines (39% of the world's supply), followed by Canada (22%) and Australia (9%) [2, 3, 28].

1.3.3 Supply and Marketplace

The uranium marketplace has three main areas: (a) uranium, (b) the processing market, and (c) the unique aspects of purchasing fuel by nuclear plant operators. Uranium markets are primarily dominated by buyers from large utility companies that operate nuclear power plants, although a small number of brokers and traders also operate. The sellers are uranium producers and a few other entities that hold uranium inventories in various forms. The uranium production industry is international in scope, with a small number of companies operating in relatively few countries. Eight companies provide approximately 83% of the estimated global uranium mine production of 70 million kg of U_3O_8. By 2016, over 430 commercial nuclear power reactors were operating, with more than 60 new reactors under construction. Uranium consumption has increased from 34 million kg U_3O_8 in 1980 to an estimated 72.5 million pounds or more and is expected to grow at an average annual growth rate of about 3% in the 2020s.

The uranium market was shaken by the March 2011 nuclear accident at Japan's Fukushima crisis-driven Daiichi nuclear power station. Immediately after the accident, several ad hoc energy policies affected worldwide short- and long-term energy security. The accident led to a loss of public acceptance of nuclear power. It led countries like Germany and Italy to shut down some nuclear reactors immediately and abandon plans to build new plants. However, contrary to some predictions, the Fukushima Daiichi accident has not dramatically changed the outlook for nuclear energy. With some exceptions, support for nuclear energy remains strong globally, driven by rising electricity demand and concerns about the link between carbon emissions and climate change [29, 30]. More than 60 reactors were under construction, and at least 150 new nuclear projects were in the licencing or advanced planning stages by the end of 2016. Nevertheless, the Fukushima Daiichi accident is a global reminder of the need to prioritize safety in nuclear energy.

1.4 Major Uranium Deposits of the World

Uranium deposits are distributed all over the world. Figure 1.7 shows the global distribution of uranium as mapped in 1806. The radiation decay is too negligible to affect the distribution map due to the long half-life of uranium. Thus, the glocal distribution remains the same.

About 8,070,400 metric tonnes of uranium reserves globally are currently identified as economically viable and distributed in 53 countries (see Table 1.6). As of 2024, Kazakhstan, Canada, Russia, Australia, and Namibia were the top five world

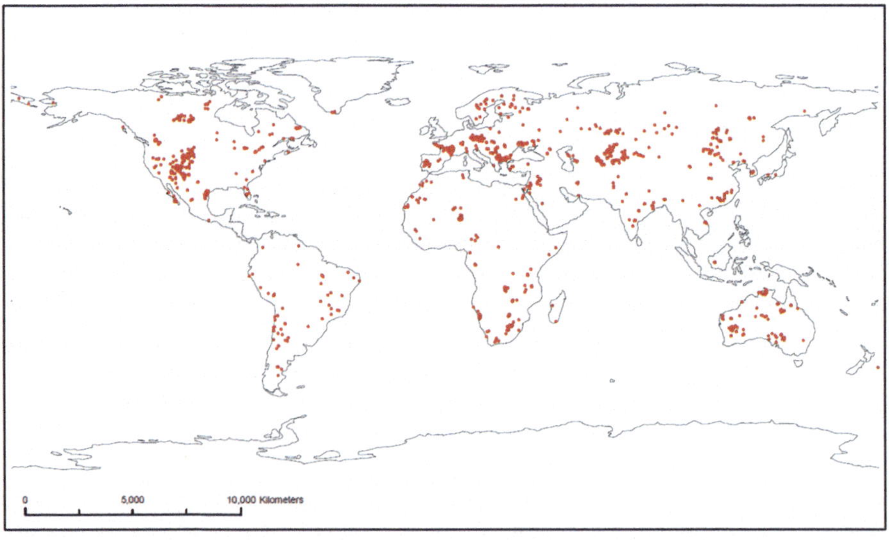

Fig. 1.7 Global distribution of uranium deposits, mapped in 1806. The figure is reproduced here with permission from the IAEA. (Source: IAEA, UDEPO [3])

Table 1.6 Global uranium reserves by country as of 2019

Continent	Country	Reserves (metric tonnes)
South America	Argentina	39,800
South America	Brazil	276,800
South America	Chile	1400
South America	Paraguay	3600
South America	Peru	33,400
Oceania	Australia	2,049,400
North America	Canada	873,000
North America	Mexico	5000
North America	United States	101,900
Europe	Czech Republic	119,200
Europe	Finland	1200
Europe	Germany	7000
Europe	Greece	7000
Europe	Denmark (Greenland)	114,000
Europe	Hungary	13,500
Europe	Italy	6100
Europe	Portugal	7000
Europe	Romania	6600
Europe	Slovakia	15,500
Europe	Slovenia	9200
Europe	Spain	28,500
Europe	Sweden	9600
Europe	Ukraine	186,900
Asia/Europe	Russia	661,900
Asia/Europe	Turkey	13,600
Asia	China	269,700
Asia	India	30,480
Asia	Indonesia	8400
Asia	Iran	7500
Asia	Japan	6600
Asia	Jordan	52,500
Asia	Kazakhstan	969,200
Asia	Mongolia	143,500
Asia	Uzbekistan	132,300
Asia	Vietnam	3900
Africa	Algeria	19,500
Africa	Botswana	87,200
Africa	Central African Republic	32,000
Africa	Chad	2400
Africa	Democratic Republic of Congo	2700
Africa	Egypt	1900

(continued)

Table 1.6 (continued)

Continent	Country	Reserves (metric tonnes)
Africa	Gabon	5800
Africa	Malawi	14,300
Africa	Mali	8900
Africa	Mauritania	24,500
Africa	Namibia	504,200
Africa	Niger	439,400
Africa	Senegal	1100
Africa	Somalia	7600
Africa	South Africa	447,700
Africa	Tanzania	58,200
Africa	Zambia	31,000
Africa	Zimbabwe	1400
	Total	**8,070,400**

producers. However, about two-thirds of global uranium production from mines comes from Kazakhstan, Canada, and Australia.

Although Australia has the largest uranium reserves, with about 1,684,100 metric tonnes (i.e., 28% of the world's uranium resource), Kazakhstan was the leading global uranium producer. It had the second-highest global uranium reserves, estimated at 815,200 metric tonnes, accounting for 13% of total global uranium resources. Australia was the fourth-largest uranium producer in the world, after Namibia and before Russia.

Canada is the second-largest uranium producer and the third-largest uranium country by reserves. It has an estimated 588,500 metric tonnes, representing 10% of the world's uranium resources. Namibia and Russia accounted for 8% of global uranium (i.e., 470,100 metric tonnes and 480,900 metric tonnes, respectively). Namibia was the third global producer of uranium.

References

1. Meinrath, G. (1999). Aquatic Chemistry of Uranium: A review focusing on aspects of environmental chemistry. *Freiberge On-line Geoscience, 1*, 1–112.
2. Association, W. N. (2025). World Nuclear Association London, United Kingdom. https://world-nuclear.org/
3. IAEA. (2018). *World Distribution of Uranium Deposits (UDEPO)* (2016th ed.). International Atomic Energy Agency.
4. Peterson, J., et al. (2007). In Argonne National Laboratory (Ed.), *Human Health Fact Sheet* (Vol. 4). Argonne National Laboratory, EVS in collaboration with U.S. Department of Energy.
5. Mkandawire, M. (2013). Biogeochemical behaviour and bioremediation of uranium in waters of abandoned mines. *Environmental Science and Pollution Research, 20*, 1–28. https://doi.org/10.1007/s11356-013-1486-3
6. Golovich, E., Wellman, D., Serne, R., & Bovaird, C. (2011). *Summary of Uranium Solubility Studies in Concrete Waste Forms and Vadose Zone Environments*. U.S. Department of Energy.

7. Giannardi, C., & Dominici, D. (2003). Depleted uranium, the longterm effects of the US bombing of Iraq. *Ponte, 59*, 227–238.
8. Giannardi, C., & Dominici, D. (1999). Side-effects of depleted uranium: Assessments for the war in Kosovo. *Ponte, 55*, 118–129.
9. Howard, M. (2001). Diversity of propaganda (the use of depleted uranium in former Yugoslavia). *Index Censorship, 30*, 156–160.
10. Ketterer, M. E., Wetzel, W. C., Layman, R. R., Matisoff, G., & Bonniwell, E. C. (2000). Isotopic studies of source of uranium in sediments of Ashtabula River, Ohio, USA. *Environmental Science Technology, 34*, 1.
11. Konstantinou, M., Demetriou, A., & Pashalidis, I. (2007). Adsorption of hexavalent uranium on dunite. *Global NEST Journal, 9*, 229–236.
12. Meinrath, A., Schneider, P., & Meinrath, G. (2003). Uranium ore and depleted uranium in the environment, with reference to uranium in the biosphere from Erzgebirge/Sachsen, Germany. *Journal of Environmental Radioactivity, 64*, 175–193.
13. Miller, J. (2023). Los Alamos National Laboratory, Los Alamos. https://www.lanl.gov/
14. Rhodes, R. (2012). *The making of the atomic bomb: 25th anniversary edition*. Simon and Schuster.
15. Foundation, A. H. (2014). Atomic Heritage Foundation, Washington, DC. https://ahf.nuclear-museum.org/
16. IAEA. (2013). In International Atomic Energy Agency (Ed.), *Online*. International Atomic Energy Agency.
17. Association, W.N. (2025). *Current and Future Generation*. World Nuclear Association., https://world-nuclear.org/
18. IAEA. (2006). *Fundamental Safety Principles*. Nuclear Fuel Cycle and Materials Section, International Atomic Energy Agency.
19. Association, W. N. (2025). *Nuclear Power reactors*. World Nuclear Association. https://world-nuclear.org/
20. ORAU. (2004). *Radioactive consumer products*. Oak Ridge Associated Universities (ORAU) Museum of Radiation and Radioactivity.
21. Sari, S. (2023). Glazes and Glass Bodies with Uranium Content.
22. Association, W. N. (2025). *Nuclear fuel cycle*. World Nuclear Association. https://world-nuclear.org/
23. CAMECO. (2025). *Cameco Uranium 101*. Cameco Corporation. Cameco.com
24. Association, W. N. (2024). *Country Profiles*. World Nuclear Association. https://world-nuclear.org
25. Gosling, F. G., & Manhattan, P. (1994). *<<The>> Manhattan Project making the atomic bomb*. September 1994 edn, VII, 66 S. 28 cm Ill., Kart. United States Department of Energy, History Division.
26. Spence, R. (2025). *Encyclopedia Britannica*. Encyclopædia Britannica.
27. Schwartz, D. N. (2017). *The Last Man Who Knew Everything: The Life and Times of Enrico Fermi, Father of the Nuclear Age*. Basic Books.
28. IAEA. (2009). *World Distribution of Uranium Deposits (UDEPO) with Uranium Deposit Classification* (Vol. IAEA-TECDOC-1629). International Atomic Energy Agency.
29. News, W. N. (2024). *Nuclear policies*. World Nuclear Association. https://www.world-nuclear-news.org
30. IAEA. (2020). *Climate Change And Nuclear Power* (Vol. IAEAL 20-01349 110). International Atomic Energy Agency.

Chapter 2
Characteristics of Abandoned Uranium Mining Sites

Abstract This chapter examines the defining features, origins, and long-term consequences of abandoned uranium mining sites. It begins by outlining what constitutes an abandoned mine and situates these sites within the broader mine life cycle, from exploration and extraction to closure. Various types of abandoned uranium mines are described, along with both conventional and non-conventional extraction methods, associated ore-processing pathways, and the underlying chemistry that drives uranium recovery. This chapter also explores accompanying non-ore processes that contribute to environmental alteration. Reasons for mine abandonment are analysed, highlighting economic, regulatory, and operational drivers. Finally, this chapter discusses the legacy of these sites, identifying key factors that shape their environmental and social impacts and detailing the persistent issues, such as contamination, landscape degradation, and community risks, that define abandoned uranium mining areas.

Keywords Mine life cycle · Uranium extraction methods · Ore processing · Mine closure · Contamination · Mining impacts · Uranium ore chemistry · Post-mining landscapes

2.1 An Abandoned Mine

For some reason, a shared legacy is that most mining companies either cease to exist or become untraceable after mine closure. Consequently, until recently, most closed mine sites were left during mining and processing operations, frequently in an excavated state, without any restoration or reclamation. Such mining and processing sites are called abandoned mines. In addition, there are mines for which the owner cannot be found, or who is financially unable or unwilling to clean them up. Such abandoned mines are sometimes called orphaned mines. As time passed, land leases lapsed, records were lost, and the original ownership of the mines became unclear. In short, no one is held accountable for waste sites. Further, abandoned mines are frequently orphaned, leaving governments to close and rehabilitate them.

M. Mkandawire, *Fate and Effects of Uranium in Abandoned Uranium Mines*,
https://doi.org/10.1007/978-3-032-10447-2_2

This chapter aims to put the legacy of abandoned uranium mines and processing sites into perspective. Therefore, it starts with the ideal process for closing mining activities at a site, then explains why most mining companies fail to close their operations properly, and finally discusses the specifics of uranium-abandoned mines. Ore extraction and processing methods are detailed to understand the complexity and characteristics of uranium-abandoned mines. The different characteristics and behaviour of uranium sites determined the fate and effect of uranium in these abandoned mining sites.

2.2 Background to Abandoned Mine

2.2.1 Mine Life Cycle

Mining is a temporary use of land, with the active life lasting from a few years to several decades, depending on the resource [1–3]. Therefore, it is evident that once the resource is depleted, the mining operation ceases. A typical mining project goes through different phases, collectively known as a mine life cycle. A mine life cycle has the following stages: exploration, discovery, development, production, and reclamation (Fig. 2.1).

Exploration aims to gather information about the potential presence of a mineable orebody. It usually begins with research to select target areas, followed by geological mapping and many geochemical and geophysical surveys. Eventually, something of value is discovered—mineral resources of interest. The mine development stage includes feasibility, geoscience, and engineering studies. Thus, very few discovered mineral deposits became producing mines because they had to pass economic feasibility tests. Once it is determined that the resource is economically feasible, permits, leases, licences, and environmental impact assessments are required. Upon favourable outcomes and obtaining all approvals, the mine construction begins. This is the most expensive phase of the mining cycle. Then the production phase begins, including extraction, milling, and processing of raw materials. The time a mine is in production depends on the amount and quality of the mineral in the deposit and the profitability of the operation.

Finally, the mine closes, and the land has to be returned to its pre-disturbance condition. Ideally, mine site reclamation and environmental protection should begin at the start of a project and continue after closure. There must be closure and reclamation plans, but progressive reclamation is recommended throughout the life of the mine. If well reclaimed, the former mine site can add significant value to land in communities for recreational purposes and future development.

Fig. 2.1 Life cycle for mineral resource development. (Modified from an unknown source)

2.2.2 Types of Abandoned Uranium Mines

There is no generalized classification of abandoned uranium mines. However, for this book, abandoned uranium mines have been classified by defining issues related to mining techniques and types, namely:

(a) Abandoned moon cast (*aka* open pit) uranium mining sites.
(b) Abandoned underground uranium mining sites.
(c) Abandoned in situ leaching mining sites.

Despite these classifications used in this book, the issues related to abandoned uranium mines are further defined by other factors specific to regional, climatic, and even socio-economic and socio-political conditions.

2.2.3 Conventional Uranium Ore Extraction

There are several mining procedures for extracting uranium from its ore deposits. However, the most frequently used are open-pit, underground, and in situ leaching (ISL), also known as in situ recovery (ISR) borehole mining [4–6]. Orebody size, location, orientation, rock quality, mineral distribution, site location, and infrastructure influence the selection of the mining method and the overall mine development plan. Generally, orebodies are either vein type, massive, or tabular in shape, and both the shape and ore thickness influence the mining method used. Vein-type orebodies are usually dipped steeply. Uranium orebodies are often narrow and irregular.

2.2.3.1 Open-Pit Mining

When orebodies lie close to the surface, the overburden (overlying rock) and waste rock are excavated and removed until the ore or mineral-bearing rocks are reached and extracted [7–9]. In most cases, this creates a large pit or borrows from the surface. This is commonly referred to as open-pit mining of uranium, open-cast or open-cut mining, and moon-cast mining. This surface mining technique is appropriate, mainly if the ore deposit is relatively large and there is little overburden. When designing open-pit mines, several factors are considered. For instance, the angle of the open-pit walls depends on the arrangement and characteristics of the overlying material and the depth to the ore [5, 10]. These determine the strength of the wall to support a safe slope.

Further, they dictate how much waste overburden must be removed before production can begin [10, 11]. Another consideration in designing open-pit mines is the first "bench" (see Fig. 2.2), as each bench is always smaller than the last. Consequently, the dimensions of the initial bench dictate the final depth of the open pit. The ratio of waste rock to ore mined is known as the stripping ratio. It is critical for determining the economic feasibility of exploiting a particular ore deposit, whether by open-pit mining or by underground mining [12, 13]. In most cases, this stripping ratio is high for the first bench and decreases steadily for each successive bench. Open-pit uranium mines are only economically feasible when the cost of mining the waste rock does not exceed the value of the ore [14].

2.2.3.2 Underground Mining

Underground mining is usually employed when the orebodies are relatively deep. The technique involves constructing access shafts and tunnels to the ore deposits [5, 15]. The shaft can be sunk vertically or inclined. Both vertical and inclined shafts are equipped with hoists and headframes, structures at the top of the shafts that enclose and operate the hoists used to transport ore and mine personnel [16]. The general structure of the underground mine is presented in Fig. 2.3. Usually, a

Fig. 2.2 (**a**) A type of open-pit mine structure; (**b**) an example: the Rössing open-pit uranium mine in Namibia (Source: Ikiwaner; reproduced under GNU Free Documentation License v1.2)

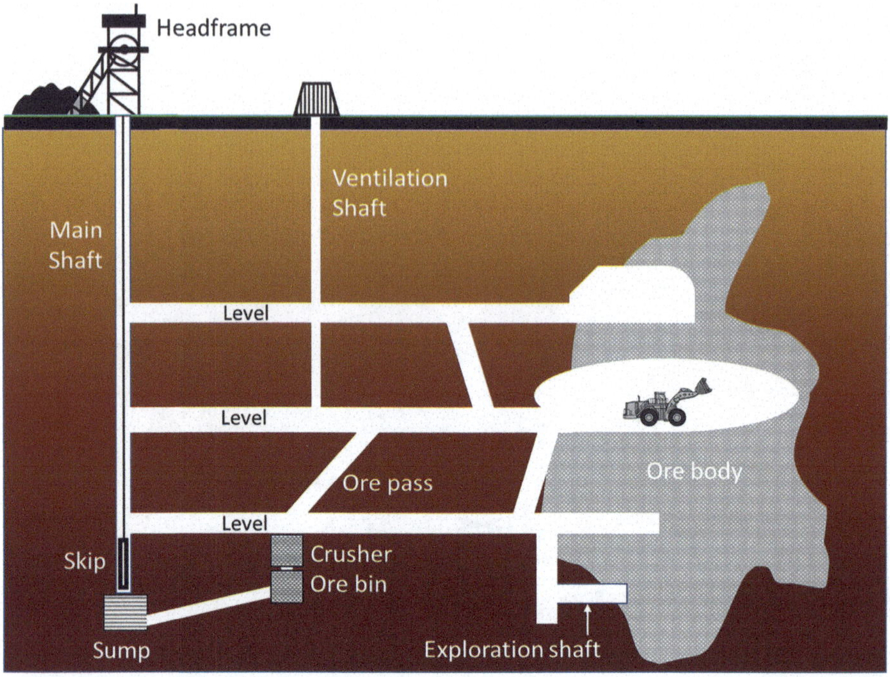

Fig. 2.3 Simplified and generalized illustration of an underground mine with a vertical shaft. The shaft mine has compartments for a skip hoist, elevators (cages), and service facilities, such as ventilation ducts

secondary or tertiary inclined opening is driven downward to connect levels and used for haulage and is called a ramp. A ramp driven into the earth, usually a spiral, is called a decline, while a horizontal opening driven into the side of a hill or mountain is called an adit. Ramps usually spiral downward so that rubber-tired mobile equipment has access to the mine [5, 15, 16]. In some cases, ramps are driven straight to accommodate conveyor belts. Horizontal or level mine workings are called crosscuts and drifts, while vertical access workings are called raises or winzes [17].

Underground mining can be done using different techniques, depending on the ore deposit's depth, shape, and surrounding geological conditions. For instance, the cut-and-fill technique is used in steeply dipping or irregular ore zones, where the mineral deposit is contained in a rock mass with good to moderate stability [16, 17]. It involves removing the ore in horizontal slices starting from a bottom cut and advancing upward, allowing the stope boundaries to be adjusted to follow irregular mineralization [18]. Although cut-and-fill mining is relatively expensive, it minimizes ore loss and dilution. The drift-and-fill technique is used when the ore zone is too wide for a single cut [15, 16, 18]. As in cut-and-fill mining, the ore is removed after blasting, and the resulting space is backfilled with fill material.

The shrinkage-stopping technique is used for steeply dipping orebodies. Ore is extracted in horizontal slices, starting at the stope bottoms and advancing upward [15, 16]. Most of the blasted rock remains in the stope to provide a working platform for miners drilling holes in the roof, and it also helps keep the stope walls stable [16, 18]. Room-and-pillar mining is commonly done in flat or gently dipping orebodies, accessed by horizontal drilling advancing along a multifaced front, forming empty rooms behind the producing front [15, 18]. The usual result is a regular pattern of rooms and pillars. Once the rooms are mined, the pillars can be mined, allowing the roof to collapse [15, 16, 19]. Block caving is a large-scale mining technique used to mine massive orebodies, involving a complex excavation system beneath the orebody. The orebody is drilled and blasted above the undercut, and ore is removed through the accessway [15, 16, 19]. As the ore is removed from the collection areas, subsequent caving provides continuous ore [19]. Extensive rock bolting and concrete lining are required to keep the openings intact, and if caving stops and removal of ore continues, a large void may form that can have the potential for sudden and massive collapse.

2.2.3.3 Combined Underground and Open-Pit Mining

There are instances in which near-surface deposits extend to considerable depth. The extraction of such ore deposits involves a few common steps for open-pit and underground mining. Depending on the chosen strategy, initial ore extraction can occur underground or in an open pit, followed by transportation to the processing facility [7, 8, 11]. In this manner, the most acute problem is determining the optimal transition depth between open-pit and underground mining, or vice versa. In most cases, this depth is calculated based on the block economic values of open-pit and underground methods and the net present value (NPV) attained through mining [19–22]. After an underground area has been mined, it is often necessary to backfill it with waste material. This can occur immediately or be delayed until the stope is thoroughly mined out.

2.2.3.4 In Situ Leaching Mining

In situ leaching (ISL), also known as solution mining or in situ recovery (ISR) mining, extracts uranium from the ore-bearing minerals by dissolving with a leaching solution underground and then pumping pregnant solution to the surface for uranium recovery in much the same way as in any other uranium processing plant (Fig. 2.4). The general process involves: (1) injection of a solution, lixiviant (typically containing water mixed with oxygen and hydrogen peroxide, as well as sodium carbonate or carbon dioxide) through a series of wells into the orebody to dissolve the uranium; (2) pumping the pregnant liquor of lixiviant and uranium in a series of recovery wells to a processing plant, where the uranium is extracted from the solution through an ion-exchange process; and (3) further purifying, concentrating, and drying to produce a material, which is called "yellowcake" because of its yellowish

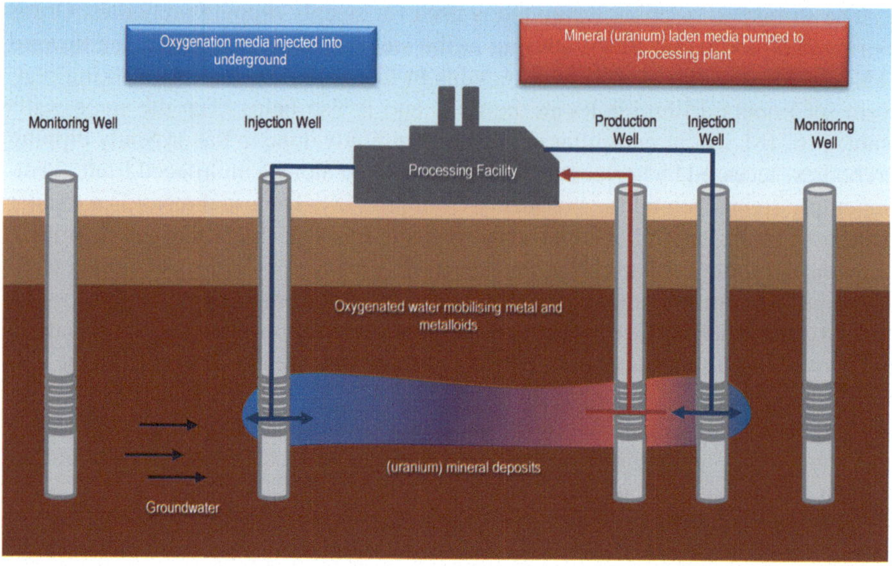

Fig. 2.4 Generalized in situ leaching (ISL) mining technology is often used in uranium extraction. The extracted media can be acidic or alkaline, depending on the deposits' geological background. Initially, native groundwater is extracted from the host aquifer, followed by injection of uranium-complexing reagents (acid or alkaline) and an oxidant (hydrogen peroxide or oxygen) into the wellfield. The leach liquors pass through the ore to oxidize and dissolve the uranium minerals in situ. Modified from USEPA [23]

colour. The yellowcake is sent to a uranium conversion facility for processing into fuel for nuclear power reactors and other uses. Chapter 6 details ISL mining and related issues in ISL-abandoned mines and processing sites.

The design of ISL wellfields varies widely depending on local conditions, including permeability, sand thickness, deposit type, ore grade, and distribution [24, 25]. The pattern consists of injection wells that introduce leach solution into the orebody and extraction wells with submersible pumps that deliver a pregnant solution to the processing plant. Wells are typical of standard water bores [24, 26]. Due to the low capital costs of conventional mining, ISL can often be more effective in mining low-grade uranium deposits.

2.2.4 *Non-conventional Uranium Extraction Techniques*

2.2.4.1 Heap Leaching

When the ore is very low-grade (typically below 0.1% U), heap leaching is used to recover uranium from the ore-bearing rock. The techniques involve stacking crushed ore into a heap, usually 5–30 metres high, on an impermeable pad, and irrigating it with an extracting solution to leach uranium over many weeks to months. The solution can be either acidic or alkaline. Most ore is barren rock or other minerals that remain undissolved during leaching. These reliable waters, known as tailings, are separated from the uranium-rich solution, usually allowing them to settle out. Then, the pregnant liquor is collected and treated to recover uranium, as in ISL, typically using an ion-exchange or solvent-extraction system. Recovery rates are generally 50–80%. After the material ceases to yield significant uranium, the depleted material can pollute, so it is usually securely emplaced back into mined-out pits to affect both surface and groundwater. The advantage of heap leaching is that the ore does not need to be finely ground, water consumption is low, and remediation is simplified, avoiding tailings impoundment.

2.2.4.2 Byproduct Uranium Recovery

Uranium ore may sometimes coexist in low quantities with another precious mineral. In this case, uranium recovery may occur when another metal (e.g., gold, copper, or nickel) is the primary product, with uranium recovered as a minor byproduct. The objective may be to recover uranium for its own sake or to remove it for product purity or environmental reasons, such as in the production of phosphoric acid fertilizer or copper [27, 28].

2.2.4.3 Secondary Sources of Uranium

Obtaining uranium from secondary sources is recycling uranium by recovering it, for example, from the decommissioning of nuclear warheads. Since 2000, this ex-military material has accounted for 13% of global uranium requirements. Other uranium sources include government and utility stockpiles, as well as depleted uranium from historical enrichment, which has been re-enriched using more efficient processes. Finally, a little comes from recycled uranium from reprocessing used fuel [29].

2.2.4.4 Uranium Extraction from Seawater

The world's oceans contain 4.5 billion tons of uranium, over a thousand times more than the terrestrial supply. However, the uranium concentration of approximately 3.3 parts per billion in the ocean is too low to be economically viable unless very efficient extraction technology for recovering uranium from seawater could be developed [30, 31]. Additionally, seawater contains trace amounts of almost every naturally occurring element, so high selectivity is necessary for efficient extraction. Nevertheless, interest in extracting uranium from seawater has been growing since the 1960s, as the world's oceans could provide a nearly limitless fuel source for nuclear reactors. Further, oceanic extraction would present fewer perceived environmental problems than terrestrial mining, including no toxic mining waste, less machinery and pollution, and fewer uranium processing steps [32].

Initially, the uranium extraction from seawater was based on phosphates and other uranyl extractants used in a liquid-liquid extraction ion-exchange method—the PUREX process. However, the most advanced and promising method of extracting and concentrating uranium from seawater involves polymeric sorbents containing the amidoxime binding moiety, such as glutarimide-dioxime. So far, these sorbents have shown success at the pilot plant scale. The extraction process is based on coordination chemistry, exploiting the functional groups on typical polymer sorbents and small-molecule analogues of these ligands (Fig. 2.5).

The affinity of uranium dictates the whole process towards the amidoxime substrates, which depends on the aqueous chemical equilibria of uranium (as UO_2^{2+}) in seawater. At the chemical equilibria, UO_2^{2+} in seawater likely forms stable complexes, $M_m(UO_2)(CO_3O3)_3^{(2m-4)}$ (aq) (where M = Ca or Mg and m = 0–2). These binary and ternary species dominate the chemistry of uranium in seawater. Apart from uranium, amidoximes are used as chelator ligands for several other metals. Consequently, selectivity for uranium over other metals, particularly vanadium, remains problematic [32]. However, moieties like glutarimide-dioxime form stable complexes with UO_2^{2+} · even in carbonate. Other materials that have extracted uranium well from seawater include hydrogels, nanostructured ceramics, porous aromatic frameworks, chitin-based materials, computationally optimized proteins, and combinatorial peptoid-based ligands.

2.2.5 *Ore Processing*

Once the uranium ore-bearing rocks have been excavated in the mine, they are processed to obtain concentrated uranium, known as yellowcake (U_3O_8), with 75–85% insufficiently high purity. A generalized flow diagram for uranium processing is shown in Fig. 2.6. The first step in uranium ore processing involves mechanical crushing and grinding the rock containing uranium ore. The ground materials are then subjected to a hydrometallurgical process using chemicals and solutions to extract uranium from the ore matrix. In the case of high-grade ore (i.e., uranium content >0.10%), the uranium is extracted from the resulting liquor using organic solvents or ion-exchange resins. Then, the extracts are washed and filtered, precipitated, and dried to produce yellowcake. For low-grade ores (< 0.10%), ore pretreatment or beneficiation is performed immediately after crushing the rock into sizable particles to reduce the amount of ore requiring chemical treatment.

The particles are aggregated in an agglomerator using water and acid to enhance the permeability and stability of the heaps. The ore is heaped on the leach pads, where an acid solution percolates through the heap for about 3 months. The uranium-bearing solution drains from the heap and is collected, from which uranium is extracted using a solvent. Generally, very few operations have used beneficiation

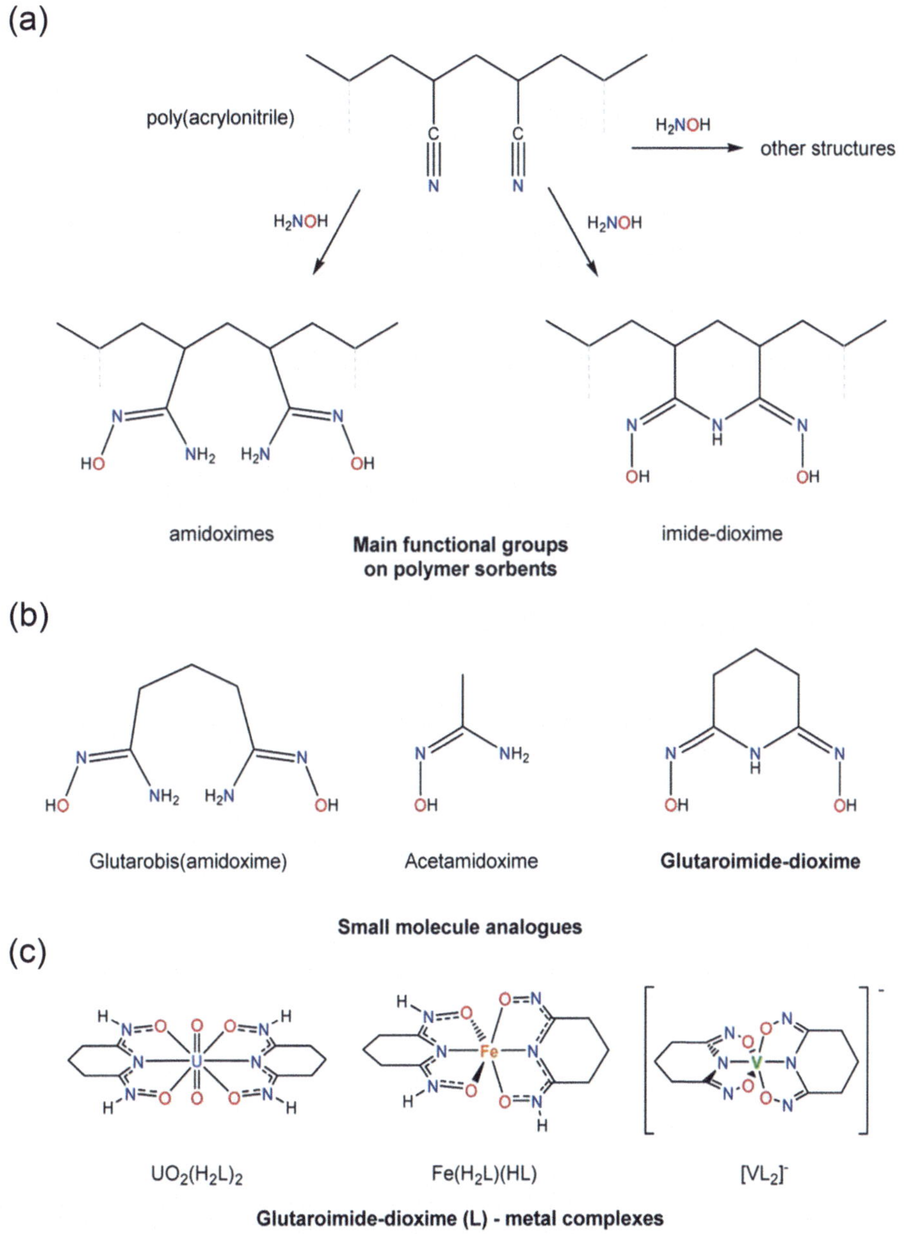

Fig. 2.5 (**a**) Synthesis and structure of polymer sorbents used for uranium extraction from seawater, (**b**) small-molecule analogues used as ligands for solution studies, notably glutarimide-dioxime, and (**c**) metal complexes of glutarimide-dioxime [33]

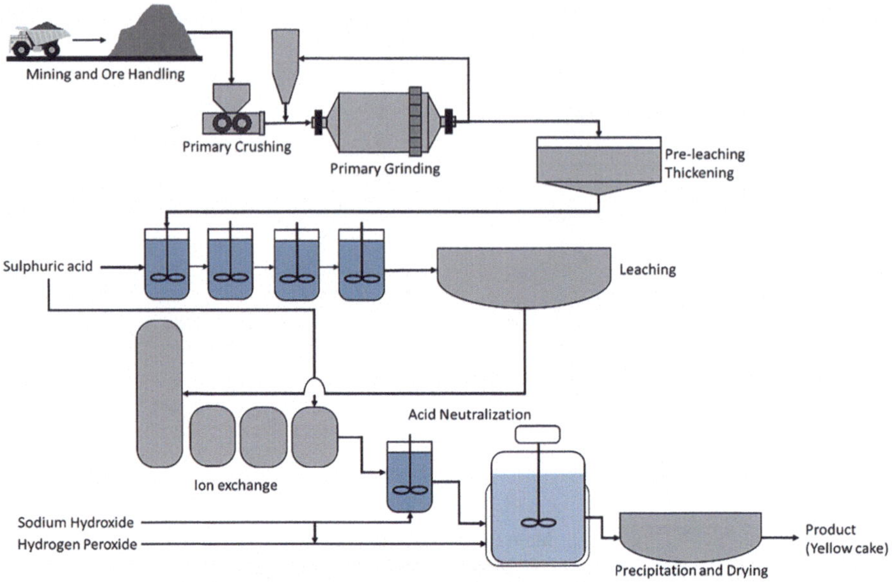

Fig. 2.6 The uranium processing flow diagram shows the unit process steps, from ore produced by an open-pit or underground mine to yellowcake production

processes that concentrate uranium by removing gangue constituents, because the value of the uranium losses is often higher than the cost of processing the whole ore.

In brief, open-pit and underground uranium mining involve physically excavating ore from the earth, followed by crushing, grinding, and chemical processing to extract uranium. In contrast, in situ leaching (ISL), as illustrated in Fig. 2.7, involves injecting a leaching solution directly into underground uranium-bearing formations through wells, dissolving the uranium in place. The uranium-laden solution is then pumped to the surface for processing. Unlike conventional mining, ISL minimizes surface disturbance, reduces the volume of tailings, and reduces worker exposure to radiation.

2.2.6 *Ore Extraction Chemistry*

The uranium extraction chemistry is summarized in Fig. 2.8. When the crushed and ground ore, or the underground ore in ISL mining, is leached with sulfuric acid, the ore's UO_2 is oxidized to UO_3. The UO_3 reacts further to form the uranyl ion UO_2^{2+}, as follows:

$$UO_3 + 2H^+ \rightarrow UO_2^{2+} \tag{2.1}$$

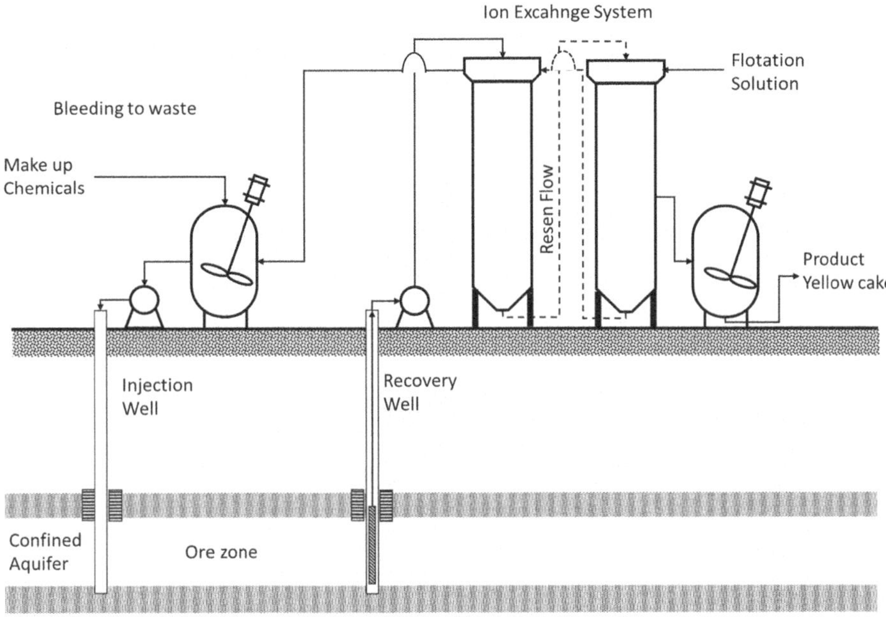

Fig. 2.7 Generalized flow chart for in situ leaching operations

$$UO_2^{2+} + 3SO_4^{2-} \rightarrow UO_2\left(SO_4\right)_3^{4-} \tag{2.2}$$

When carbonate leaching is used, a soluble uranyl tricarbonate ion ($UO_2\left(CO_3\right)_3^{4-}$) is formed:

$$UO_2^{2+} + 3CO_3^{2-} \rightarrow UO_2\left(CO_3\right)_3^{4-} \tag{2.3}$$

This can then be precipitated with an alkali (e.g., sodium or magnesium diuranate). Next, the uranium in the solution is recovered using either a resin/polymer ion-exchange or a liquid ion-exchange system. Next, the uranium is stripped from the resin/polymer with a strong acid, chloride, or nitrate solution. The pregnant solution produced by the stripping cycle is then precipitated by adding ammonia, hydrogen peroxide, caustic soda, or caustic magnesia.

For solvent extraction, tertiary amines are used in a kerosene diluent, and the phases move counter-currently:

$$2R_3N + H_2SO_4 \rightarrow \left(R_3NH\right)_2 SO_4$$

$$2\left(R_3NH\right)_2 SO_4 + UO_2\left(SO_4\right)_3^{4-} \rightarrow \left(R_3NH\right)_4 UO_2\left(SO_4\right)_3 + 2SO_4^{2-}$$

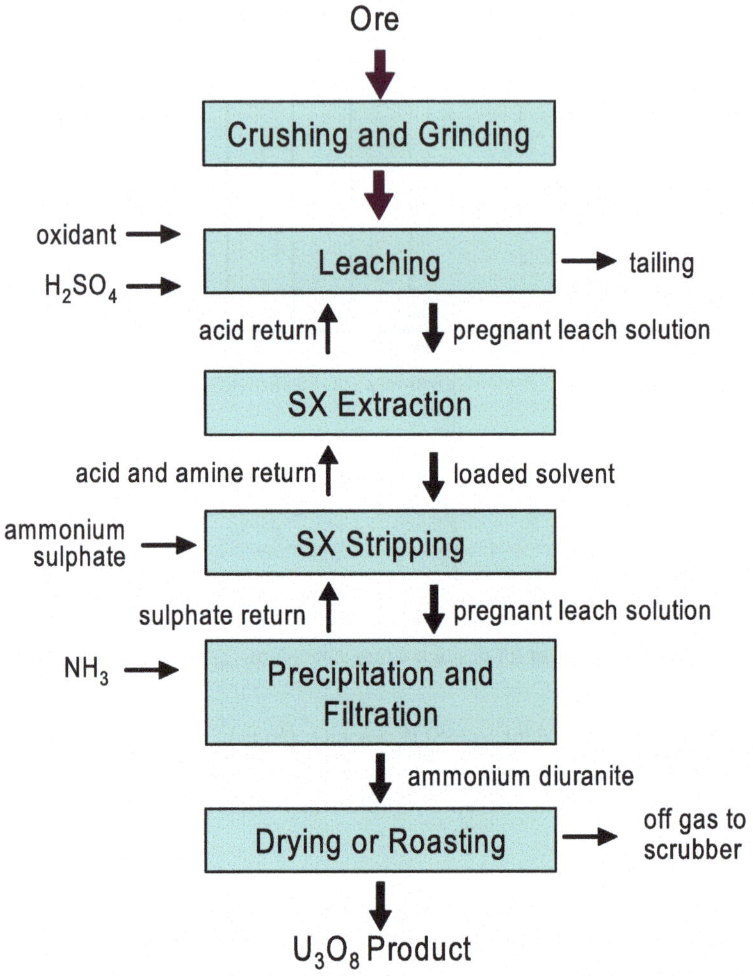

Fig. 2.8 The uranium extraction chemistry from ore mining to yellowcake production. In ISL mills, uranium recovery is very similar, without the need for crushing and grinding [34]

"R" is an alkyl (hydrocarbon) grouping with a single covalent bond. The loaded solvents may then be treated to remove impurities. First, cations are removed at pH 1.5 with sulfuric acid, and then anions are removed with gaseous ammonia. The solvents are then stripped in a countercurrent process using ammonium sulfate solution:

$$\left(R_3NH\right)_4 UO_2\left(SO_4\right)_3 + 2\left(R_3NH\right)_2 SO_4 \rightarrow 4R_3N + \left(NH_4\right)_4 UO_2\left(SO_4\right)_3 + 2H_2SO_4$$

Precipitation of ammonium diuranate is achieved by adding gaseous ammonia to neutralize the solution (though in earlier operations, caustic soda and magnesia were used):

$$2NH_3 + 2UO_2\left(SO_4\right)_3^{4-} \rightarrow \left(NH_4\right)_2 U_2O_7 + 4SO_4^{2-}$$

The diuranate is then dewatered and roasted to yield U_3O_8, the form in which uranium is marketed and exported.

Peroxide products can be dried at ambient temperatures to produce a product containing about 80% U_3O_8. Ammonium or sodium diuranate products are dried at high temperatures to convert them into uranium oxide concentrate (U_3O_8), the yellowcake, which is usually khaki and contains about 85% uranium by mass [34, 35]. The U_3O_8 is only mildly radioactive.

2.2.7 Accompanying Non-ore Extraction Processes

Accompanying main mining and processing activities are water treatment, waste rock and tailings disposal, and processing effluents. Generally, environmental and hydrological conditions, including the depth of a mine, whether underground or open-pit, determine the need for removing water—mine dewatering—to keep the groundwater level low and the mine workable without submersion. In addition, ore extraction leaves a large amount of waste rock because the uranium content in the rocks is very low. Further, after uranium recovery in processing, solid waste is called tailings, consisting of everything in the ore except the extracted uranium. Thus, tailings contain the primary radioactive materials (e.g., 230th and ^{226}Ra) and toxic metals (e.g., $^{207,\ 208,\ \text{and}\ 210}Pb$) that are remnants of the uranium decay series.

Further, toxic metalloids, such as arsenic, are prevalent in uranium tailings depending on the uranium ore type. Moreover, liquid effluents used in uranium processing and extraction, mainly from open-cast and underground mining, are usually highly acidic and contaminated, as are the tailings. Hence, the process effluents require a distinct management approach as well.

2.2.7.1 Water Treatment from the Mine Dewatering Process

The water removed to keep the mine dry requires treatment and exceptional management before use. Modern uranium mining and processing plants recycle the water from mines, while excess water that cannot be recycled must be treated to meet regulatory requirements before being released to the receiving environment. Water treatment from the mine dewatering process depends on the uranium recovery process, chemicals used, and ore contaminants. The water treatment process is multistep, neutralizing effluents, immobilizing metals, and diminishing uranium and radionuclide content. In some cases, contaminated mine water may contain minor metals that require additional treatment technologies, such as reverse osmosis or nanofiltration. After treatment, the water is discharged, recycled to plant operations, or sent for further treatment. Water management within a mining project starts with characterizing all potential water sources, their possible uses, and potential contamination issues. This includes a site water balance analysis, a plant water balance analysis that assesses water flows and quality, and identifies water recycling options.

2.2.8 Mine Closure Process

Most resource extraction industries have an end when the resources are depleted. Thus, during ore prospection, it is known that the mine will close at a given time, and in most countries, it is obligatory to plan the mine closure. The mine closure process consists of four main steps. Figure 2.9 is a flow diagram of the process involved in a mine closure. The first step is a shutdown, which stops production and massively reduces the labour force, retaining only a few to permanently shut down the mining equipment. The second step is decommissioning, which involves removing the mining processing facilities and equipment, draining pipelines, and repurposing or demolishing buildings. The third step is remediation/reclamation, which aims to return the land and watercourses to an acceptable standard of productive use, ensuring that any landforms and structures are stable and any watercourses are of acceptable water quality. Finally, post-closure monitoring programs are instituted to assess the effectiveness of reclamation measures and identify any corrective actions where needed.

Mine closure planning has two distinct components, namely: (a) On-site planning, which is mostly about environmental rehabilitation and returning the landform to a reasonable condition for alternative land use, and (b) Off-mine closure plan,

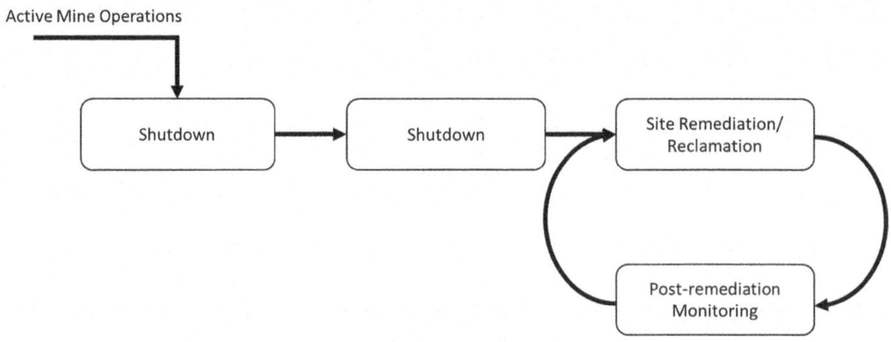

Fig. 2.9 An ideal mine closure process consists of proper planning for shutdown, followed by actual decommissioning, reclamation, and repurposing of the mine for other land use. The sites should be continuously monitored, and remediation revisited and redesigned as long as risks exist

which looks at the socio-economic issues of the post-mining era, mostly about the loss of livelihood. Also, mines require long-term care and maintenance after mine closure, including ongoing treatment of mine water discharge, periodic monitoring and maintenance of tailings containment structures, and ongoing monitoring of any remediation technologies, such as constructed wetlands.

2.2.9 Reasons for Abandonment

Factors which contribute to the abandonment of a mine include economic events such as a decrease in the price of minerals or an unforeseen disaster leading the mining company to bankruptcy and lack of finances, making it unable to complete the proper mine closure process; preference to pay creditors to the restoration of the mining sites by liquidators during mining company bankruptcy; and small-scale mining operations by artisanal or illegal miners. The concept of mine closure planning is relatively new to the mining industry, introduced in the 1970s. Before the 1970s, there were few requirements to clean up mine sites due to a lack of reclamation policies or regulations. Nowadays, most countries have developed mine reclamation policies and regulations, but enforcement may be ineffective or penalties for non-compliance are minimal. This is common in developing countries with insufficient law enforcement resources. On the one hand, regulations and policies are deliberately ignored to attract foreign investment in the mining sector to support economic development. Some abandoned mines are a result of political unrest, social conflict, or political instability, leading to the unscheduled halt of mining operations.

2.3 Legacy of Abandoned Uranium Mines

It is common to see that most uranium mining sites worldwide are abandoned or orphaned. Usually, the owners are either nowhere to be found, financially unable, or unwilling to clean up the former mine site. A new paradigm leading to abandoned mines is that most uranium mining companies are moving their operations to developing countries while running away from stringent regulations and laws in developed countries. These mining companies are capitalizing on the desperation of most developing countries to woo foreign investment to improve their economic situations, while lacking the resources to enforce laws and regulations. One common incentive is to give tax and environmental act compliance holidays. Sometimes, developing countries may have robust laws and regulations but lack the resources to enforce them. Consequently, the uranium mining companies may leave the countries without proper closure. Incidents of corruption and war are contributing to the situation.

2.3.1 Factors Defining Abandoned Uranium Mining Sites

The general characteristics of abandoned uranium mines are, inter alia, left infrastructure (buildings and open mine shafts), disturbed landscapes, waste rock and tailings heaps, unaddressed chemical and radiation contamination, and, in some cases, the formation of acid-rock drainage. Some abandoned uranium mines were reclaimed to the standards of the day but were still considered abandoned. Sometimes, the methods are inadequate to prevent environmental contamination and do not conform to modern remediation and reclamation standards. However, the specific characteristics of uranium-abandoned mines depend on the mining and processing methods used for ore extraction, site hydrogeological and geological conditions, and weather conditions (see Fig. 2.10). Nevertheless, all attributes of abandoned uranium mines contribute to the legacy of safety, health, and environmental hazards.

2.3.2 Issues of Abandoned Uranium Mines

With their associated physical, environmental, and public safety concerns, abandoned mines are a legacy of past uranium mining operations. Generally, the expected environmental consequences associated with abandoned mine sites include altered landscape, new pits, shafts and adits, land no longer useable due to loss of soil or soil contamination, spoil heaps covering the land, abandoned tailings disposal facilities, contaminated aquatic sediments, subsidence, derelict work sites with compacted and polluted soil, burning coal waste dumps and workings, and changes in vegetation [36–38].

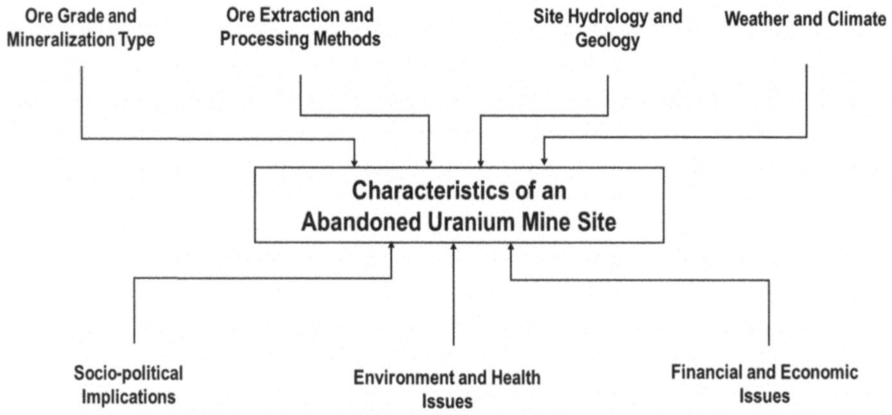

Fig. 2.10 Factors and issues that characterize the abandoned uranium mine and processing sites

Specifically, abandoned mines can pose health, safety, financial, and environmental liabilities for communities, the mining industry, and governments. These liabilities are passed on to current stakeholders by former mine owners, who may no longer exist. Further, the abandoned uranium mining site may pose severe risks due to the gradual deterioration of its features, including tailings storage facilities and underground workings.

Nevertheless, mining legacies can also be positive, such as providing clean drinking water from dewatering operations, training opportunities for workers, reusing mining infrastructure, or investing in the local economy [39]. Consequently, this section discusses the issues that characterize abandoned uranium mining sites. The discussion is highly generalized, as abandoned uranium mines vary in size and danger, ranging from shafts and test pits to large, extensive sites.

2.3.2.1 Physical Issues

Although the physical issues related to abandoned uranium mines are site-specific, common public health and safety hazards transcend the geopolitical spheres and sites. Here, some of the most common issues which can include abandoned infrastructure, accessible mine openings (i.e., unblocked pits, shafts, and tunnels), hazardous mine wastes, and ground surface instability are discussed below:

Abandoned Infrastructure

When a mining site is abandoned, most of the infrastructure is left as it was during operations. Similarly, old equipment and mining machines are left on the site without being disassembled. Like other abandoned mines, uranium mines pose physical hazards, including mine shafts and vertical entries (aka adits). These adits have some physical concerns in abandoned uranium mines, including visual impacts and stability issues. For instance, accidents related to vertical openings or deteriorating structures are the most common cause of death and injury in abandoned underground uranium mines. In underground mines, lethal concentrations of explosive and toxic gases such as methane, carbon monoxide, and hydrogen sulfide can accumulate in underground passages. Rockfalls and cave-ins from adits or pit walls can be a safety hazard. Unsafe structures include support timbers, ladders, cabins, and other related features. These may seem safe, but they may easily crumble under a person's weight due to weathering. In most cases, abandoned mines flood, and shallow water can conceal hazards, such as tunnels and sharp objects.

Disturbed Landscapes

Except for ISL, most abandoned uranium mining and processing sites are also known as landscapes and are left disturbed, with mine wastes and tailings unconsolidated into well-protected heaps [40, 41]. Numerous small waste stockpiles and borrow pits, adits, and shafts are left unattended. Although such scenarios are declining in modern uranium mining, abandoned mines with disturbed land and unprotected slopes are susceptible to erosion due to uncontrolled surface drainage, which can remove soils and loose waste particles. They may destabilize large areas and transfer pollution to previously unpolluted environments. These physical impacts of abandoned mines, such as slope instability and contamination of soils by acid drainage and other metals released from waste piles, often result in the loss of productive land. Further, most abandoned mines are used for dumping industrial and household waste, both legally and illegally, exacerbating contamination at the sites.

Waste Rock and Tailings Piles

The uranium content of the ore is often between only 0.01% and 0.2%. Therefore, large amounts of ore must be mined to get the uranium, resulting in massive excavations and large heaps of waste rock. Depending on the geological background, rock heaps can be waste that may produce acid mine drainage (AMD) due to sulfide-containing minerals like pyrite, or they may be clean or stable waste placed on the surface without special consideration. Uranium tailings are a waste byproduct (tailings) of uranium mining. In mining, raw uranium ore is brought to the surface and crushed into fine sand. The valuable uranium-bearing minerals are then removed via heap leaching with acids or bases, and the remaining radioactive sludge, called "uranium tailings," is stored in massive impoundments. A tonne (907 kg) of ore yields one to five pounds (0.45–2.3 kg) of uranium, depending on the uranium content of the mineral. Uranium tailings can retain up to 85% of the ore's original radioactivity [42–44].

Pit Lakes

Large-scale open-pit uranium mining has left a permanent legacy on the landscape, creating thousands of lakes by flooding abandoned open-cut mines. Many of these lakes are toxic and radioactive, posing risks to adjacent communities and ecosystems [7–9, 45, 46, 47]. Pit lakes form when open-pit mining operations are discontinued, and dewatering ceases. The remaining pit fills with ground, surface, and rainwater [7, 8, 46]. These are mine pits; thus, pit lakes have high depth-to-surface-area ratios with relatively flat bottoms and steep sides. The morphologies of most modern mine pit lakes resemble those of asteroid or crater lakes rather than co-occurring lakes, which tend to be shallower and more nutrient-rich [7, 48].

2.3.2.2 Environment and Health Issues

Environmental and health concerns in abandoned uranium mines are generally of two types: chemical pollution and radiation. Typical environmental problems stem, among other things, from uranium radionuclide and metal contamination of surface and groundwater, stockpiled waste rock and mill tailings piles, and compost.

Chemical Pollution Issues

Water is the most frequently contaminated resource from abandoned uranium mines, and the main conduit of contamination impacts the abandoned site—it is the main exposure pathway for pollutants. In addition, elevated levels of metals, suspended sediment, acidity, hydrocarbons, and brine leaching threaten surface and underground water quality and aquatic habitats.

1. Acid-Rock Drainage or Mine Drainage (ARD/AMD): The most severe environmental concerns at abandoned mines are acid mine drainage (AMD) from underground workings, open-pit mine faces, waste rock piles, and tailings storage areas that were left exposed to the elements or inadequately reclaimed, resulting in the contamination of water with dissolved metals and acidity. AMD occurs when surface or groundwater flows from or over abandoned mine features that contain sulfide mineralization. Discharge from adits or open pits, and surface flow over and seepage through sulfide-rich waste rock and tailings, can produce acid drainage. Acid drainage begins when iron sulfide materials are exposed to air and water. The exposed, relatively insoluble sulfide materials are oxidized to soluble sulfuric acid and iron compounds. In turn, sulfuric acid dissolves other metals, such as aluminium, copper, zinc, and cadmium. Although these constituents can occur naturally in water at trace levels, their concentration can increase substantially through acid drainage driven by hydrologic and weathering processes.

   Generally, Fe-Mn (oxy)hydroxides and silicate minerals are significant uranium-hosting minerals, and a substantial fraction of uranium exists as adsorbed ions on minerals. Thus, during AMD formation, uranium is quickly released and controlled by multiple interactive processes, including the dissolution of waste rock or tailings containing remnants of U-bearing minerals, uranium desorption from mineral surfaces, the formation of aqueous uranium complexes, and reductive precipitation of uranium. Consequently, abandoned uranium mines can produce AMD for over 100 years, posing significant risks to surface and groundwater. In addition, AMD lowers the pH of the surrounding surface water, making it corrosive and unable to support many aquatic organisms and vegetation. Humans may also be affected by consuming water and food containing metal.
2. Mine Waste and Tailing Leachates: The remnants of materials left behind after the desired uranium is extracted are often piled on-site while a mine is in opera-

tion and left behind when the mine is abandoned. Most of the mine waste tailings may be inert rock, just as the raw ore was before uranium extraction. However, physically altering, grinding up, mixing with large amounts of water, and exposing to atmospheric oxygen substantially alter chemical behaviour. When rain runs off the piles, these mine waste and tailings leach harmful toxins like lead, mercury, and arsenic into the nearby receiving environment. Further, most uranium mining has occurred in extensive mineralized areas, and natural runoff may contain elevated levels of metals, metalloids, and radionuclides.

3. Pit-Lake Outflows: The water quality varies among the abandoned mining pit lakes, a function of surrounding geology and catchment interactions, including connections to groundwater and rivers, riparian vegetation, and mine discharge. Thus, open pits, which remain after uranium mining operations cease, can form meromictic lakes that develop conditions suitable for the containment of dissolved and colloidal metals, spanning the full spectrum from alkaline to acidic, fresh to saline, and toxic to non-toxic [46, 49, 50]. Most uranium ore-bearing sites are rich in sulfides, particularly pyrite (FeS_2); when mining exposes these materials to water and air, oxidation accelerates, producing acidity that leaches metals from surrounding rocks, creating acid mine drainage. Thus, due to their size and potential interaction with the broader receiving environment through groundwater and surface-water outflows, abandoned uranium mining pit lakes can substantially impact the environment.

Radiation Exposure Issues

Abandoned uranium mines pose the added threat of radiation exposure to the list of health concerns [51–53]. Mining uranium ores by underground and surface methods produces large and small amounts of bulk waste material containing naturally occurring radioactive components. It becomes exposed to the environment through mining activities [54].

Uranium minerals are always associated with other radioactive elements in the ore, such as radium and Radon, which arise from the radioactive decay of uranium in the ground over millions of years. Thus, although uranium is barely radioactive, the ore, especially if it is very high-grade, poses significant radiation hazards in abandoned uranium mines. About 95% of the ore radioactivity is from the ^{238}U decay series, totalling about 150 kBq/kg in ore with 0.1% U_3O_8. The ^{238}U series has 14 radioactive isotopes in secular equilibrium. Thus, each represents about 11 kBq/kg, irrespective of the mass proportion [55–57].

When the ore is processed, the ^{238}U and the smaller masses of ^{234}U (and ^{235}U) are removed. The balance becomes tailings and, at this point, has about 85% of its original intrinsic radioactivity. With the removal of most ^{238}U, two short-lived decay products in the uranium decay series, namely 234th and ^{234}Pa, soon disappear, leaving the tailings with a little over 70% of the radioactivity of the original ore after several months. 230th, which decays with a half-life of 77,000 years to ^{226}Ra, which then decays to ^{222}Ra, becomes the controlling long-lived isotope in most abandoned uranium mines.

When radium undergoes natural radioactive decay, one of the products is radon gas. Since radon and its decay products (daughters) are radioactive, the tailings still pose a risk of emitting radon gas and causing poisoning. Therefore, during the operational life, water often covers the tailings dam material to reduce surface radioactivity and radon emission [58, 59]. On completion of the mining operation, it is normal for the tailings dam to be covered by two metres of clay and topsoil with enough rock to resist erosion. This reduces gamma radiation and radon emanation rates to levels near those typically experienced in the orebody region and vegetation cover [60, 61].

If uranium tailings are stored aboveground and allowed to dry out, the radioactive sand can be carried great distances by the wind, entering the food chain and water bodies. The danger posed by such sand dispersal is uncertain at best, given its dilution effect. Tailings are typically neutralized with lime and other additives, such as barium chloride, to stabilize radium content and prevent metal contaminants from causing environmental contamination [62–66]. Then, they are thickened to reduce water content and deposited in an impoundment facility. One concern with tailings impoundments is the potential for radon gas release.

Therefore, radiation risks remain high in most abandoned mines, are long-term, and are not dependent on uranium availability. This implies that abandoned uranium mines emit substantial amounts of radon gas, a natural decay product of radium, a leading source of ionizing radiation. A colourless, odourless gas is formed from radium's radioactive decay; radon emits alpha particles during decay. It is not radon itself that is a health concern, but the radioactive products it decays into. Being a gas, radon is simply the vehicle by which members of the uranium decay chain can enter the lungs. Radon is not a concern outside the body since the alpha particles it emits cannot penetrate the skin.

2.3.2.3 Socio-Economic Issue

As in other mining industries, the primary socio-economic concern with abandoned uranium mines is the loss of employment and business activities due to unscheduled closures. However, the other socio-economic concerns arising from abandoned mines mostly stem from physical and environmental problems. These include the safety hazards caused by abandoned mines that usually result in losing lives.

Contamination of both the ground and surface waters by abandoned uranium mines impacts the aquatic systems. Contamination may be due to radionuclide, metal, and metalloid releases resulting from acid mine drainage, waste, and tailing material leaching. Acid drainage from mines alters pH, driving away or killing aquatic life. A manifestation of acid drainage is ochre, also known as a yellow boy, an orange-coloured slime caused by metals precipitation, often smothering aquatic life. Thus, community livelihoods that depend on fishing and other water resources are disturbed, leading to immigration. Furthermore, restrictions on legitimate water users may also find the water unsuitable for irrigation, livestock watering, and industrial or potable water supply.

Another potential socio-economic impact from abandoned uranium mines is radiation and dust from the mine waste and tailings. Thus, populations are usually dispersed due to chemical and radiation exposure epidemiology. Further, the disturbed landscape is so ugly that no tourists are attracted to it. Although not common among abandoned uranium mines, mines abandoned by a company may become targets for legal or illegal mining by small-scale miners [67–69]. In some cases, this involves expanding the previous mine workings; in others, it consists of reworking the waste materials to extract any remaining metal. In some cases, miners may use hazardous processing chemicals to extract metals, harming their health and the environment. This uncontrolled mining can affect ecosystems and water resources.

2.3.2.4 Financial Implications

In most cases, owners of abandoned mines cannot be traced, or the companies that operated the mines are liquidated immediately after operations close. Consequently, public funds are required to rehabilitate the abandoned mine sites. Further, it becomes evident that governments are forced to rehabilitate abandoned uranium mines when there are no identifiable owners or the owners have no funds to pay for rehabilitation. Further, responsibility for the rehabilitation work and the sites' long-term care ultimately lies with the government. Consequently, abandoned mines have financial implications for the public. Further, economic repercussions lead to financial losses.

2.4 Conclusion

This chapter has examined abandoned uranium mines in the most general terms. At the same time, it fully recognized that the condition and impact of each abandoned uranium site are site- and geopolitically specific. Hence, sometimes, these discussions on abandoned uranium may diverge from reality on the ground in some countries, while the issues and concerns are shared and similar at a high level.

Conventional mining involves removing mineralized rock (ore) from the ground, breaking it up, and treating it to extract the desired minerals. Uranium mining poses special issues because of the associated radiation and chemical risks. Uranium itself is only slightly radioactive. However, radon, an inert radioactive gas, is released into the atmosphere in small amounts during mining and crushing. Radon occurs naturally in most rocks—minute traces of it are present in the air, which we all breathe, and it is a significant contributor to the natural radiation dose we all receive.

Thus, as uranium is not much in waste and tailings, the daughter elements and secondary contaminants may be a greater concern. Further, uranium's ore content in rocks is generally very low, and uranium mining generates substantial waste. In situ leaching (ISL), also known as solution mining or in situ recovery (ISR) in North America, involves leaving the ore in place and recovering the minerals by dissolving them and pumping the pregnant solution to the surface for recovery. Consequently, there is little surface disturbance and no tailings or waste rock generated. However, the orebody needs to be permeable to the liquids used and not contaminate groundwater away from the orebody.

Thus, abandoned uranium mines will continue to impact the environment and the communities in which they are located unless acceptable mine closure practices are implemented.

References

1. Environment, N. (Ed.). (2009). *Environmental assessment branch policy and corporate services division* (Vol. 17). NS Environment.
2. Kivinen, S. (2017). Sustainable post-mining land use: Are closed metal mines abandoned or re-used space? *Sustainability, 9*, 1705.
3. Limpitlaw, D., & Briel, A. (2014). Post-mining land use opportunities in developing countries - a review. *Journal of the Southern African Institute of Mining and Metallurgy, 114*, 899–903.
4. Kim, S. S., et al. (2016). Advanced remediation of uranium-contaminated soil. *Journal of Environmental Radioactivity, 164*, 239–244. https://doi.org/10.1016/j.jenvrad.2016.08.005
5. Seidel, D. C. (1981). Extracting uranium from its ores. *IAEA Bulletin, 23*, 24–28.
6. Coral, T., et al. (2018). Microbial communities associated with uranium in-situ recovery mining process are related to acid mine drainage assemblages. *Science of the Total Environment, 628-629*, 26–35. https://doi.org/10.1016/j.scitotenv.2018.01.321
7. Blanchette, M. L., & Lund, M. A. (2016). Pit lakes are a global legacy of mining: An integrated approach to achieving sustainable ecosystems and value for communities. *Current Opinion in Environmental Sustainability, 23*, 28–34. https://doi.org/10.1016/j.cosust.2016.11.012
8. Castro, J. M., & Moore, J. N. (2000). Pit lakes: Their characteristics and the potential for their remediation. *Environmental Geology, 39*, 1254–1260. https://doi.org/10.1007/s002549900100
9. Strømman, G., et al. (2013). Uranium activity ratio in water and fish from pit lakes in Kurday, Kazakhstan and Taboshar, Tajikistan. *Journal of Environmental Radioactivity, 123*, 71–81. https://doi.org/10.1016/j.jenvrad.2012.05.014
10. Poniewierski, J.. (2018). (ed Deswik) 16 (Deswik, Brisbane, Australia).
11. Hustrulid, W. A., & Kuchta, M. (2006). *Open pit mine planning and design*. CRC Press.
12. Senes, C. L. (2008). *Alberta Environment* (Vol. 57). Government of Alberta.
13. Committee on Uranium Mining in Virginia & Committee on Earth Resources National Research Council. (2012). *Uranium Mining in Virginia: Scientific, technical, environmental, human health and safety, and regulatory aspects of uranium mining and processing in Virginia* (Vol. 342). National Academy of Sciences.
14. NEA. (2014). In Nuclear Energy Agency (Ed.), *Organisation for economic co-operation and development* (Vol. NEA No. 7062). OECD.
15. Morin, M. A. (2001). *Underground Hardrock Mine Design and Planning— A System's Perspective* PhD thesis, Queen's University.
16. Brazil, M. et al. (2004). *Optimisation in the design of underground mine access.*
17. Vergne, J. D. L. (2008). *Hard rock miner's handbook.* Stantec Consulting Ltd.
18. Puhakka, T., & Tamrock. (1997). *Underground drilling and loading handbook.* Tamrock.
19. Harraz, H. (2010). *Underground mining methods.*
20. Diddle, B., Agioutantis, Z., Moradi, A., Roghanchi, P., & Schafrik, S. 59th U.S. rock mechanics/Geomechanics symposium.

21. Chen, H., et al. (2025). Uranium contamination mediating soil and ore microbial community assembly at four mining sites, South China. *Frontiers in Microbiology, 16*, 2025. https://doi.org/10.3389/fmicb.2025.1553072
22. Yusuf, I. (2024). *Uranium biokinetics and internal dosimetry*. Linköping University Electronic Press.
23. EPA. (2017). *2017 notice of proposed rulemaking* (Vol. 40 CFR Part 192 2). United States Environmental Protection Agency.
24. Li, G., & Yao, J. (2024). A review of in situ leaching (ISL) for uranium mining. *Mining, 4*, 120–148.
25. IAEA. (1989). *In situ leaching of uranium: Technical, environmental and economic aspects*. International Atomic Energy Agency.
26. Shen, N., Li, J., Guo, Y., & Li, X. (2020). Thermodynamic modeling of in situ leaching of sandstone-type uranium minerals. *Journal of Chemical & Engineering Data, 65*, 2017–2031. https://doi.org/10.1021/acs.jced.9b01152
27. IAEA. (2025). *Recovery of uranium from phosphate ores* (Vol. 41). International Atomic Energy Agency.
28. IAEA. (1993). *Uranium extraction technology* (Vol. 380). International Atomic Agency.
29. Awwad, N. (2021). In N. Awwad (Ed.), *Nuclear power plants—the processes from the cradle to the grave*. IntechOpen.
30. Guidez, J., & Gabriel, S. (2016). Extraction of uranium from seawater: A few facts. *EPJ Nuclear Science and Technology, 2*, 10.
31. Bardi, U. (2010). Extracting minerals from seawater: An Energy analysis. *Sustainability, 2*, 980.
32. Endrizzi, F., Leggett, C. J., & Rao, L. (2016). Scientific basis for efficient extraction of uranium from seawater. I: Understanding the chemical speciation of uranium under seawater conditions. *Industrial & Engineering Chemistry Research, 55*, 4249–4256. https://doi.org/10.1021/acs.iecr.5b03679
33. Parker, B. F., Zhang, Z., Rao, L., & Arnold, J. (2018). An overview and recent progress in the chemistry of uranium extraction from seawater. *Dalton Transactions, 47*, 639–644. https://doi.org/10.1039/C7DT04058J
34. Mkandawire, M. (2013). Biogeochemical behaviour and bioremediation of uranium in waters of abandoned mines. *Environmental Science and Pollution Research, 20*, 7740–7767. https://doi.org/10.1007/s11356-013-1486-3
35. Mkandawire, M., & Dudel, E. (2010). *Environmental behaviour of uranium in closed mining sites*.
36. IAEA. (1997). *Closeout of uranium mines and mills: A review of current practices*. International Atomic Energy Agency.
37. Rapantova, N., Licbinska, M., Babka, O., Grmela, A., & Pospisil, P. (2013). Impact of uranium mines closure and abandonment on groundwater quality. *Environmental Science and Pollution Research International, 20*, 7590–7602. https://doi.org/10.1007/s11356-012-1340-z
38. Shaoping, H., Xincai, C., Jiyan, S., Yingxu, C., & Qi, L. (2008). Particle-facilitated lead and arsenic transport in abandoned mine sites soil influenced by simulated acid rain. *Chemosphere, 71*, 2091–2097.
39. Brueckner, M., Spencer, R., Knowles, S., & Paull, M. (2021). Mining legacies—Broadening understandings of mining impacts. *The Extractive Industries and Society, 8*, 100950. https://doi.org/10.1016/j.exis.2021.100950
40. Hubbard, B. E., Gallegos, T. J., & Stengel, V. (2023). Mapping abandoned uranium mine features using Worldview-3 imagery in portions of Karnes, Atascosa and live oak counties, Texas. *Minerals, 13*, 839.
41. Schramm, L. (2012). Cleaning-up abandoned uranium mines in Saskatchewan's North. *Bulletin of the Canadian Nuclear Society, 33*, 17–23.
42. Lahrouch, F., et al. (2021). Uranium retention on iron oxyhydroxides in post-mining environmental conditions. *Chemosphere, 264*, 128473. https://doi.org/10.1016/j.chemosphere.2020.128473

43. Vriens, B., et al. (2018). Localized sulfide oxidation limited by oxygen supply in a full-scale waste-rock pile. *Vadose Zone Journal, 17*, 180119. https://doi.org/10.2136/vzj2018.06.0119
44. Lefebvre, R., Hockley, D., Smolensky, J., & Gélinas, P. (2001). Multiphase transfer processes in waste rock piles producing acid mine drainage: 1: Conceptual model and system characterization. *Journal of Contaminant Hydrology, 52*, 137–164. https://doi.org/10.1016/S0169-7722(01)00156-5
45. Miller, G. C., Lyons, W. B., & Davis, A. (1996). Peer reviewed: Understanding the water quality of pit lakes. *Environmental Science & Technology, 30*, 118A–123A. https://doi.org/10.1021/es9621354
46. von Gunten, K., et al. (2018). Biogeochemistry of U, Ni, and as in two meromictic pit lakes at the Cluff Lake uranium mine, northern Saskatchewan. *Canadian Journal of Earth Sciences, 55*, 463–474. https://doi.org/10.1139/cjes-2017-0149
47. (!!! INVALID CITATION !!! 9-11).
48. Cockell, C. S., & Lee, P. (2002). The biology of impact craters , a review. *Biological Reviews, 77*, 279–310. https://doi.org/10.1017/S146479310100584X
49. Von Gunten, K. (2019). *Biogeochemistry of meromictic pit lakes and permeable reactive barriers at the Cluff Lake uranium mine* PhD thesis, University of Alberta.
50. Oremland, R. S., et al. (2000). Bacterial dissimilatory reduction of arsenate and sulfate in meromictic mono Lake, California. *Geochimica et Cosmochimica Acta, 64*, 3073–3084.
51. Antunes, S. C., Pereira, R., & Gonçalves, F. (2007). Acute and chronic toxicity of effluent water from an abandoned uranium mine. *Archives of Environmental Contamination and Toxicology, 53*, 207–213.
52. Dudel, M. M. A. E. G. (2009). In G. H. Wolfe (Ed.), *Uranium: Compounds, isotopes and applications*. Nova Science Publishers.
53. Mkandawire, M. (2005). *Fate and effect of uranium and arsenic in surface waters of abandoned uranium mines: Studies with Lemna gibba L.* Shaker.
54. Beddow, H., Black, S., & Read, D. (2006). Naturally occurring radioactive material (NORM) from a former phosphoric acid processing plant. *Journal of Environmental Radioactivity, 86*, 289–312.
55. Association, W. N. (2025). *Nuclear fuel cycle*. World Nuclear Association. https://world--nuclear.org/
56. Association, W. N. (2025).World Nuclear Association London, United Kingdom. https://world-nuclear.org/
57. Noubactep, C., Sonnefeld, J., Merten, D., Heinrichs, T., & Sauter, M. (2006). *Effects of the presence of pyrite and carbonate minerals on the kinetics of the uranium release from a natural rock* (Vol. 270). Springer.
58. Lottermoser, B. G., & Ashley, P. M. (2005). Tailings dam seepage at the rehabilitated Mary Kathleen uranium mine, Australia. *Journal of Geochemical Exploration, 85*, 119–137.
59. Association, W.N. (2017). *Nuclear fuel cycle*. World Nuclear Association. https://world--nuclear.org/
60. IAEA. (2014). In International Atomic Energy Agency (Ed.), *Online* (Vol. NW-T-3.6). International Atomic Energy Agency.
61. NEA. (2014). In Nuclear Energy Agency (Ed.), *Online* (Vol. 7062). OECD Publishing.
62. Abdelouas, A. (2006). Uranium mill tailings: Geochemistry, mineralogy, and environmental impact. *Elements, 2*, 335–341. https://doi.org/10.2113/gselements.2.6.335
63. Webb, G. (2004). Warm sands: Uranium mill tailings policy in the atomic west. *Journal of the West, 43*, 95–95.
64. Chenoweth, W. L. (2004). Warm sands: Uranium mill tailings policy in the atomic west. *Technology and Culture, 45*, 866–868.
65. Landa, E. R. (2004). Uranium mill tailings: Nuclear waste and natural laboratory for geochemical and radioecological investigations. *Journal of Environmental Radioactivity, 77*, 1–27.
66. Landa, E. R. (1992). *Field studies of radon in rocks, soils, and water* (Vol. 10). CRC Press.

67. Fabianova, E., Szeszenia-Dabrowska, N., Kjaerheim, K., & Boffetta, P. (1999). Occupational cancer in central European countries. *Environmental Health Perspectives, 107*, 279–282.
68. Hoffmann, W., Kranefeld, A., & Schmitz-Feuerhake, I. (1993). Radium-226-contaminated drinking water: Hypothesis on an exposure pathway in a population with elevated childhood leukemia. *Environmental Health Perspectives, 101*, 113–115.
69. Sevc, J., et al. (1993). A survey of the Czechoslovak follow-up of lung cancer mortality in uranium miners. *Health Physics, 64*, 355–369.

Chapter 3
Uranium in Mine Waste Rock and Tailings Piles

Abstract This chapter presents the generalized view of uranium dynamics in waste rock and tailing piles. The behaviour varies from one mine to another, depending on several site-specific characteristics, including geology, hydrogeology, geochemistry, climate, and vegetation. The geochemical dynamics also change with time and the nature of their exposure to atmospheric conditions. At the same time, the geochemical and physical characteristics of waste from one operation to another are broadly similar, indicating that uranium production facilities worldwide share many common characteristics and problems. Similarly, the uranium dynamics are present in their waste rock and tailing piles.

Keywords Uranium mine waste · Tailings management · Geochemical processes · Hydrological processes · Geotechnical characteristics · Tailings ponds and heaps · Uranium plume migration · and Abandoned uranium sites

3.1 Wastes of Uranium Production

Waste is mainly generated during the mining and beneficiation of uranium ore. The uranium production system and the waste generated are summarized in Fig. 3.1. Regardless of the mining method (open pit or underground), the ore-bearing rock is crushed, milled, and leached in a uranium mill. The chemical plant extracts uranium from ore. The mill is typically located near the mines to minimize transportation costs. Thus, the main types of waste are waste rock generated during mining and tailings generated during uranium milling.

M. Mkandawire, *Fate and Effects of Uranium in Abandoned Uranium Mines*,
https://doi.org/10.1007/978-3-032-10447-2_3

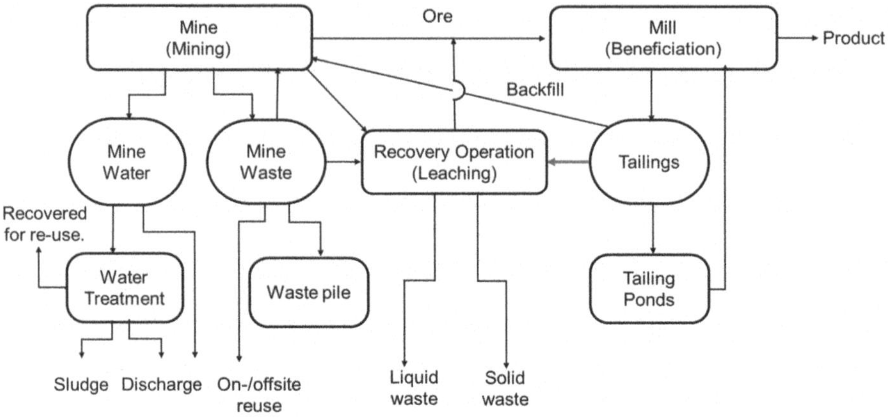

Fig. 3.1 The sources and types of waste in a uranium production process

3.1.1 Waste Rock

Waste rock is material removed from the mine to access the orebody. They may contain rock and target minerals in low concentrations for economic recovery [1, 2]. Waste rock includes granular, broken rock that ranges from fine sand to large boulders, depending on the nature of the formation and the mining methods employed. In open pits or open-cast mining, waste rock includes the overburden (i.e., soil, vegetation, and rocks above the orebody) [1, 2].

One of the legacies of uranium mining is the creation of vast heaps of waste rock. The uranium ore content is often between 0.1% and 0.2% [3]. As a result, large quantities of rock are excavated, from which minimal amounts of uranium minerals are extracted. The waste rock can be separated into mineralized and non-mineralized rock based on its mineral content. Waste rock with a U_3O_8 concentration greater than 0.03% is deemed mineralized waste rock in uranium mining. Non-mineralized waste rocks have concentrations of U_3O_8 less than 0.03%. Such waste rock is categorized as either clean or potentially acid-generating based on its content of acid-producing minerals like pyrites (FeS_2).

The characteristics of the waste rock are primarily influenced by the geological, hydrological, and climatic conditions of the mining site. Uranium ore deposits can occur in igneous, sedimentary, and metamorphic rocks [4]. Climate conditions influence the geochemical processing of waste rocks. Although they may usually have little to no uranium concentration, waste rock can contain radionuclides (mainly thorium-230 and radium-226, along with their associated decay products) and toxic metals, such as nickel, copper, arsenic, molybdenum, selenium, and cadmium. Therefore, waste rock may pose serious environmental hazards. For instance, uranium content and related radioactivity in ore-bearing rocks are shown in Fig. 3.2.

The rate of waste rock accumulation depends significantly on the mining method and the ore-to-waste ratio. Until the 1960s, uranium was predominantly mined from

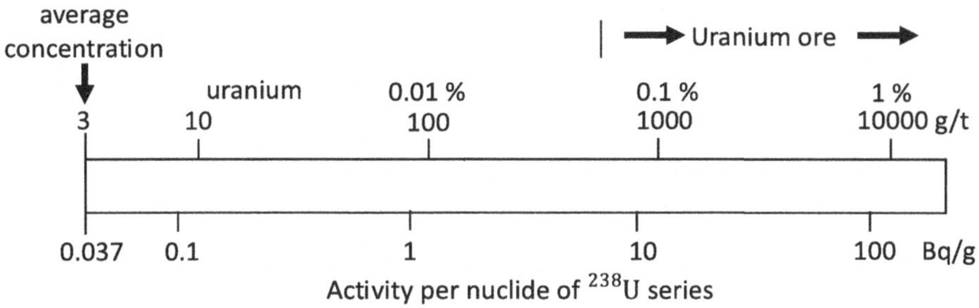

Fig. 3.2 Uranium content and related radioactivity in ore-bearing rocks. (Adapted from https://www.wise-uranium.org)

near-surface ore deposits in open-pit mines. Currently, most uranium mines operate underground and in situ leaching, which do not produce large quantities of waste rock [5].

3.1.2 Tailings

Uranium is primarily recovered from ores through hydrometallurgical processes. The process involves several separation and extraction steps. In open-pit and underground mining, the ore rocks are brought to a mill, crushed, and ground before leaching agents are added to dissolve the uranium. The leaching agent extracts uranium from the ore, along with several other constituents, including molybdenum, vanadium, selenium, iron, lead, and arsenic. The pure triuranium octaoxide (U_3O_8) is then separated from the leaching solution using solid-liquid extraction, solvent extraction, and ion exchange. The extraction agents include, but are not limited to, ammonia, chlorine, hydrochloric acid, sulfuric acid, and cyanide [6]. The waste material produced and the ore residues after U_3O_8. The materials that have been extracted are tailings, also known as slimes, which, when dried, have the consistency of fine sand. Thus, uranium mill tailings are the residual waste from the extraction of uranium from uranium ore.

Uranium tailings contain significant concentrations of radioactive decay products from the uranium chains, a range of metals, and other compounds. Since only uranium is extracted, all other uranium decay daughter elements remain in the tailings at their original radioactivity. The most important are thorium-230, radium-226, radon-222, and the daughter isotopes of radon decay, including polonium-210, because of their short half-lives. The radionuclides in uranium tailings emit 20–100 times as much gamma radiation as natural background levels on deposit surfaces. Depending on the extraction process, some residual uranium may remain in the tailings. Consequently, uranium tailings account for about 85% of the total activity of uranium ore. It is estimated that the total activity in the tailings remains at approximately 75% of that in the ore, a constant level that has persisted for over 10,000 years [1, 7–9].

Since only uranium is extracted, the tailings still contain a spectrum of all the naturally occurring non-radioactive and radioactive elements found in uranium ore, as well as the chemicals used in the extraction process. The granular-to-slime constituency of uranium tailings makes them readily leachable, erodible, or collapsible under different conditions. The reduction in size of ore rock to sand grains increases the exposure area of metal minerals to leaching-favourable environmental conditions. Then, with the tailing grains already being exposed to the extraction chemicals, the leaching kinetics of the metal are enhanced. Further, the tailings may contain sulfidic minerals that are prone to generating acid mine drainage (AMD). The formation of AMD facilitates the further release of many ore mineral constituents, which may be toxic, such as certain metals [10].

3.1.3 Other Wastes

3.1.3.1 Sludge from Liquid Effluent Treatment

Liquid effluents are generated throughout the uranium production cycle and post-production, including processes such as crushing, grinding, leaching, and precipitation that use water and chemicals. A common practice is to treat effluents before releasing them into the environment. Thus, large quantities of contaminated sludge are generated and become waste and byproducts of uranium mining and processing.

The generated effluents contain radioactive and non-radioactive elements and compounds. The radioactive constituents of the effluent may include uranium and radium (^{226}Ra) and other water-soluble daughter radionuclides. The non-radioactive contaminants include nickel, arsenic, manganese, magnesium, molybdenum, selenium, fluorides, and sulfates, which come from the leaching of the minerals. Further, the influence may also contain various uranium (VI) stripping extraction agents, such as ammonium carbonate ($(NH_4)_2CO_3$), ammonium chloride (NH_4Cl), oxygenated water (H_2O_2), ammonia (NH_3), hydrochloric acid (HCl), sulfuric acid (H_2SO_4), and nitric acid (HNO_3) [6]. All these contaminants end up in the effluent treatment sludge.

3.1.3.2 Scarps, Machinery, and Infrastructure

Besides waste rock and tailings, uranium mining and processing sites also generate other solid wastes, including obsolete equipment parts, chemical supply containers, scrap metals, and scrap lumber. Most of these scraps are contaminated with radioactivity. After mining operations cease, materials from dismantled and demolished equipment, masonry, mine headframes and hoist rooms, sheds, warehouses, and other buildings, such as concrete, glass, wood, and metals, may also constitute potentially radioactively contaminated waste.

The mining waste may also be removed by blasting and excavation machines, loaders and dump trucks, abandoned rail tracks, and hand tools. Like most industries, the uranium mining and processing site has a damp environment, where most containers for supplies, including uranium extraction agents, trash, oil, solvents, and chemicals, are stored. The dumping site would contain waste from mining support systems, such as facility repair workshops, where vehicles are serviced or repaired, as well as domestic trash and solid waste from human activities.

3.2 Waste Rock Piles

Waste rock piles are substantial accumulations of non-economic rock material—waste rocks—that are displaced during mining operations. These waste rocks are heterogeneous, containing a mix of particle sizes and lithologies, ranging from fine sand to large boulders, and are mineralized. The mineralized waste rock consists of low-grade ore with minimal value and insignificant concentrations of secondary minerals. Most mineralized waste rocks are associated with metal sulfide (Me_xS_y) minerals like pyrite (FeS_2) and other chalcopyrites oxidize and degrade when exposed to oxygen and water. Consequently, the material can generate acid or release contaminants at rates that could negatively impact the environment if left exposed on the surface indefinitely. As a result, during the restoration of abandoned mining and processing sites, waste is consolidated into waste rock piles (WRP) to manage or prevent mine drainage and pollution migration from the sites to receiving environments.

3.2.1 The Waste Rock Piles Profiles

Waste rock piles (WRP) are man-made landscapes of non-economic rock material displaced during mining operations, commonly referred to as waste rock, overburden, and other materials unsuitable for further processing. These WRPs can cover large areas spanning several square kilometres and can reach heights of up to 150 metres, or even 400 metres in the case of large mountaintop mines. Small mines typically have piles that range in height from 10 to 30 metres. The WRPs in surface mining operations can exceed 100 metres and contain over a billion cubic metres of material.

The conceptual profile of the WRP is shown in Fig. 3.3. The topmost layer is a vegetation cover, primarily consisting of shallow-rooted plants, preferably those with fibrous roots. Plants with long taproots are strongly discouraged. This vegetation serves to stabilize the topsoil and prevent erosion. The vegetation layer grows on clean soil, which can be up to 30 cm thick. Beneath the clean topsoil is a layer of

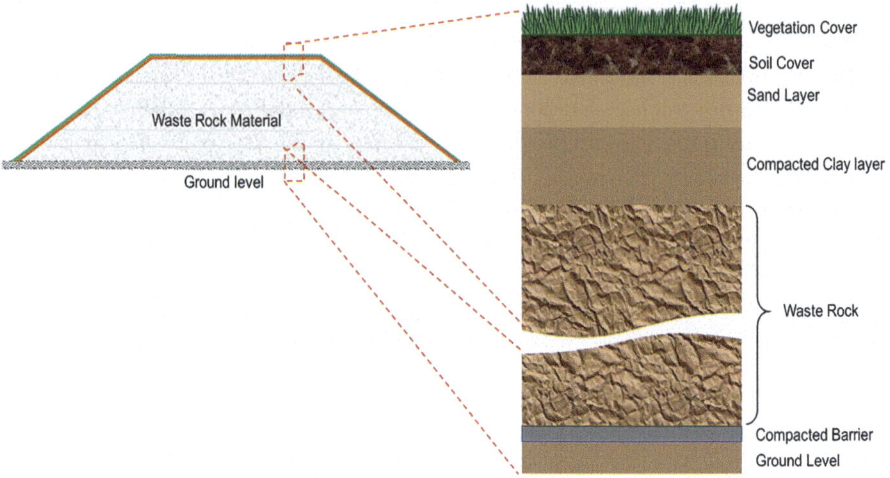

Fig. 3.3 (**a**) A generalized conceptual profile of WRP

compacted clay material ranging from half a metre to a metre thick, which controls air ingress and surface water infiltration into the inner WRP, where the waste rock is located. Some WRPs are designed with a synthetic geotextile or HDPE cover between the topsoil and the clay layer.

The rock materials in WRPs are heterogeneous, consisting of a diverse mix of particle sizes and rock types, ranging from fine sand to large boulders. Some rock fragments contain reactive minerals, particularly sulfides, which can lead to acid mine drainage (AMD) and the leaching of metals. In the context of uranium mining, the rock fragments may contain high levels of daughter radionuclides resulting from uranium decay, posing a significant risk of radon gas emanation.

The primary goal of constructing the WRPs is to limit the exposure of mineralized rocks to water and oxygen, thereby preventing the oxidation of sulfide minerals and the subsequent formation of acid mine drainage (AMD). Additionally, WRPs aim to reduce radiation exposure from radionuclides and minimize radon gas emissions at uranium mining sites. Consequently, the design of a WRP is influenced by the quantity of potential acid or alkaline-forming materials present, as well as the site's hydrology, geological characteristics, and prevailing weather and environmental conditions (Fig. 3.4). Furthermore, the type of mining operation and the site's topography also contribute to the design and construction methods employed for the WRP.

Figure 3.5 shows various WRP constructions across different topographies. Typically, WRPs are often constructed in horizontal layers known as benches. The benches are designed with compacted layers and cover systems to control surface water infiltration and oxygen ingress while influencing drainage patterns. Inert materials or coverings are extensively used to minimize water infiltration and oxygen exposure.

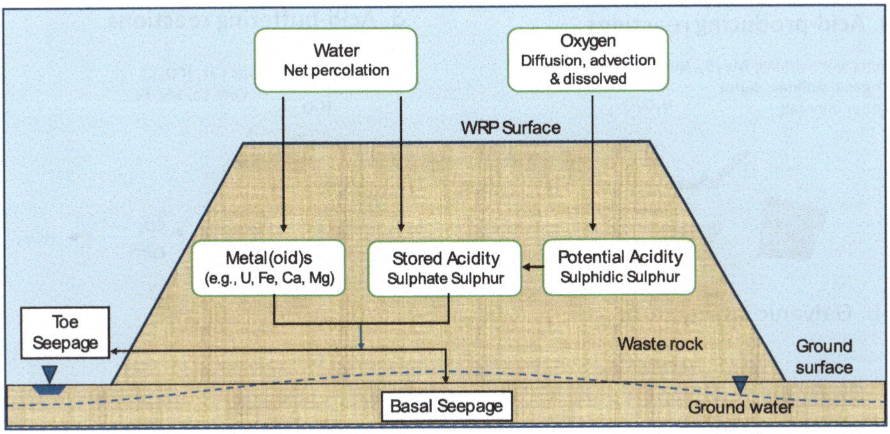

Fig. 3.4 A summary outline of the various components and interactions involved in the geochemical processes occurring in a waste rock pile. These processes can be categorized into three key elements: (1) the ingress of water and oxygen from the atmosphere, (2) the contaminant loads, including rock acidity and metal leaching, within the waste rock pile, and (3) the loading of contaminants into the surrounding environment, affecting both groundwater and surface water

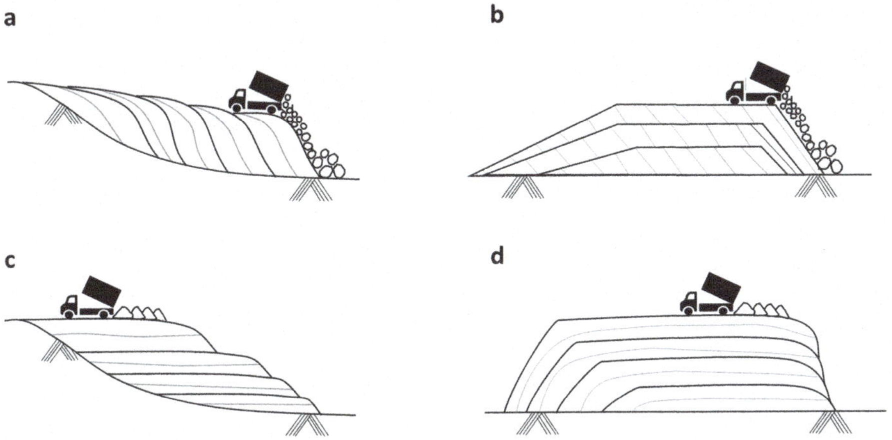

Fig. 3.5 Construction of mine waste dumps: (**a**) top-down on inclined topography, (**b**) top-down on horizontal topography, (**c**) bottom-up on inclined topography, and (**d**) bottom-up on horizontal topography. Modified from Zevgolis (2018) [11]

3.2.2 Geochemical and Hydrological Processes

The geochemical processes occurring in waste rock piles are driven by the interaction of water and oxygen with the minerals present in the waste rock fragments. The resulting geochemistry can vary, leading to acidic, alkaline, or neutral conditions, which depend on the mineral composition of the waste rocks. Figure 3.6

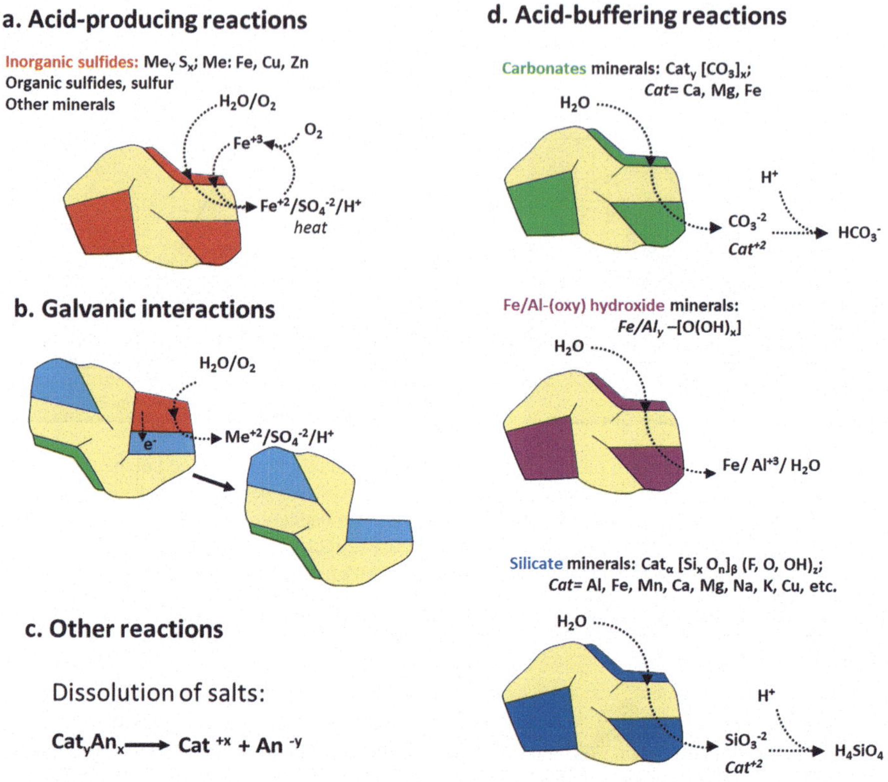

Fig. 3.6 Geochemical reactions occurring in waste rock piles include: (**a**) The oxidative dissolution of sulfide minerals, which is a primary mechanism contributing to acid generation; (**b**) galvanic interactions that can enhance the preferential dissolution of sulfide minerals; (**c**) various reactions that dissolution of metals and other solutes without directly impacting the drainage pH; and (**d**) the dissolution of carbonates, iron/aluminium-(oxy)hydroxides, and silicates, which may consume protons and thereby provide a net buffering effect.(Adapted from Vriens et al. [12])

summarizes key geochemical reactions that may occur in the mine waste rock pile, resulting in different drainage qualities.

3.2.2.1 Acid and Metal Producing

Formation of acid in waste rock piles occurs due to the oxidative dissolution of sulfide minerals present in mineralized rock fragments. The sulfide minerals are mostly in the form of chalcophile metals, commonly represented as generalized metal sulfides (Me_xS_y), with pyrite (FeS_2) being the most abundant. Other significant minerals include chalcopyrite ($FeCuS_2$), covellite (CuS), sphalerite (ZnS), pyrrhotite (FeS), arsenopyrite (FeAsS), and galena (PbS).

The oxidation of sulfidic minerals involves the transfer of multiple electrons from each sulfur atom to a suitable aqueous oxidant. This chemical process generates a variety of intermediate sulfur species, such as polysulfides, thiosulfate, and elemental sulfur, each of which plays a crucial role in the overall transformation. For example, the oxidation of FeS_2 proceeds through an electrochemical mechanism, where electron transfer occurs at specific surface sites of the mineral. This process can be illustrated by the following chemical reaction:

$$FeS_2 + 72O_2 + H_2O \rightarrow Fe^{2+} + 2H^+ + 2SO_4^{2-}$$

Under oxic conditions, dissolved ferrous Fe^{2+} rapidly oxidizes into ferric Fe^{3+}, which can act as an additional oxidant:

$$Fe^{2+} + 14O_2 + H^+ \rightarrow Fe^{3+} + 12H_2O$$

$$FeS_2 + 14Fe^{3+} + 8H2O \rightarrow 15Fe^{2+} + 16H^+ + 2SO_4^{2-}$$

The Fe^{3+} ions break metal sulfide bonds more effectively than protons. As a result, the Fe^{3+} reaction accelerates the oxidation of FeS_2, thereby creating a significant autocatalytic feedback. FeS_2 is naturally abundant in many chalcophile metal ores, and Fe^{3+} also acts as an oxidant for other metal sulfides (e.g., $CuFeS_2$, ZnS, and PbS). The kinetics of abiotic sulfide oxidation by Fe^{3+} allow for much faster sulfide oxidation than oxygen alone. The ferrous Fe^{3+} is very insoluble under oxic and near-neutral conditions and readily precipitates as Fe^{3+}-(oxy)hydroxide to release additional acidity. The chemical reaction is:

$$Fe^{3+} + 3H_2O \rightarrow Fe(OH)_3 + 3H^+ \tag{1}$$

As a result, the overall reaction equation for pyrite oxidation is often presented as:

$$FeS_2 + 154O_2 + 72H_2O \rightarrow Fe(OH)_3 + 2H_2SO_4$$

As previously described, abiotic sulfide oxidation reaction rates tend to be relatively slow, even when facilitated by the catalytic presence of Fe^{3+}. However, these oxidation reactions are significantly accelerated by the activities of various iron- and sulfur-oxidizing bacteria that harness the energy released during oxidation. Notable chemolithotrophic bacteria involved in sulfide oxidation include both acidophilic and neutrophilic species from genera such as *Thiobacillus*, *Leptospirillum*, and *Ferrobacillus*, including ferrooxidans and thiooxidans.

3.2.2.2 Acid-Buffering Reactions

Some WRP have high abundant of carbonate minerals (e.g., calcite, dolomite, and ankerite), Al- and Fe-(oxy)hydroxides (e.g., gibbsite and ferrihydrite, respectively), and silicate minerals (e.g., feldspars, chlorites, smectites, micas, and amphiboles) in waste rock, which can dissolve and consume protons and introduce alkalinity that offsets acidity produced by sulfide oxidation. The simplified reaction mechanism involved in the dissolution of these alkaline-producing minerals is illustrated as follows:

1. The dissolution of carbonate minerals follows the reactions:

$$CaCO_3 + H^+ \rightarrow Ca^{2+} + HCO_3^-$$

2. The dissolution reaction of Al- and Fe-(oxy)hydroxides occurs as:

$$[Fe/Al]OOH + 3H^+ \rightarrow [Fe/Al]^{3+} + 2H_2O$$

3. The dissolution reaction of silicate minerals occurs as:

$$CaSiO_3 + 2H^+ + H_2O \rightarrow Ca^{2+} + H_4SiO_4$$

And/or.

$$CaAl_2Si_2O_8 + 8H^+ \rightarrow Ca^{2+} + 2Al^{3+} + 2H_4SiO_4$$

The dissolution reactions of carbonates, oxides, and silicates that neutralize acids accelerate as the pH decreases, in line with their solubility characteristics. This process creates what is known as a buffering sequence for acid-rock drainage:

1. Carbonates buffer acidity until they are typically depleted at $pH < 6$.
2. Aluminium- and iron-(oxy)hydroxides begin to dissolve and buffer acidity at pH levels below 4 to 5 and 3 to 4, respectively.
3. The dissolution of aluminosilicates contributes significantly to the acid-buffering capacity only at $pH < 3$.

When carbonates dissolve, they neutralize protons through the carbonate equilibrium. However, the dissolution of iron-rich carbonates, such as siderite ($FeCO_3$), introduces dissolved ferrous iron into the system. This releases surplus protons upon oxidation and hydrolysis, reducing the overall neutralizing effect.

Acid-buffering minerals dissolve simultaneously with the oxidation of sulfides when they are naturally co-located in the waste rock matrix. Additionally, these minerals may be intentionally mixed, blended, or added to sulfide-rich waste material as part of an active or passive strategy for acid-rock drainage remediation.

The kinetics of carbonate dissolution are typically rapid compared to sulfide oxidation, while the dissolution of silicates is orders of magnitude slower. In addition to consuming protons, the dissolution of carbonate, oxide, or aluminosilicate minerals introduces metals, such as Ca, Mg, Mn, Al, and Fe, into the leachate, which affects the geochemical processes in the WRP.

3.2.2.3 Physical Transport Processes

The migration of drainage from waste rock sources is intricately linked to the movement of gas, water, and heat within the porous waste rock piles. Aqueous transport primarily governs the final export of solubilized solutes from the waste rock, while gas transport influences weathering rates by regulating the supply of oxygen, which is essential for the oxidation of sulfides. The oxidation of sulfides generates heat that moves through temperature gradients, affecting various processes, including pore gas density and flow. This, in turn, helps sustain elevated weathering rates within the waste rock. Variations in hydrogeological properties result from the physical heterogeneity of waste rock piles, which, in turn, affect the movement of water, air, and heat.

Aqueous Transport Process

The movement of water within WRPs exhibits considerable variability across sites, contingent on factors such as climate conditions, internal structural composition, and the hydrogeological characteristics of the WRP [13, 14]. At the macroscale, the hydrological dynamics of a WRP are predominantly influenced by surface water percolation and groundwater flow, both of which contribute to fluctuations in internal water storage. Ideally, the internal water storage dynamics of a WRP should be minimally influenced by interactions with groundwater and percolating surface water; therefore, such piles should be constructed on relatively impermeable or lined surfaces to mitigate groundwater interaction.

Furthermore, WRPs are engineered with compacted subsurface cover layers specifically designed to reduce surface water percolation, particularly during intense precipitation, thereby safeguarding the reactive materials below. The compaction of these cover materials facilitates surface runoff and encourages the formation of ponding events. However, when cover systems fail or become overwhelmed, the portion of precipitation that does not runoff or evaporate may infiltrate the unsaturated portions of the waste rock pile. This infiltration process triggers significant changes in internal water storage and drives downward water movement, culminating in the formation of a wetting front. This wetting phenomenon subsequently triggers drainage driven by gravity, ultimately leading to base seepage.

At the pore scale, the flow dynamics of water within WRPs are primarily governed by infiltration processes occurring within the interstitial pores. The porosity of WRPs varies substantially, depending on the particle-size distribution of the waste materials, which ranges from boulders exceeding 1 m in size to sub-millimetre clay fractions. Consequently, hydraulic conductivity plays a crucial role in shaping flow dynamics, which can differ markedly across the poorly sorted, heterogeneous matrix of WRPs. Infiltration is typically characterized by two distinct components: (a) A slower matric flow component, wherein water translocates under the influence of capillary and gravitational forces and (b) a more rapid macropore flow component, distinguished by preferential pathways facilitating expedited channelized flow.

Gas Transport Process

Gas movement within the waste rock pile (WRP) primarily results from pressure gradients caused by oxygen consumption during sulfide oxidation and the production of CO_2 from the dissolution of carbonates. This leads to pore gas pressures and compositions that deviate from atmospheric conditions. Variations in pore gas composition and pressure initiate gas transfer through both diffusion and advection, which in turn affects the distribution of internal pore gases and the overall outward transport of CO_2, as compared to the inward replenishment of O_2. The role of dissolved oxygen transport in percolating water is generally minimal due to its low solubility in water.

In low-permeability waste rock systems, gas transport is predominantly by molecular diffusion. Additionally, pressure gradients can induce advective gas transport within porous media. Both diffusive and advective gas transfer are significantly influenced by the characteristics of the waste rock, including its permeability, degree of porewater saturation, and spatial distribution of heterogeneity.

Heat Transport

The internal temperatures in waste rock piles can be tens of degrees above ambient levels. This is attributed to sulfide oxidation, an exothermic process. For example, the oxidation of pyrite can release between 1000 and 1500 $kJ \cdot mol^{-1}$ [15–17]. This heat can be transferred convectively from reactive "hotspots" within these waste rock piles, which is responsible for the so-called chimney effect. This process involves a combination of conduction (i.e., heat diffusion) and advection (i.e., bulk fluid flow), with radiative transfer and viscous dissipation typically considered negligible in waste rock environments.

The heat generated by the oxidation process induces natural convection, driven by buoyancy forces, in waste rock piles. This results in a change in the density of the pore gas, such as water vapour. Buoyant force, also known as upthrust, is the upward force exerted by a fluid (i.e., liquid or gas) on a submerged object. The buoyant force results from the pressure difference between the top and bottom of the submerged object. The temperature gradients created by this internal heating initiate upward convection, counteracting the downward diffusive transport of oxygen into the waste rock pile [18, 19]. The interplay of convective heat and gas transfer is responsible for high rates of weathering in the waste rock piles.

3.2.3 Geophysical and Geotechnical Characteristics

3.2.3.1 Properties and Implications

Waste rock piles possess important geotechnical properties, including grain-size distribution, porosity, and hydraulic conductivity, which significantly influence their stability and the potential for water infiltration and acid mine drainage (AMD). These geochemical characteristics are reflected in various physical attributes, including macroscale features such as the size and morphology of waste rock stockpiles, the distribution of particle sizes within the stockpiles, and the rock's thermal conductivity. Furthermore, microscale properties, particularly the mineral surface areas, play a significant role in determining the behaviour of waste rock structures.

The substantial spatial variability in mineralogy and disparities in particle size and porosity engender pronounced variations in hydrogeological and geochemical properties within the waste rock piles (WRP). The diversity of particle sizes and their distribution critically influence the hydrogeological characteristics of the pile, where the formation of porosity, defined as the voids or spaces between rock particles, substantially affects the waste rock's capacity for water and air storage as well as flow dynamics. For instance, larger particles can facilitate the formation of macropores, which directly affect water flow and infiltration rates.

Moreover, voids and pores within the waste rock are paramount in determining its ability to store and transmit water, thereby influencing infiltration rates and water flow patterns within the WRP. The interplay among grain-size distribution, porosity, and other relevant factors is integral to understanding the hydraulic conductivity of the WRP. Hydraulic conductivity, the material's capacity to transmit water, may exhibit significant variability throughout a waste rock pile, thereby affecting contaminant residence times and migration pathways (Fig. 3.7).

Key physical properties, such as bulk density and particle density, which pertain to the density of the solid rock particles, are crucial considerations. Bulk density is defined as the dry weight of a specific volume of waste rock material, including both solid particles and voids. The interrelationships among porosity, grain size, and their distribution are vital in influencing the shear strength of the waste rock pile (WRP). Shear strength, which reflects the waste rock's capacity to resist deformation and failure, is essential for maintaining slope stability and overall integrity of the pile. Moreover, these factors also affect specific weight, defined as the waste rock's weight per unit volume, a key parameter for assessing overall stability.

The shear strength and specific weight of the WRP are instrumental in determining the friction angle, which represents the angle at which a slope of waste rock would fail under its gravitational influence. This angle is a key consideration in overall stability assessments. Cohesion, the force that binds particles together, also contributes to the stability evaluation of waste rock piles.

Stability calculations involve determining the likelihood of failure due to sliding or collapse. Such evaluations are typically conducted using the limit equilibrium method (LEM) or the finite-element method (FEM) to determine the factor of safety

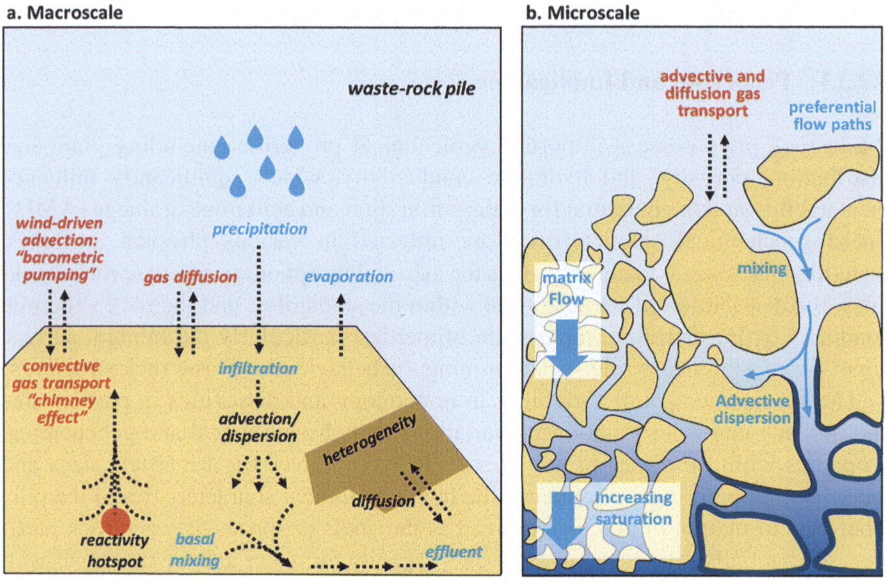

Fig. 3.7 Relevant aqueous (blue) and gaseous (red) mass transport processes in waste rock piles: (**a**) macroscale and (**b**) pore scale or microscale processes (Source: Vriens et al.) [12]

(FS) of the waste rock structure. These methodologies provide a framework for understanding the critical factors that govern the stability of waste rock piles in geotechnical engineering contexts.

3.2.3.2 Factors Influencing Geotechnical Properties

The geotechnical and geophysical properties of waste rock piles (WRP) are significantly influenced by factors, including geometry, material properties, construction methods, and external loads, such as rainfall and seismic activity. Specifically, the external geometry, which includes slope angle and the presence of benches, plays a critical role in slope stability. Waste rock piles with a uniform slope and a single bench are generally less stable compared to those that feature benches, which help to distribute stress and reduce the risk of slope failure.

Internally, the structure of the waste rock pile, shaped by construction techniques and the type of waste rock, affects water infiltration, porewater pressure, and overall hydrogeological behaviour. Parameters such as grain-size distribution, the composition of fine or coarse materials, and the dumping method (e.g., end-dumping or push-dumping) can significantly influence water flow patterns and contaminant movement within the pile.

Additionally, rock type and mineralogy can affect its geotechnical properties, including susceptibility to weathering and the potential to generate acid mine drainage (AMD). The construction methods used for dumping and compacting the waste rock can alter its internal structure and geotechnical attributes. Over time, environmental factors such as freeze-thaw cycles and rainfall can alter the geotechnical properties of waste rock. The amount of water infiltrating the waste rock pile is crucial, as it can significantly affect stability and the likelihood of acid mine drainage (AMD).

3.3 Uranium Mine Waste Rock Piles

To comprehensively understand the dynamics and behaviour of uranium in waste piles long after a uranium mine has been abandoned, it is essential to grasp the key characteristics of an abandoned uranium mine, as discussed in the previous chapter. Moreover, it is crucial to thoroughly understand the features of a waste rock pile resulting from uranium mining. By correlating the attributes of an abandoned mine with those of waste rock piles, valuable insights can be gained into the risks and hazards posed by these materials. This understanding is instrumental in formulating effective strategies for risk mitigation and hazard control, as well as for addressing the challenges of remediating abandoned uranium sites.

3.3.1 Characteristics of Uranium-Abandoned Sites

Most abandoned mine sites were developed decades before the environmental impacts of mining were fully understood or modern operating standards were established [20, 21]. The abandoned mines are now re-mushroomed by miners taking advantage of (a) poor and developing economies, compromising standards and legal requirements for foreign investment and (b) a lack of standards, legal monitoring, and enforcement infrastructure. When a uranium mine is abandoned, the mine owners cannot be found or are financially unable or unwilling to take responsibility.

Consequently, mining activities in most of these abandoned uranium mines never adhered to the basic principles of responsible mining. The hallmarks of abandoned uranium mines include disturbed landscapes resulting from excavation, heaps and dumps of waste (if not properly disposed of on the surface), and abandoned mining and milling infrastructure and equipment. The legacy of uranium mining is evident throughout the landscape, with numerous pit lakes, craters, and waste piles scattered throughout the mining sites [5, 22]. These landforms, abandoned infrastructure, and waste rocks are associated with physical, chemical, and radiation hazards, posing environmental, safety, human health, and economic risks.

For instance, the physical hazards associated with abandoned uranium mines often remain as unsecured vertical shafts, adits and horizontal tunnels, inclined shafts or declines, test pits and trenches, unstable rock slopes, subsidence, and caving features. If mining were responsible, the mineralized waste rock would have been deposited on impermeable pads during operations so that seepage and runoff from the piles would have been collected and processed. Even in abandoned mines, mining could have been done responsibly.

When a mining site is abandoned, the seepage and runoff are no longer treated to remove contaminants before they are released into the environment. If not treated, they can contain radioactive and toxic materials. As a result, formerly clean mining sites become contamination sources once abandoned, posing severe environmental hazards.

The mineralized waste rock contains other elements, such as toxic metals and metalloids, that coexist with uranium in the ore. Further, the mineralized waste rock may also contain traces of uranium minerals, but most uranium daughter radionuclides remain abundant. As a result, the uranium waste rock piles are radioactively dangerous to humans and the environment. Specifically, the radiation dangers increase with the increase in radium-222 (^{222}Ra) in the mine waste. The ^{222}Ra in waste rocks continuously decays into the radioactive gas radon-222 (^{222}Rn). Radon releases constitute a significant hazard from abandoned uranium mines. Several cancers, including lung cancer, are attributed to exposure to ^{222}Rn. Since radon spreads quickly with the wind, many people receive small additional radiation doses.

Another uranium daughter radionuclide of high is polonium-210 (^{210}Po). ^{210}Po is a product of the radioactive decay of uranium-238, which decays to radon-222 and then to polonium. Its half-life is 138 days, making it one of the most potent radionuclides. Nonetheless, radiation levels, especially gamma, decrease rapidly with distance from the pile.

3.3.2 Engineered Uranium Mine Waste Rock Piles

3.3.2.1 Reasons for Engineered Covers

As part of rehabilitating abandoned mining sites, waste rock, overburden, and some contaminated machinery left on the surface are systematically consolidated into waste heaps and covered. These engineered covering systems are known as engineered mine waste rock piles. Like many mining operations, the engineered uranium waste rock piles are primarily designed to prevent water and air ingress, as their interaction with the waste rock can lead to oxidation of sulfide-containing rocks, resulting in acid and metalliferous drainage (AMD). The AMD is commonly generated from waste rocks at uranium mines, as is the case with other abandoned mines. On average, sulfidic waste rock from abandoned uranium mines accounts for 60–80% of the AMD discharge.

3.3.2.2 Recommended Uranium Mine WRP Designs

The engineered cover systems for uranium mine waste rock piles are also designed to minimize porosity and permeability, effectively blocking radiation penetration and limiting radon gas release. Logically, conventional end-dumping of waste rock is strongly discouraged in the engineering cover system for uranium mine rock waste. End-dumping is a technique for creating waste rock dumps by placing material from haul trucks on the surface in layers, which creates angle-of-repose pathways for air and water.

Consequently, strategic construction practices such as base-up, layered, and compacted (BULC) are recommended for managing uranium mine waste rock piles. The construction of BULC waste rock piles employs geochemical engineering principles. The technique involves constructing the waste rock piles by initially using paddock dumping to place the waste rock, followed by initial compaction with a dozer to create a uniform layer. Each flattened layer should ultimately have a thickness between 1 and 3 metres. Some key benefits of BULC construction methods are summarized in Tables 3.1 and 3.2.

3.3.2.3 Special Consideration for Uranium WRP

Like most waste rock piles, thinner layers enhance adequate compaction, reducing optimal permeability and decreasing air entry. For uranium mine rock wastes, they additionally increase the retardation of argon gas emanating. Each compacted thin lift manages surface runoff. The thin-layer construction techniques also allow for the integration of other AMD management measures into the design of BULC waste rock piles. For instance, pore gas and water quality monitoring should be implemented to verify operational performance and manage radon emanation in the recommended uranium mine rock waste.

Radon originates from the decay of uranium, a naturally occurring element found in most rocks. The waste rock from uranium mines typically contains uranium decay daughter elements that have advanced in decay towards radon. As a result, waste rock piles produce and release high levels of radon gas. When radon is released from a waste rock pile, it can accumulate in the nearby air, potentially leading to elevated radiation exposure if inhaled, particularly for individuals living or working near the site. Therefore, uranium mine waste rock piles should always be covered. For example, a layer of soil or other materials can seal the pile, reducing radon release. Additionally, unique ventilation systems should be installed to extract radon gas from the pile and safely disperse it. Regularly monitoring radon levels around the waste rock pile through the pole is essential to assess potential exposure risks.

3.3.2.4 Other General Considerations

Most waste rock contains sulfides, often pyrite (FeS_2), which are typically left on the ground surface, especially in abandoned mines, before being consolidated into waste rock piles. Surface waste rock dumps containing loose-dumped, potentially acid-forming sulfidic waste rock essentially act as "oxidation reactors" because they have a ready supply of oxygen and function as a "sponge" for rainfall infiltration,

Table 3.1 Types of uranium deposit rocks

	Description	
Rocks	Associated minerals	Distributions
Intrusive	Alaskite, granite, pegmatite, and monzonite	Rössing and Husab (Namibia), Kvanefjeld (Greenland), Bancroft area (Canada), and Palabora (South Africa)
Granite-related	Vein deposits in granite, adjacent meta-sedimentary rocks, and disseminated mineralization in granite	Some of the orebodies in Europe, Canada, and Australia
Polymetallic iron-oxide breccia complex	The deposit contains iron, copper, uranium, gold, silver, rare earth elements (mainly lanthanum and cerium), and fluorine Haematite-rich breccias host all the economic copper-uranium mineralization	The majority of Australian uranium resources
Volcanic-related	These uranium deposits occur in and around volcanic calderas, within acid to intermediate volcanic rocks, and are associated with faults and shear zones	China (Xiangshan), Kazakhstan, Mongolia (Dornod and Gurvanbulag), the Russian Federation (Streltsovka caldera, a significant occurrence), Peru, and Mexico
Metasomatite	Consists of unevenly disseminated uranium in structurally deformed rocks that were affected by sodium and potassium metasomatism	Elkon District (Russia), the Lagoa Real-Caetite District (Brazil), Novokonstantinovskoye and those near Zheltye Vody (Ukraine), Valhalla and Skal (Australia), Michelin (Canada), and Lianshanguan (China)
Collapse breccia pipe	These occur in circular, vertical collapse structures filled with coarse fragments and a fine matrix of sediment that has been penetrated (a) The collapsed pipes range in diameter from 30 to 200 metres and are up to 1000 metres deep (b) Uranium mineralization is primarily located within permeable sandstone breccias in the pipe (c) The principal uranium mineral is pitchblende	Near the grand canyon (the USA), notably in the Arizona strip

Table 3.1 (continued)

Rocks	Description	
	Associated minerals	Distributions
Metamorphite	This type of uranium deposit occurs in: (a) Metasediments and metavolcanic rocks are unrelated to granite (b) Zones of calcium-rich alteration within Proterozoic metamorphic rocks (c) Marble-hosted phosphate	Forstau (Austria), Shinkolobwe deposit (Democratic Republic of Congo), Rozna (Czech Republic), Jaduguda (India), Kokshetau District (Kazakhstan), port radium (Canada), and Mary Kathleen uranium/rare earth deposit near Mount Isa, Queensland (Australia)

Table 3.2 Some key advantages of the BULC waste rock pile

Feature	Key benefits
Base-up construction	(a) Prevents the creation of preferential pathways for air and water, reducing AMD generation and acidity discharge.
	(b) Allow waste rock masses to behave more like tailings deposits, where controlling air entry is standard practice, even before installing a cover system.
	(c) Enhancing the geotechnical stability of waste rock storage requires focusing on surface water control and runoff.
Thin-lift configurations	(a) Create unstructured horizontal layers to prevent the formation of preferential pathways
	(b) Improve the potential for adequate compaction to reduce porosity and permeability
Compaction of thin layers	(a) Reducing the potential for air entry, AMD generation, and water discharge is essential
	(b) Lower water discharge rates result in longer residence times for waste rock porewater, promoting carbonate-neutralization reactions and allowing silicate-neutralization reactions to play a more prominent role in managing acid mine drainage (AMD)
	(c) Stringent air entry control can be achieved by strategically placing occasional thinner layers with better compaction (e.g., <1 m).
	(d) Adding water to optimize compaction can further reduce air entry.

which collects and flushes oxidation products, leading to the generation of acid mine drainage (AMD).

The design of mine waste rock piles can incorporate multiple thin layers (e.g., 0.5 m) of densely compacted waste rock. These layers can be strategically arranged within a PAF waste rock pile to distribute responsibility for pollution control throughout the entire pile, rather than relying on a single layer or cover system.

Additionally, thin layers of engineered oxygen-consuming materials, such as sulfide-bearing waste rock combined with excess carbonate, can be strategically distributed throughout a PAF waste rock pile to create multiple internal barriers that effectively exclude oxygen. Alkalinity-generating layers, including limestone, dolomite, and caustic magnesia, are deliberately incorporated to function as passive alkalinity addition layers, acid water neutralization zones, and oxygen entry control zones, especially when the carbonate is ultra-fine-grained. Similarly, highly reactive sulfidic waste can be applied in layers with controlled thickness to aid in heat dissipation and reduce the risk of spontaneous combustion.

3.3.3 Uranium Behaviour in Waste Rock Piles

The primary factors influencing uranium behaviour in waste rock piles, particularly concerning its migration and fate, include oxygen and water infiltration. Under oxidizing conditions, uranium mobility in the waste rock piles depends on a series of geochemical reactions involving the dissolution of sulfide and carbonate minerals after the waste rock is exposed to oxygen and water. These reactions ultimately govern solution pH, redox, aqueous U complexation, and sorption site availability. First, sulfide minerals in the presence of atmospheric O_2 and water undergo oxidation, which produces hydrous ferric oxide and sulfuric acid. The pyrite oxidation reaction is:

$$4FeS_2 + 15O_2(aq) + 14H_2O \rightarrow 4Fe(OH)_3 + 8SO_4^{2-}(aq) + 16H^+(aq)$$

The dissolution of carbonate minerals buffers acidity. The reaction of carbonate mineral dissolution by acid attack is as follows:

$$CaCO_3 + 2H^+(aq) \rightleftharpoons Ca^{2+}(aq) + H_2CO_3(aq)$$

Dissolved inorganic carbon and calcium released in the reaction become available to complex uranyl and enhance its mobility through the reaction:

$$UO_2^{2+}(aq) + x\,Ca^{2+}(aq) + 3HCO_3^-(aq) \rightleftharpoons Ca_xUO_2(CO3)_3^{2x-4} + 3H^+(aq); (x = 1\,\mathrm{or}\,2)$$

The release of calcium can, in theory, enhances uranium mobility by competing with uranium for sorption sites on hydrous ferric oxide. Conversely, the oxidation of sulfide minerals forms hydrous ferric oxide, which reduces uranium mobility by increasing the availability of sorption sites.

Several conditions affect the dissolution of uranium from the rock fragments: (a) the creation of acidic or alkaline environments within the pile, (b) the site's hydrogeological characteristics, and (c) the natural weathering processes that impact the rocks. The mobility of uranium is governed by pH, redox conditions, aqueous complexation, interactions with mineral surfaces, and the dissolution or precipitation of minerals within the waste rock piles.

3.3.3.1 Influence of Acid Formation

When the ore contains the mineral pyrite (FeS_2) and uranium deposits, the pyrite is often abundant within the mine waste rock pile. When this pyrite is exposed to air and water, an oxidation reaction occurs, producing sulfuric acid (H_2SO_4). This reaction primarily drives acid mine drainage (AMD) and acid formation.

Mine waste rock piles create ideal conditions for acid formation due to several factors:

(a) They have the right weathering conditions, including temperature, humidity, and exposure to air, all of which influence oxidation rates.
(b) Rock fragmentation, as smaller particles have a larger surface area, increases oxidation.

Additionally, the presence of bacteria can accelerate oxidation, further increasing acid production.

AMD can be pretty acidic. Upon contact with the rock fragments in the waste rock piles, solid-state uranium is converted to a soluble form under conditions with a pH < 2. When solid uranium is in the U(VI) form of UO_3, it reacts directly with the acid in AMD. In this context, the key reaction process governing the mobilization of uranium from the associated waste rocks is:

$$UO_3 + 2H^+ \rightarrow UO_2^{2+} + H_2O$$

Under the influence of H^+, U(VI) transforms from a solid state to the most soluble uranyl ion (UO_2^{2+}) form. The uranyl ion combines with sulfate, ultimately existing in the productive solution as complexes like UO_2SO_4.

When waste rocks predominantly contain uranium in the U(IV) form, uranium leaching proceeds in two steps. The U(IV), mostly present as UO_2 in the rocks, oxidizes to U(VI), enhancing its reactivity and subsequently reacting with the acid to form soluble complexes. The reaction requires an oxidizing agent. High levels of oxygen (O_2) in the waste rock pile create oxidizing conditions, which favour the reaction. The fundamental principle of the oxidation process is that ions act as electron transfer mediators to facilitate the transformation of uranium from the U(IV) to the U(VI) form. The oxidation process, in the presence of oxygen as the oxidizing agent, is as follows:

$$2UO_2 + O_2 \rightarrow 2UO_3$$

The oxidation of UO_2 and subsequent reaction with an acid can be combined and represented as:

$$UO_2^{+} + 2H^+ \rightarrow UO_2^{2+} + H_2O$$

Uranyl species are highly soluble and mobile, allowing them to migrate easily within piles and potentially contaminate the surrounding environment. However, iron generated from pyrite oxidation plays a crucial role.

Ferrous iron is produced when all the oxygen is consumed during pyrite oxidation. This scenario is not typical in waste rock piles, where much of the iron is oxidized to ferric ions. This oxidation process results in the formation of hydrous ferric bodies in the presence of water. Hydrous ferric oxides serve as significant sorbents for uranyl in mine waste, effectively sequestering uranyl through a bidentate inner-sphere complex at pH levels ranging from 3 to 9. Uranyl species are highly soluble and mobile, allowing them to migrate easily within piles and potentially contaminate the surrounding environment.

To this point, the assumption has been that waste rock piles have low carbonate content. In the case of high carbonate, the carbonate undergoes neutralization reactions involving the carbonate reacting with the AMD as follows:

$$2H^{+} + CO_3^{2-} \rightarrow H_2O + CO_2$$

High carbonate conditions in waste rock piles may occur naturally in geological settings where uranium deposits are present, such as sandstone, hydrothermal veins, karst cavities, and stratified carbonate rocks. However, these high carbonate conditions can also arise from management practices that involve the addition of excess limestone, specifically, calcium carbonate (calcite) or the double carbonate of calcium and magnesium (dolomite), during the consolidation of waste rock piles.

3.3.3.2 Influence of Alkaline Formation

Alkaline formation in a mine waste rock pile occurs when the rock contains naturally occurring alkaline carbonate or silicate minerals, such as calcite (calcium carbonate), dolomite (calcium magnesium carbonate), or certain types of feldspar (containing sodium or potassium). When exposed to air and water, these alkaline minerals weather and dissolve, releasing alkaline ions into the solution. This process raises the pH level of the surrounding water, creating an alkaline environment.

Grain size affects the release of alkaline ions from waste rock piles: smaller particles weather more quickly, leading to faster release. Other influencing factors include temperature and rainfall, which can accelerate the weathering process and increase alkalinity. Additionally, the initial chemistry of the water that infiltrates the waste pile can significantly impact the rate and extent of alkalinity formation.

When alkalinity is high ($pH > 9$), uranium becomes soluble. The mobilization of uranium in alkaline conditions has low kinetics and high-energy consumption. Thus, waste rock piles in warm regions with uranium-bearing sandstone are mostly developed under these conditions. The behaviour of uranium in waste rock piles with alkaline formation is primarily dictated by the type of inorganic carbon (i.e., carbonates) formed and dissolved in water after weathering, complemented by the availability of oxygen and other oxidizing conditions within the pile. In situations where the predominant constituents are carbonate and bicarbonate, U(VI) reacts with carbonate ions to form uranyl tricarbonate ions, as in the reaction:

$$UO_3 + 3CO_3^{2-} + H_2O \rightarrow UO_2\left(CO_3\right)_3^{4-} + 2OH^-$$

For bicarbonate ions, the reaction result is uranyl dicarbonate ions. The reaction is:

$$UO_3 + 2HCO_3^- \rightarrow UO_2\left(CO_3\right)_2^{2-} + H_2O$$

In cases where the alkaline media reacts with U(IV), the reaction for carbonate ions and bicarbonate ions produces uranyl dicarbonate ions ($UO_2\left(CO_3\right)_2^{2-}$) and uranyl tricarbonate ions ($UO_2\left(CO_3\right)_3^{4-}$), respectively. However, the reaction step differs. The mono-carbonate reaction is as follows:

$$UO_2 + 2CO_3^{2-} + 0.5O_2 + H_2O \rightarrow UO_2\left(CO_3O3\right)_2^{2-} + 2OH^-$$

For bicarbonate, the reaction is:

$$UO_2 + 3HCO_3^- + 0.5O_2 \rightarrow UO_2\left(CO_3O3\right)_3^{4-} + H_2O + H^+$$

Both $UO_2\left(CO_3\right)_2^{2-}$ and $UO_2\left(CO_3\right)_3^{4-}$ are soluble complexes that can easily migrate within and from the waste rock pile to contaminate receiving environments.

Uranium waste rock piles containing significant amounts of pyrite (FeS_2) and other sulfides can experience accelerated oxidation of pyrite in alkaline environments due to the presence of carbonate or bicarbonate ions. This process leads to the formation of various soluble iron-carbonate complex compounds, including $FeHCO_3^-$, $FeCO_3$, $Fe(CO_3)(OH)^-$, and $FeCO_3^{2-}$. These complexes play a crucial role in immobilizing uranium as precipitates and in promoting the nucleation of secondary mineralization.

3.3.3.3 Influence of Neutral Condition Formation

The presence and abundance of sulfide minerals, such as pyrite and chalcopyrite, along with neutralizing minerals like carbonates and silicates in the waste rock, significantly influence the potential for neutral rock drainage (NRD). NRD occurs when water interacts with the minerals in the waste rock pile, primarily oxidizing sulfide minerals and generating acidity that is then neutralized by naturally occurring carbonate or silicate minerals within the rock. As a result, the discharge exhibits a neutral pH but contains elevated levels of dissolved metals.

The NRD has a pH ≥ 6 and is typically characterized by high alkalinity (i.e., as $CaCO_3$) and dissolved calcium. As discussed in the previous section, NRD formation in uranium waste rock piles can occur naturally when the waste rocks contain both acid-forming and alkaline minerals. It can also be induced by placing a significant amount of alkaline rocks, such as limestone, during the consolidation of waste into waste rock piles.

The behaviour of uranium in NRD-forming waste rock piles is influenced by the presence of bicarbonate ions (HCO_3^-) in weakly acidic water, which complexes with uranium to form uranyl ions. The release of uranium from waste rock is explicitly linked to metal-carbonate species, particularly calcium, leading to the formation of weakly sorbing calcium-carbonate-uranyl aqueous complexes. In some instances, magnesium-carbonate-uranyl may also play a role. This phenomenon contributes to increased uranium release from carbonate-bearing rocks. When these rocks contain a high amount of sulfide minerals, even larger quantities of uranium are released, as sulfuric acid generated during the oxidation of sulfide minerals enhances the dissolution of carbonate minerals, promotes calcium release, and facilitates the formation of calcium-carbonate-uranyl aqueous complexes.

However, a considerable amount of uranium is re-immobilized in iron oxide-rich materials via surface complexation. This surface complexation is sensitive to the concentrations of dissolved calcium and pCO_2 under NRD conditions. The uranyl quocations (UO_2^{2+}) can adsorb onto clays, organic matter, and hydrous ferric oxides in the waste rock piles. However, uranyl sorption is considerably weakened by the formation of the calcium-carbonate-uranyl aqueous complexes (i.e., $Ca_2UO_2(CO_3)_3$ and $CaUO_2\left(CO_3\right)_3^{2-}$) under neutral to alkaline pH conditions due to competition for sorption sites. Analogous Mg-uranyl-carbonate species also exist, although their influence on U mobility is less understood.

3.3.3.4 Influence of Gases: Coexistence of CO_2 and O_2 Formation

Carbon dioxide can be produced in waste rock piles, primarily through the oxidation of sulfides. During this process, sulfur-containing minerals in the rock interact with oxygen and water, forming sulfuric acid. The sulfuric acid then reacts with carbonate minerals present in the rock, releasing carbon dioxide gas.

The presence of oxygen, whether it enters through ingression or is trapped within waste rock piles, interacts with the carbon dioxide generated during sulfide oxidation. This interaction results in the formation of a CO_2-O_2 dissolution medium by dissolving infiltrated water or groundwater. This CO_2-O_2 medium is highly effective for dissolving uranium from the waste rocks. In this context, O_2 is an oxidizing agent, while CO_2 is a complexing agent.

During the process of extracting uranium from the waste rocks, carbon dioxide dissolves to create carbonic acid (H_2CO_3), as illustrated in the reaction:

$$CO_2 + H_2O \rightleftharpoons H_2CO_3$$

Carbonic acid subsequently undergoes dissociation to form bicarbonate ions (HCO_3^-) within the solution, as represented by the following reaction:

$$H_2CO_3 \rightleftharpoons H^+ + HCO_3^-$$

In parallel, molecular oxygen facilitates the oxidation of a substantial fraction of U(IV) present in the waste rocks, converting it to its hexavalent form through the following reaction:

$$2UO_2 + O_2 \rightarrow 2UO_3$$

Subsequently, U(VI) may form soluble uranyl dicarbonate ions ($UO_2(CO_3)_2^{2-}$) through the following reaction:

$$UO_3 + 2HCO_3^- \rightarrow UO_2(CO_3)_2^{2-} + H_2O$$

Alternatively, some UO_2 may directly react with O_2 and HCO_3^- to generate uranyl tricarbonate ions ($UO_2(CO_3)_3^{4-}$) following the reaction:

$$UO_2 + 3HCO_3^- + 0.5O_2 \rightarrow UO_2(CO_3)_3^{4-} + H_2O + H^+$$

In addition to the fundamental reactions associated with the dissolution of U(IV) and U(VI), the CO_2-O_2 medium also triggers various water-rock interactions. When carbon dioxide dissolves in water, it forms H_2CO_3, HCO_3^-, and H^+, resulting in an increased concentration of H+ and a consequent rise in groundwater acidity. This shift enhances the dissolution of carbonate minerals in uranium deposits, yielding bicarbonate species that are readily complexed with uranium minerals. Interactions with carbonate minerals may increase the porosity and permeability of the pore media within the waste rock pile, thereby facilitating uranium movement.

The CO_2-O_2 medium can also initiate complex reactions, forming uranyl-carbonate ternary complexes of U(IV). This phenomenon occurs because U(IV) readily reacts with ligands such as oxygen, nitrogen, and sulfur under specific conditions, forming coordination bonds. Carbon dioxide further enhances this process by facilitating the formation of uranyl-carbonate ternary complexes. Notable examples include species such as $MUO_2(CO_3)_3^{2-}$ and $M_2UO_2(CO_3)_3$, where M^{2+} represents divalent cations like Mg^{2+}, Ca^{2+}, Sr^{2+}, and Ba^{2+}.

These uranyl-carbonate ternary complexes significantly reduce the tendency of U(VI) complexes to adsorb onto the charged surfaces of iron (hydro)oxides and clay minerals in the presence of carbonate. This reduction promotes the movement of uranium-bearing solutions. In summary, the formation of U(IV) complexes is influenced by carbon dioxide and oxygen, which play a crucial role in the migratory behaviour of U(VI) complexes.

3.3.3.5 Influences of Microbial Activity

Microbial activity plays a crucial role in either the dissolution of uranium from rocks or its immobilization in water, depending on whether the environment within uranium waste rock piles is oxygen-rich (oxic) or oxygen-depleted (anoxic). Like soil and groundwater environments, uranium waste rock piles serve as habitats for both aerobic and anaerobic microorganisms. The activities of these microbes can significantly alter the pH and redox conditions in the system, thereby affecting uranium behaviour.

Oxidizing Microbial Activity

The influence of microbial activity on the behaviour of uranium and related metals in the waste rock pile varies based on whether aerobic or anaerobic microbes are dominant. For example, interactions between solid uranium and aerobic microorganisms often result in oxidation of uranium from the U(IV) state to U(VI).

The dissolution of uranium from waste rocks is facilitated by the activities of acidophilic and autotrophic prokaryotes that oxidize iron or sulfur. Microorganisms such as *Acidithiobacillus ferrooxidans*, *Leptospirillum* sp., and *Ferroplasma acidiphilum* thrive in acidic environments, where they extract energy through chemolithotrophic processes that oxidize ferrous iron to ferric ions and reduce inorganic sulfur compounds. They obtain energy by oxidizing a variety of sulfur compounds, including sulfide, elemental sulfur, and thiosulfate. Figure 3.8 summarizes the mechanism and processes involved in the bio-oxidation of uranium. Chemolithotrophs are organisms that derive energy from the oxidation of inorganic compounds rather than organic material. Alongside *Acidithiobacillus* and *Leptospirillum*, other notable acidophiles include *Sulfobacillus* and *Ferrimicrobium*, both of which are recognized for their ability to thrive in acidic conditions and oxidize inorganic substances.

The bio-oxidation of iron and sulfur by acidophilic microorganisms involves specific enzymes. For example, *Acidithiobacillus ferrooxidans* produces tetrathionate hydrolase, which facilitates the hydrolysis of tetrathionate. They also utilize a chain of membrane-bound cytochromes to mediate the oxidation of Fe^{2+} to Fe^{3+}. Additionally, other strains of *Acidithiobacillus* possess a cytoplasmic enzyme, sulfur-oxidizing reductase, which plays a role in the oxidation of elemental sulfur.

Acidophilic consortia of Fe- and S-oxidizing microorganisms create an acidic environment in waste rock piles, thereby promoting oxidation. Under the acidic conditions established by these microbial communities, U(VI) is transformed into soluble uranyl ions, which are a key factor in the dissolution of UO_3 from the rocks. The reaction involved is as follows:

$$UO_3 + 2H^+ \rightarrow UO_2^{2+} + H_2O$$

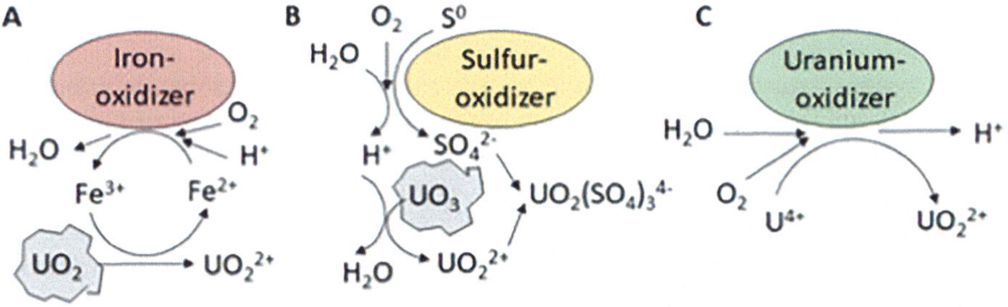

Fig. 3.8 Mechanism of bio-oxidation of uranium from U(VI) to U(VI) through interaction with acidophilic (**a**) Fe- and (**b**) S-oxidizing microorganisms. (**c**) is a general depiction of the process in the presence of an oxidizer from the processes in (**a**) and (**b**). (Modified from Kaksonen et al. [23])

The trivalent iron ions within the microbial consortia act as oxidizing agents, driving a reaction that converts solid UO_2 into soluble uranyl ions. The reactions involved are:

$$UO_2 + 2Fe^{3+} \rightarrow UO_2^{2+} + 2Fe^{2+}$$

Sustaining acidophilic Fe- and S-oxidizing microorganisms in the waste rock pile requires a consistent availability of dissolved O_2 and CO_2. This highlights the combined effect of CO_2 and O_2 in the processes that release uranium from the waste rock pile. Oxygen triggers the formation of soluble ferric iron from Fe^{2+} and the generation of sulfuric acid from reduced sulfur compounds, including elemental sulfur. The reaction plays a crucial role by continuously supplying trivalent iron ions, which serve as effective oxidizing agents in the oxidation of UO_2 to uranyl ions, thereby promoting their dissolution. The reaction that produces ferric ions is:

$$Fe^{2+} + O_2 + 4H^+ \rightarrow 4Fe^{3+} + 2H_2O$$

The process involved in generating sulfuric acid is:

$$2S^0 + 3O_2 + 2H_2O \rightarrow 2SO_4^{2-} + 4H^+$$

The dissolution of sulfides generates additional H^+, thereby sustaining the acidic environment that facilitates both uranium dissolution and the proliferation of acidophilic bacteria. This highlights the critical role of pyrite-rich waste rock in the pile, which, under the influence of microorganisms and acidic conditions, releases Fe^{2+}, SO_4^{2-}, and H^+. As a result, the dissolved species are continually converted into Fe^{3+}, thereby enhancing the bioleaching process by supplying the additional iron ions necessary for the oxidation of U(IV). The reaction can be outlined as follows:

$$FeS_2 + 14Fe^{3+} + 8H_2O \rightarrow 15Fe^{2+} + 2SO_4^{2-} + 16H^+$$

Reductive Microorganisms

Anoxic zones in waste rock piles are primarily inhabited by anaerobic microorganisms, fungi, and bacteria, which significantly influence the migration of labile uranium. These microbes interact with uranium through four key mechanisms: bioreduction, biosorption, biomineralization, and bioaccumulation. The interaction of microbes with uranium is summarized in Fig. 3.9. Specifically, anaerobic microbes are responsible for the bioreduction of uranium, resulting in biotransformation and biomineralization.

Uranium Bioreduction Mechanism

In the uranium bioreduction process, bacteria transform soluble hexavalent uranium (U(VI)) into less soluble tetravalent uranium (U(IV)). This occurs through microbial metabolism, where bacteria use U(VI) as an electron acceptor in the absence of oxygen. Various prokaryotes, especially iron-reducing and sulfate-reducing bacteria, utilize organic electron donors to generate electrons for this reduction. The resulting transfer of electrons leads to the reduction of the uranyl ion to uraninite (UO_2). Iron-reducing bacteria such as *Geobacter uraniireducens* and *Geobacter daltonii* are involved in this process, while *Desulfovibrio desulfuricans* can also utilize U(VI) as an electron acceptor.

The process involves enzyme catalysis. Figure 3.8 illustrates the mechanism involved. In summary, enzymatic uranium reduction can occur directly or indirectly in the cytoplasm, periplasm, at the outer membrane, or extracellularly in the bacterial cell. Generally, cytochromes play a critical role in the reduction process. Located

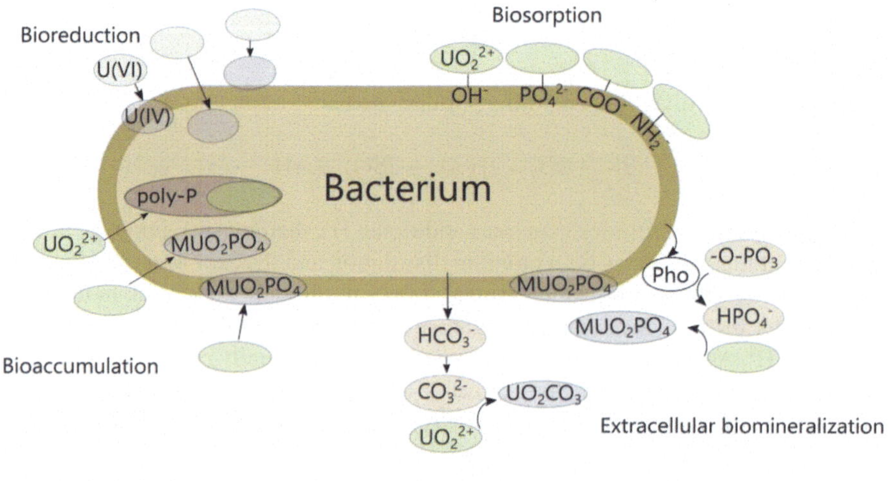

Fig. 3.9 Bacterial interactions with uranium. Uranium minerals are typically presented as UO_2CO_3 or MUO_2PO_4, where M represents a metal ion. Pho, phosphatase; poly-P, polyphosphate granules. (Modified from Rogier et al. [24])

on either the subsurface, periplasmic, or outer surface in various bacteria, cytochromes serve as significant intermediate electron carriers (in the absence of hydrogen). In some bacteria, outer-membrane cytochromes can act as terminal uranium reductases. However, the precise electron transfer pathways involved in uranium reduction remain unclear. The process requires enzyme catalysis, as illustrated in Fig. 3.10, which shows the underlying mechanism. In essence, enzymatic uranium reduction can occur directly or indirectly within the cytoplasm, periplasm, outer membrane, or extracellularly in bacterial cells. Cytochromes are generally crucial to the reduction process, as they are located on the subsurface, periplasmic, or outer surface of various bacteria and act as essential intermediate electron carriers in the absence of hydrogen. In certain bacteria, outer-membrane cytochromes may function as terminal reductases for uranium. However, the exact pathways involved in electron transfer for uranium reduction remain unspecified [24].

Biomineralization of Uranium

Another influence of microbes on uranium behaviour in the waste rock pile is the biomineralization of uranium. Uranium biomineralization is the process by which microorganisms induce the precipitation of uranium into insoluble minerals. This process involves interactions between uranium and microbial enzymes, microbe-associated ligands, such as phosphate, carbonate, or hydroxide, and cell-surface interactions, ultimately leading to the formation of uranium-phosphate precipitates or other minerals. Biomineralization can occur both on the cell surface

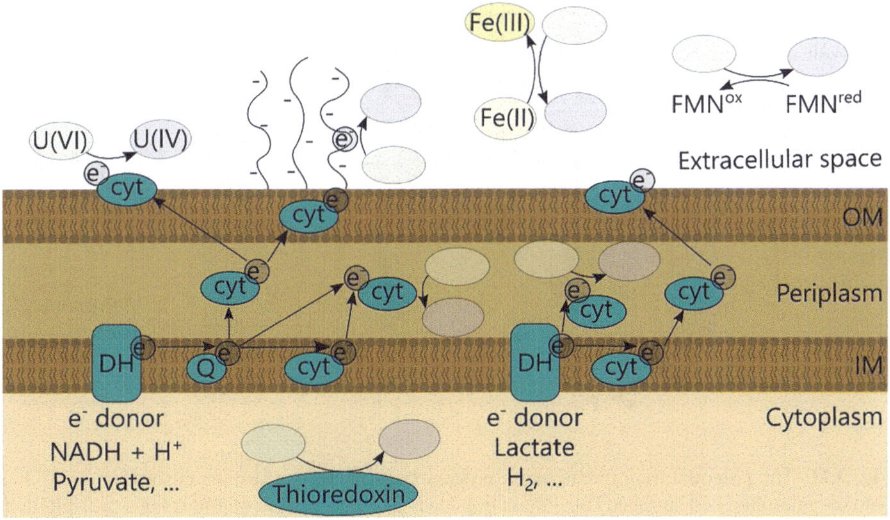

Fig. 3.10 Bacterial uranium reduction mechanisms. *DH* Dehydrogenase, *Q* quinone, *cyt* cytochrome, *FMN* flavin mononucleotide. (Modified from Rogier et al. [24])

(extracellular) and within the cells (intracellular). The tentative mechanisms involved are schematically illustrated in Fig. 3.11.

The mechanism in question involves microbes that produce enzymes, specifically phosphatases, which convert organic phosphorus substrates, such as polyphosphates, into inorganic phosphate. This inorganic phosphate then leads to the precipitation of uranium. Notably, around 80% of soil microbes can cleave organophosphates through their phosphatase activity. Some of the key species involved include *Serratia*, *Proteus*, *Bacillus*, *Arthrobacter*, and Streptomyces, as well as various fungi. For example, *Serratia* sp. converts exogenous β-glycerophosphate to phosphate, thereby precipitating uranium. At the same time, *Pseudomonas aeruginosa* degrades its polyphosphates to phosphate, leading to the formation of uranyl phosphate on cell membranes. Additionally, *Enterobacter* sp. promotes the intracellular mineralization of U(VI) to uramphite ($NH_4UO_2PO_4 \cdot 3H_2O$) under neutral conditions, facilitated by alkaline phosphatase located in the periplasmic space.

Microbes produce ligands, such as phosphate, carbonate, and hydroxide, that bind to uranyl ions (U(VI)). These ligands, in conjunction with cell surface components like carboxylate, amino, and amide groups, create nucleation sites for the precipitation of uranium minerals. For instance, *Ankistrodesmus* sp. captures uranium on its cell surface through complexation with carboxylate, amino, and amide

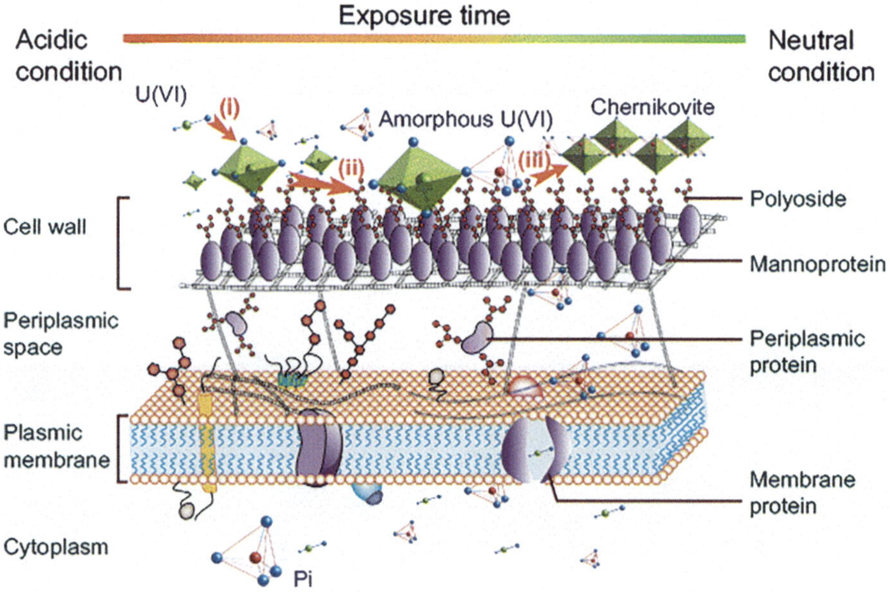

Fig. 3.11 The potential biomineralization process begins with the surface adsorption of UO_2^{2+} onto the microbial cell surface. This interaction leads to the formation of amorphous precipitation when U(VI) reacts with the released phosphate (Pi) and carbohydrates. Subsequently, the amorphous U(VI) precipitation transforms into the crystalline chernikovite extracellularly. Proposed by Shen et al. [25]

groups, thus facilitating the nucleation of insoluble uranium minerals. Certain species, such as *Pseudomonas* sp., can immobilize uranium by utilizing acrylamide, carbonyl, and phosphate as the primary functional groups that interact with uranium. Other species, such as *Shewanella putrefaciens*, form complexes with uranium and phosphoryl- and nitrogen-containing groups on their proteins within extracellular polymeric substances [26].

The most common biomineralized products are uranium-phosphate minerals. Table 3.3 lists some of the common uranium-phosphate minerals found in various waste rock piles worldwide. Amorphous uranium-phosphate complexes convert into crystalline nano-minerals. However, other minerals, including carbonates and silicates, form with these ligands. In some cases, other uranium minerals, such as uraninite (UO_2) or compreignacite ($K_2[(UO_2)_6O_4(OH)_6]\bullet 8H_2O$), can also form.

In certain waste rock piles, elevated levels of uranium co-precipitate with calcite. However, in undersaturated environments, calcite can significantly influence uranium mobility. Dissolved calcium ions and carbonate from calcite can form a complex with U(VI), resulting in the creation of Ca_2UO_2.

Evidence of uranium biomineralization with silicates is provided by the presence of sodium boltwoodite in specific uranium waste rock piles. When sodium boltwoodite [$Na(UO_2)(SiO_3OH) \cdot 1.5H_2O$] interacts with phosphates and silicates, U(VI) precipitates as sparingly soluble minerals, including metaschoepite ($UO_3{\cdot}2H_2O$) and various phosphate-silicate compounds [27].

3.3.3.6 Influence of Weathering Processes

Weathering, the breakdown of rocks and minerals through physical and chemical processes, is accelerated in the waste rock pile by exposure to water, oxygen, and temperature fluctuations. In addition to these environmental factors, geochemical and mineralogical reactions significantly influence the weathering rates of waste rock, which, in turn, affect the quality, quantity, and behaviour of uranium within the piles.

As rocks weather, uranium-bearing minerals, including those found in magmatic and hydrothermal systems, are released, potentially increasing uranium mobility. However, the weathering process also leads to the formation of secondary uranium

Table 3.3 Some major uranium-phosphate mineral products of uranium biomineralization are found in different uranium waste rock piles

Mineral phase	Chemical formula
–	HUO_2PO_4
–	$NH_4UO_2PO_4$
Autunite	$Ca(UO_2)_2(PO_4)_2 \cdot 6H_2O$
Ankoleite	$K_2(UO_2)_2(PO_4)_2 \cdot 6H_2O$
Uramphite	$(NH_4)(UO_2)PO_4 \cdot 3H_2O$
Chernikovite	$H_2(UO_2)_2(PO_4)_2 \cdot 8H_2O$

minerals, such as uranyl phosphates (e.g., autunite), which are commonly found within the clay fraction of the waste rock pile.

A notable phenomenon arising from the oxidative weathering of uranium-containing rocks in waste rock piles is the trapping of uranium by natural gels. These gels, composed of amorphous silica, crystalline hydrated U-hydroxides, and silicates, consist of intimate mixtures of short-range ordered aluminosilicates and hydrous ferric oxides. In silicon/aluminium-rich gels, aluminium is six-fold coordinated to oxygens, while silicon tetrahedra are mostly isolated, as seen in Al-rich allophane. Al-rich allophane is a clay mineral characterized by a high aluminium (Al) to silicon (Si) ratio, typically around 2. Iron-rich gels likely result from hydrolysis and oxidation of Fe^{2+} in solution, combined with the presence of silicon and aluminium during the precipitation of Fe^{3+}.

When the waste rock pile contains a high proportion of U-mineralized granite, its weathering can lead to the formation of gels. In addition to uranium, this weathering releases silicon, aluminium, and iron as major components. For example, gels with zircon compositions are responsible for trapping mobile uranium in the acidic waters of waste rock piles.

During the weathering process, uranium is mobilized in the form of soluble uranyl (UO_2^{2+}). Uranyl can readily form complexes with anions, such as carbonate, phosphate, sulfate, fluoride, chloride, and silicate; at elevated pH, it can also form polynuclear species. In the first step, uranyl is complexed by low-Z elements, such as silicon and aluminium. In the second step, it becomes trapped within hydrous ferric oxides during iron precipitation.

The mobility of uranium is significantly influenced by sorption, co-precipitation, and U-trapping processes, leading to secondary mineralization in a waste rock pile. As uranium migrates, it can react with other elements, forming new secondary minerals, including uranyl phosphates like autunite and torbernite. These phosphates are often present in the clay fraction of the waste rock. The retention of U in the matrix at the base of the waste rock pile, and the underlying paleo-soil, most likely occurs through the precipitation of (nano-) uranyl phosphates, or a combination of co-precipitation and adsorption reactions of U onto Fe (oxy)hydroxides and/or clay minerals.

3.3.3.7 Influence of the Geochemical and Hydrogeological Background.

Uranium can contaminate groundwater when water seeps and moves vertically through waste rock piles, eventually reaching the water table. The leaching of uranium from waste rocks and its subsequent migration are affected by various geochemical and hydrogeological factors, including pH, Eh (redox potential), adsorption, sedimentation, complexation, and chelation of organic and inorganic compounds, as well as precipitation and evaporation. Additionally, these factors influence geochemical processes involving radionuclides, metals, and metalloids that coexist with uranium in ore.

3.4 Tailing Piles

Tailings are waste materials produced during ore processing to extract valuable products, such as uranium. Chemically, the tailings contain diluted ore and various pollutants that are mixed during milling. Most mine tailings are generally deposited underwater to prevent them from oxidizing, provide shielding from radiation emanating from the tailings, and stop dust from blowing off the surface of the tailings.

Specifically, uranium tailings are a waste byproduct of uranium mining and processing. Raw uranium ore is brought to the surface and crushed into fine sand during mining. The uranium-bearing minerals are then removed through heap leaching with acids or bases, and the remaining radioactive sludge, known as uranium tailings, is stored in large impoundments or tailings management facilities (TMFs). These impoundments can be either tailing heaps or ponds.

3.4.1 Tailing Heaps and Ponds

The TMFs are designed to isolate the tailings from the surrounding environment for an extended period. The TMFs are typically located within mined-out open pits or in surface impoundments that utilize natural features and artificial barriers, such as dams, to contain the tailings. They are designed to divert groundwater and surface water away from the tailings, preventing contamination and acid mine drainage.

3.4.1.1 Tailing Heaps

A tailings heap is a large mound of solid tailings, typically constructed in layers or lifts, forming a ridge-like structure. The tailings are deposited onto the heap and may be compacted or organized to ensure stability. Generally, the tailings in a heap have low water content, as they are often partially dewatered prior to placement. While tailings heaps may resemble waste rock piles, they specifically contain byproducts and waste materials derived from the extraction and processing of ores. Consequently, it is crucial that the heap design effectively prevents landslides or other structural failures.

3.4.1.2 Tailings Pond

A tailings pond is a basin designed to hold the liquid slurry of mining tailings. It is created by constructing a dam or embankment, and the tailings slurry is pumped into the pond, allowing the solid particles to settle. Subsequently, the water can be treated for potential recycling or safely discharged into the environment.

Essentially, a tailings heap is a solid pile, while a tailings pond is a basin containing liquid. Both methods involve complex geochemical processes in tailings ponds, including interactions among water, air, and the minerals in the tailings. Similar to waste rock piles, the geochemical processes in tailings heaps and ponds primarily involve the oxidation of sulfide minerals, which release metals (e.g., Cu, Pb, Zn, As, Cd, Co, and Ni) and increase acidity, often leading to acid mine drainage (AMD). These processes are influenced by factors such as redox potential, water saturation, and pH in the tailings. A more detailed discussion on the geochemical processes and mechanisms involved can be found in the section on waste rock piles above.

Tailings materials are generally associated with the highest risk of acid mine drainage (AMD) due to their highly sulfidic content, ultra-fine-grained characteristics, and presence in large, partially unsaturated deposits. These attributes are inherently tied to the reactive nature of tailings. The elevated sulfide content and the reactive nature of the tailings lead to rapid oxygen consumption within the pore spaces. This oxygen depletion can slow down or even halt the oxidation of sulfides throughout the tailings heap. Oxygen replenishment can occur solely through atmospheric exchange at the air-tailings interface, on the one hand.

On the other hand, the grain size of most tailings limits this process to diffusion, which impacts only the uppermost 100–2000 mm [28]. Consequently, sulfide oxidation in tailings can be inherently self-limiting. This may be contrary to waste rock piles.

As water flows through the tailings, the pH may rise due to mineral buffering, such as carbonates, or dilution. This increase in pH can cause metal hydroxides and other secondary minerals to precipitate, potentially trapping some dissolved metals. Metal ions can also be adsorbed onto the surfaces of different minerals, such as iron oxides, further affecting their mobility. As water evaporates, dissolved salts become more concentrated, which can lead to the precipitation of salts like gypsum. The process can change the overall chemistry of the tailings and influence the solubility of other elements. Therefore, in arid or semi-arid climates, evaporation can significantly influence the behaviour of tailings ponds.

The geochemical processes in tailings ponds can continue for decades, even centuries, after mine closure. The long-term stability of tailings deposits depends on the balance between acidity generation and buffering, the rate of metal release and sequestration, and the effectiveness of any engineered controls. Nonetheless, the geochemical processes and mechanisms in TMF are similar to those in WRP, as discussed in detail above.

3.4.2 *Uranium Tailings*

3.4.2.1 Production

Following excavation, the ore-bearing rocks are transported to the milling machine, where they are ground into fine particles. Various chemicals are then introduced as extraction agents to separate the uranium from the rock materials. The resulting byproducts and waste are referred to as uranium tailings. These tailings typically consist of a mud-like substance containing powdery gangue, which is pumped into a tailings impoundment for storage and retention. Once dried, the tailings have a texture similar to fine sand.

3.4.2.2 Characteristics

It is estimated that one tonne of uranium ore yields between 0.45 and 2.3 kg of uranium, depending on the mineral's uranium content. Consequently, a significant volume of waste, known as tailings, is produced during uranium ore processing. This process focuses on extracting the uranium element from the ore, resulting in tailings that retain a substantial portion of the original radioactivity, along with all uranium decay products, metals, and processing reagents [29].

Uranium tailings can retain up to 85% of the ore's initial radioactivity and contain over a dozen radionuclides. Although classified as radioactive materials of low specific activity, concerns regarding uranium tailings arise from the considerable volume of waste generated, the variety of radionuclides present, and the fact that approximately 36% of these include long-lived nuclides, posing potential long-term radiation risks [30, 31].

Among the most significant radionuclides are thorium-230, radium-226, radon-222, polonium-210, and the daughter isotopes resulting from radon decay. Radium is considered the most hazardous component of uranium tailings because it emits radon, a radioactive gas that can easily diffuse into the environment. When inhaled, airborne radon decays into a series of hazardous short-lived products.

3.4.2.3 Uranium Tailing Heaps

Uranium tailings are typically stored aboveground and allowed to dry, forming uranium tailing heaps. This method is primarily employed because uranium is extracted by leaching, and these heaps can be re-mined multiple times. The geochemical properties of these tailing heaps resemble those of waste rock piles; however, the presence of extraction agents, such as acids and ammonia, influences their dynamics. Moreover, the hydrogeology of surface-deposited uranium tailing heaps exhibits similarities to that of waste rock piles.

When these tailings dry out on the surface, radioactive dust can be carried over long distances by the wind, potentially entering the food chain and contaminating water bodies. The risks associated with this dispersal remain uncertain, primarily due to the dilution effect it causes. While the majority of the tailing mass consists of inert rock, similar to the original ore before uranium extraction, it undergoes significant physical changes: it is ground up, mixed with large amounts of water, and exposed to atmospheric oxygen, which can alter its chemical behaviour considerably.

3.4.3 Uranium Behaviour in Tailings

3.4.3.1 Geochemical Behaviour

The geochemical behaviour of uranium in tailings mirrors that observed in waste rock piles, where the oxidation of sulfur-containing minerals, such as pyrite, and the subsequent production of acid mine drainage (AMD) significantly contribute to the release of uranium from the tailing material. These tailings may contain either primary or secondary uranium minerals, or sometimes both. The mobilization of uranium from tailings is influenced by several factors, including contact time, liquid-to-solid ratio, particle size, pH, specific water chemistry conditions, and the form of uranium present.

In uranium tailing piles, water availability can both facilitate and control the formation of AMD and the release of uranium from the tailing minerals. When uranium mine tailings are submerged or oversaturated (i.e., high liquid-to-solid ratio), a water depth of 30–45 cm or more can create a gas-entrapping barrier, depriving pyrite of oxygen and preventing pyrite oxidation. Since most uranium tailings are not stored in ponds but in heaps, some of the water may drain away while other portions evaporate, eventually leaving the tailing material dry. As a result, dry uranium tailing piles are almost chemically inert until they interact with water from precipitation, groundwater, and atmospheric oxygen, similar to WRPs as previously discussed in the section above.

The activity of uranium in the tailing piles depends on the interplay between pH-Eh and microbial activity, which influences dissolution, adsorption, speciation, and co-precipitation [32, 33]. For instance, uranium activity is limited in pH below 6 due to the formation of insoluble metal hydroxide compounds. Similarly, uranium activity is high under oxidizing conditions, meaning it increases with oxidation. Further, uranium exhibits a strong adsorption capacity under oxidizing conditions. Under reducing conditions, insoluble sulfides begin to sediment, removing significant amounts of metals and metalloids as insoluble sulfides [34]. All these affect and influence the hydrological behaviour of uranium in tailing heaps (Fig. 3.12).

In the aerated zone of the tailing, the UO_2^{2+} interacts with organic and inorganic compounds, forming stable anionic complexes that are less readily absorbed, thereby influencing the uranium transportation by seeping water. When the tailing materials are rich in alkaline-earth metals, uranyl and carbonate react with the alkaline metals to form ternary complexes $MUO_2(CO_3)_3^{2-}$ and $M_2UO_2(CO_3)_3$, where M^{2+} denotes Mg^{2+}, Ca^{2+}, Sr^{2+}, and Ba^{2+}. The complexes of $MUO_2(CO_3)_3^{2-}$ and $M_2UO_2(CO_3)_3$ are simultaneously formed for Ca^{2+} and Ba^{2+}, while Mg^{2+} and Sr^{2+} form only the $MUO_2(CO_3)_3^{2-}$ complex. In low Ca^{2+}, the $CaUO_2(CO_3)_3^{2-}$ complexes dominate, but increasing Ca^{2+} leads to $Ca_2UO_2(CO_3)_3$ domination. As a result, uranium immobilization through sorption onto anion-exchange resins is inhibited by forming the neutral $Ca_2UO_2(CO_3)_3$ species.35

3.4.3.2 Hydrological Behaviour

The long-term behaviour of uranium in the tailing piles focuses on its mobility and migration. The mobility of uranium in waste rock and tailing piles depends on the mineralogical and geochemical transformations, while the migration of uranium depends on its geochemical behaviour (see the preceding section) and the hydrological characteristics of the tailing pile. The migration mainly depends on the chemical speciation of uranium in the aqueous system of the tailing, as influenced by chemistry, flow rate, and the direction of water that transports mobile uranium within and outside the tailing material. These are water ingressions into the piles from the surface, primarily atmospheric precipitation (i.e., rain and snow), which are responsible for the vertical-to-basal leaching of uranium. Additionally, mounding of groundwater due to hydraulic conductivity and capillary effects transports uranium out of the piles into the receiving environment.

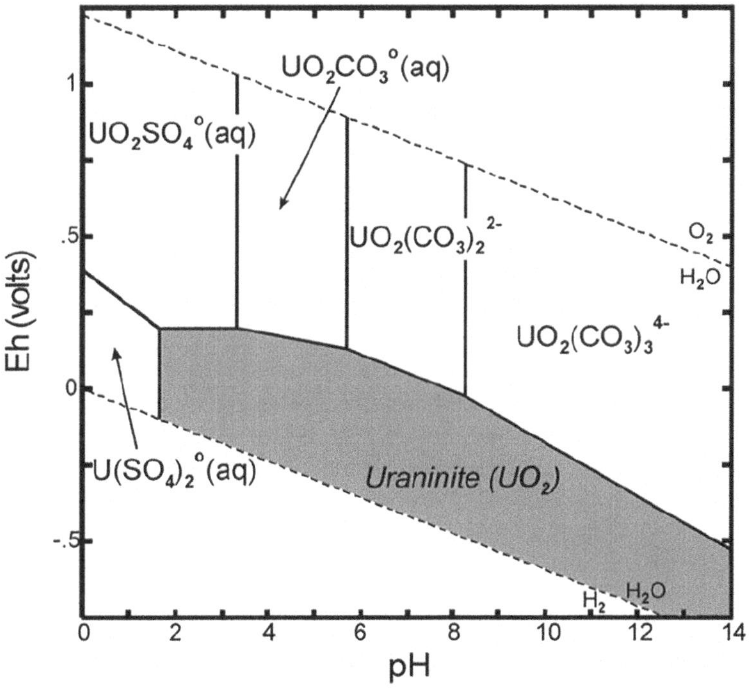

Fig. 3.12 The Eh-pH diagram illustrates the predominant aqueous species of uranium within the specified environmental conditions. The shaded region delineates the Eh-pH range characterized by increased uraninite solubility. This representation is adapted from the findings of Krupka and Serne [33]

Influence of the Circum-Arenization Process

The chemistry of the tailing material's acidity and the porewater's pH are significant factors that determine uranium migration and transport. The porosity of tailing depends on the grain size. The uranium ore processing (i.e., milling) and later weathering within the tailing piles transform the ore-bearing rock into an arena-like material. During the arenization process, the primary uranium phases, like coffinite ($U(IV)(SiO_4)_{1-x}(OH)_{4x}$) and uraninite ($U(IV)O_2$), are oxidized into readily soluble UO_2^{2+} readily mobile groups. This source of UO_2^{2+} combined with the dissolution of secondary minerals makes the tailing piles contain highly mobile uranium species [35, 36].

However, several reactive mechanisms between the mobile species and the mineral and chemical constituents of the tailing control the uranium species. The main mechanisms of uranium immobilization in the tailing piles are co-precipitation, complexation, and encapsulation. Uranium tends to accumulate in certain levels of clay minerals in the tailing, due to being absorbed on precipitation of Fe, Al, and OM or P-Al-Fe-Si colloids.

Influence of Hydroxyapatite

Specifically, the uranium migration in the form UO_2^{2+} species is controlled by sorption onto mineral phases, such as phyllosilicates and ferric oxyhydroxides, and by precipitation of secondary minerals, such as uranyl phosphates. For example, uranium migration is controlled by the presence of soddyite (i.e., $(UO_2)_2SiO_4 \bullet 2H_2O$). Furthermore, uranyl phosphates are efficient traps for uranium sequestration. Conversely, UO_2^{2+} forms soluble complexes with various anionic species, including carbonate, phosphate, or organic matter, which tend to limit hydrolysis and increase mobility [32, 35].

Over 90% of uranium immobilization is due to the formation of calcium uranyl phosphate; however, increasing carbonate concentration significantly inhibits uranium immobilization due to the competition between carbonate and phosphate. Tailing piles typically contain a substantial amount of calcium and carbonates. Some are added artificially as crushed limestone to neutralize the tailings when they are AMD-generating. Others are naturally occurring from the geosphere, where soluble inorganic phosphates are ubiquitous. The replacement reactions are spontaneous processes that occur at solid-liquid interfaces between calcium carbonate minerals and phosphate, resulting in the formation of different precipitates. During the precipitate process, uranium co-precipitates into a stable, irreversibly complex hydroxyapatite, $Ca_5(PO_4)_3(OH)$, often written $Ca_{10}(PO_4)_6(OH)_2$ to denote the crystal unit cell [37].

Influence of Iron Oxyhydroxide Surfaces

The waste rock and tailings may contain substantial quantities of Al-, Mn-, and Fe-oxyhydroxide minerals, including boehmite, manganite, goethite, and lepidocrocite. These oxyhydroxide minerals play a crucial role in regulating the intestinal transport and migration of uranium within the tailings. A key process involves the immobilization of uranyl species at the solid-water interface, which is predominantly influenced by U(VI) sorption behaviour. While U(VI) tends to be highly mobile under oxidizing conditions, its species are effectively adsorbed onto the surfaces of oxyhydroxides in supergene environments, primarily through surface complexation or precipitation. For instance, uranium sorption on Al-oxyhydroxides mainly occurs via surface precipitation, whereas on Mn- and Fe-oxyhydroxides it chiefly occurs via surface complexation.

Iron oxyhydroxide parageneses play a crucial role in regulating uranium mobility and the formation of uranyl phases in waste rock and tailing piles. The sorption process of U(VI) onto iron oxyhydroxides is primarily determined by intraparticle diffusion, which acts as the main rate-limiting factor. This sorption process is highly sensitive to pH levels but has only a minimal influence from ionic strength. Under acidic conditions, the presence of humic acid (HA) significantly enhances U(VI) sorption. Conversely, in alkaline conditions, HA inhibits sorption by forming soluble U(VI)-HA complexes. The sorption of U(VI) on iron oxyhydroxides is characterized as endothermic, favouring higher temperature conditions [38, 39].

Role of Ferrihydrite

Uranium can form ternary surface complexes with phosphates or organic matter on ferrihydrite. Ferrihydrites are short-range order minerals that appear as spherical nanoparticles and are thermodynamically unstable, gradually converting into more stable iron oxides, such as goethite or haematite, over time. However, substances such as organic compounds, phosphates, or silicates can act as crystallization inhibitors, stabilizing ferrihydrite and slowing its transformation into more stable mineral forms (Fig. 3.13). In addition, ferrihydrite boasts a high sorption capacity owing to its high surface area, serving as a primary trap for uranium and related metals, metalloids, and radionuclides. It effectively retains dissolved uranyl species and metals such as chromium, arsenic, selenium, cadmium, zinc, and lead through inner-sphere sorption complexes or co-precipitation.

Phosphate ions or organic matter increase the uranyl uptake of uranyl species on ferrihydrite in oxic geochemical systems. This process is attributed to the strong surface binding of both phosphate and organic matter, as well as the subsequent formation of ternary surface complexes with uranyl during their sorption to ferrihydrite [41–44].

The crystallization of 2-line ferrihydrite (2 l-$Fe(OH)_3$(s)) into goethite and magnetite is catalysed by Fe(II). An example of the mechanism involved in the catalytic transformation of ferrihydrite is illustrated in Fig. 3.14. During the Fe(II)-catalysed transformations of both 2-line ferrihydrite and ferrihydrite, uranium that co-precipitates with silicate (referred to as silicate-ferrihydrite) exhibits different behaviour: 2 l-$Fe(OH)_3$(s) transforms uniformly into goethite, whereas silicate-ferrihydrite retains a form of ferrihydrite.

Both co-precipitated and adsorbed U(VI) initially form a mononuclear bidentate surface complex with ferrihydrite and silicate-ferrihydrite. Throughout the Fe(II)-catalysed transformation, U(VI) associated with 2-line ferrihydrite is reduced and partially incorporated into the newly formed goethite, likely as U(V). In contrast, U(VI) linked to silicate-ferrihydrite remains unreduced, maintaining its initial adsorbed characteristics. Uranium(VI) initially adsorbed to silicate-ferrihydrite demonstrates greater resistance to reductive dissolution, indicating that its mobility is at least partially reduced, when the Fe(II)-catalysed transformation of ferrihydrite-like iron oxyhydroxides is inhibited [42, 45].

In summary, the adsorption of UO_2^{2-} and P by Fe oxides is a precursor in the formation of uranyl phosphates [47]. The formation of stable uranyl phosphates dominates the limitation of uranium mobility under circumneutral pH conditions, where adsorption onto Fe(ox)hydroxides and clay minerals also hinders uranium migration [48].

3.4.3.3 Role of Pre-existing Hydroxyapatite Transformation

Hydroxyapatite is primarily a biogenic mineral found in bones. It naturally exists in phosphate rock and limestone. As a result, hydroxyapatite may already be present in uranium tailings, either as part of limestone or alongside uranium minerals in the rocks. The pre-existing hydroxyapatite transforms into autunite, hydrated calcium uranyl phosphate, $Ca(UO_2)_2(PO_4)_2 \cdot 10 - 12H_2O$.

The transformation of hydroxyapatite into autunite may occur via two distinct mechanisms, depending on the pH of the interstitial water present in the tailings. The dissolution of hydroxyapatite and the subsequent precipitation of autunite occur when uranyl-bearing solutions are sufficiently acidic to dissolve the existing hydroxyapatite. Alternatively, uranyl ions can adsorb to the surface of hydroxyapatite, leading to a substitution of Ca^{2+}. This latter mechanism occurs when the pH of

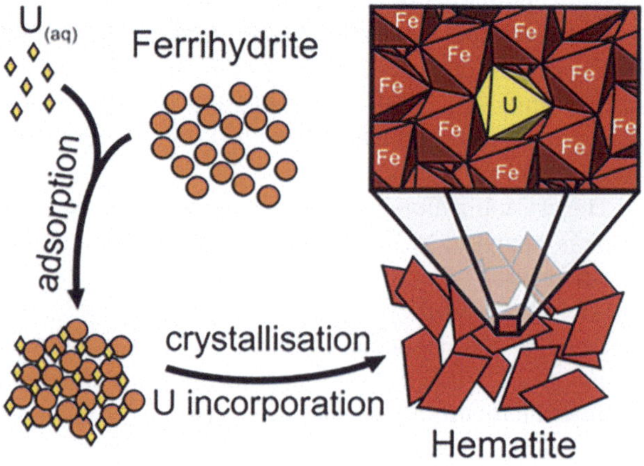

Fig. 3.13 Adsorption of uranium onto the surface of ferrihydrite, an amorphous iron oxyhydroxide, leads to incorporation of uranium into the structure of the iron oxide haematite during crystallization. (Adopted from Marshall et al. [40])

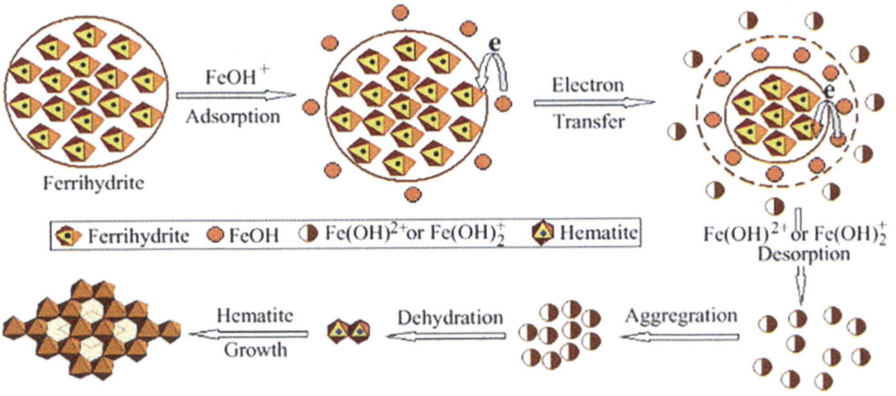

Fig. 3.14 Fe2+ catalytic transformation of ferrihydrate. (Adopted from Liu et al. [46])

the aqueous phase in the tailings approaches neutrality and when the dissolved uranyl ion concentration is low.

Influence of Processing/Extraction Reagents

Conventional mining operations employ a milling process in which ore is crushed and ground to liberate mineral particles. Subsequently, the ore is leached in tanks using sulfuric acid to dissolve uranium oxides. Ammonia is then extensively used to precipitate uranium as "yellowcakes" (ammonium diuranate) during ore processing. Consequently, the tailing waters exhibit elevated levels of ammonium, nitrate, and sulfate. While nitrate-containing reagents are typically used in limited amounts, significant nitrate concentrations in tailing water are often associated with microbial nitrification. The sulfate generally arises from the application of sulfuric acid. Still, in certain instances, it may also result from the oxidation of sulfide minerals associated with waste rock and mine tailings.

Ammonium and nitrate serve as redox regulators. The aerobic oxidation of ammonium consumes oxygen, contributing to less oxidative redox conditions. Additionally, nitrate acts as a relatively strong oxidizing agent. As such, both ammonium and nitrate play a crucial role in the mobilization and immobilization of uranium, which is sensitive to redox changes. For example, nitrate can trigger the remobilization of uranium within the reducing zones of tailing piles, particularly in the deeper sections. Moreover, the influence of ammonium and nitrate on redox conditions significantly affects dissolved organic carbon (DOC) concentration. This effective ligand stabilizes and facilitates the transport of uranium in tailing piles.

Furthermore, ammonium affects the transport and movement of uranium in tailings by changing pH levels. As a weak acid, ammonium lowers pH. The oxidation of ammonium to nitrate also releases protons, which further increases groundwater acidity. This pH change not only influences uranium speciation in solution (e.g., through pH/Eh interactions) but also plays a critical role in the adsorption of uranyl species onto solid surfaces by affecting charge interactions within the solid phase. Additionally, ammonium acts as a surface sorption competitor, competing with uranium and other metals for cation exchange sites in the tailings. Ammonia and nitrate serve as ligands for aqueous metal complexation [49].

3.5 Uranium Plume Migration

3.5.1 Plume Formation in WRP and Tailing Heaps

In uranium waste rock and tailings piles, plumes form due to acid mine drainage (AMD) and residues from extraction processes, including ammonium, sulfuric acid, and various organic acids. Uranium, in its uranyl form, is highly soluble in water, and this solubility increases under acidic conditions. Consequently, the initial formation of plumes in uranium waste rock and tailings is driven by the weathering of sulfide minerals (particularly sulfide sulfur) in the presence of oxygen. This weathering leads to the production of sulfate, sulfur, and residual extraction or leaching agents.

This process indicates the potential acidity (PA) of the piles, which represents the total acidity that can be generated when sulfide-rich waste rock interacts with water, leading to acid mine drainage or acid-rock drainage. This acidic drainage, along with processing agents, can facilitate the dissolution of uranium, other metals, and various pollutants. If these contaminants are not properly neutralized or managed, they can migrate as a plume, spreading into the surrounding environment.

The extent of plume formation is influenced by the potential acid neutralization capacity of alkaline components, such as limestone ($CaCO_3$), present in waste rock and tailing materials. The rate and extent of neutralization depend on the specific minerals present and their reactivity. Therefore, the actual plume that forms is determined by net acid generation (NAG), which reflects the remaining acidity after accounting for competing acid-producing and neutralizing reactions.

3.5.2 Plume Behaviour

The behaviour of contaminant plumes from uranium mine waste rock and tailings involves complex interactions of hydrology, geochemistry, and mineralogy. Key pathways for plume migration include groundwater, surface water, and infiltration through the unsaturated vadose zone. For effective plume migration, waste pile acidity must exceed neutralization potential, generating acidic leachate that mobilizes uranium and other contaminants.

Plume migration in groundwater can occur vertically and horizontally, influenced by hydraulic gradients, recharge rates, and aquifer permeability. Uranium species are highly mobile, with increased acidity helping to dissolve previously precipitated forms and locked minerals, particularly in tropical areas where dry periods can lead to fluctuating contaminant concentrations during initial rainfall.

Wind can aerosolize fine particulates from tailings, creating dust plumes that contaminate nearby ecosystems. The shape and extent of contaminant plumes depend on the source characteristics, aquifer properties, and hydraulic gradients. Waste materials exhibit heterogeneity, which affects plume behaviour.

Geochemical processes like mineral precipitation and adsorption can reduce contaminant concentrations. The potential for acid mine drainage (AMD) is linked to sulfide mineral types and concentrations in the waste. Environmental factors, such as precipitation and temperature fluctuations, affect AMD formation and plume migration. The storage method of waste rock and tailings also influences the levels of oxygen exposure and water infiltration, thereby prolonging contamination risks from ongoing oxidation and leaching processes.

3.6 Concluding Remarks

The behaviour of uranium in waste rock and tailing piles varies by site, shaped by local climate, geology, the type of mineral present, and the extraction methods employed. Consequently, the behaviour of uranium described in this chapter is fairly general, reflecting the average observations made across multiple abandoned uranium mining and processing locations. Thus, this chapter offers a comprehensive overview of these dynamics, which can differ notably from one site or region to another. Furthermore, the information provided is not exhaustive; ongoing research continues to enhance our understanding, and some data are evolving as advancements in research tools and methods improve our insights into uranium dynamics and behaviour at abandoned uranium sites.

References

1. Abdelouas, A. (2006). Uranium mill tailings: Geochemistry, mineralogy, and environmental impact. *Elements, 2*, 335–341. https://doi.org/10.2113/gselements.2.6.335
2. Harraz, H. (2010). *Underground mining methods*.
3. Bruneton, P., & Cuney, M. (2016). In I. Hore-Lacy (Ed.), *Uranium for nuclear Power* (pp. 11–52). Woodhead Publishing.
4. Everhart, D. L., & Mathez, M. (1951). *Geology of uranium deposits—A condensed version*. United States Atomic Energy Commission.
5. Hubbard, B. E., Gallegos, T. J., & Stengel, V. (2023). Mapping abandoned uranium mine features using Worldview-3 imagery in portions of Karnes, Atascosa and live oak counties, Texas. *Minerals, 13*, 839.

6. Morais, C. A., & Gomiero, L. A. (2005). Uranium stripping from tertiary amine loaded solution by ammonium sulfate. *Minerals Engineering, 18*, 1277–1281. https://doi.org/10.1016/j.mineng.2005.05.019
7. Landa, E. R. (2004). Uranium mill tailings: Nuclear waste and natural laboratory for geochemical and radioecological investigations. *Journal of Environmental Radioactivity, 77*, 1–27.
8. Lottermoser, B. G., & Ashley, P. M. (2005). Tailings dam seepage at the rehabilitated Mary Kathleen uranium mine, Australia. *Journal of Geochemical Exploration, 85*, 119–137.
9. Mkandawire, M. (2013). Biogeochemical behaviour and bioremediation of uranium in waters of abandoned mines. *Environmental Science and Pollution Research, 20*, 1–28. https://doi.org/10.1007/s11356-013-1486-3
10. Zhang, T., et al. (2023). A review: The formation, prevention, and remediation of acid mine drainage. *Environmental Science and Pollution Research International, 30*, 111871–111890. https://doi.org/10.1007/s11356-023-30220-5
11. Zevgolis, I. E. (2018). Geotechnical characterization of mining rock waste dumps in central Evia, Greece. *Environment and Earth Science, 77*. https://doi.org/10.1007/s12665-018-7743-5
12. Vriens, B., Plante, B., Seigneur, N., & Jamieson, H. (2020). Mine waste rock: Insights for sustainable hydrogeochemical management. *Minerals, 10*, 728.
13. Power, C., Ramasamy, M., & Mkandawire, M. (2018). Performance assessment of a single-layer moisture store-and-release cover system at a mine waste rock pile in a seasonally humid region (Nova Scotia, Canada). *Environmental Monitoring and Assessment, 190*, 186. https://doi.org/10.1007/s10661-018-6555-0
14. Lefebvre, R., Hockley, D., Smolensky, J., & Gélinas, P. (2001). Multiphase transfer processes in waste rock piles producing acid mine drainage: 1: Conceptual model and system characterization. *Journal of Contaminant Hydrology, 52*, 137–164. https://doi.org/10.1016/S0169-7722(01)00156-5
15. Schoonen, M., Elsetinow, A., Borda, M., & Strongin, D. (2000). Effect of temperature and illumination on pyrite oxidation between pH 2 and 6. *Geochemical Transactions, 1*, 23. https://doi.org/10.1186/1467-4866-1-23
16. Feng, X., et al. (2024). Electrochemical study characterizations of pyrite weathering in simulated acidic soil: Iron transformation, sulfur conversion and environmental implications. *Environmental Technology & Innovation, 35*, 103689. https://doi.org/10.1016/j.eti.2024.103689
17. Vriens, B., et al. (2018). Localized sulfide oxidation limited by oxygen supply in a full-scale waste-rock pile. *Vadose Zone Journal, 17*, 180119. https://doi.org/10.2136/vzj2018.06.0119
18. Kim, H., & Benson, C. H. (2004). Contributions of advective and diffusive oxygen transport through multilayer composite caps over mine waste. *Journal of Contaminant Hydrology, 71*, 193–218. https://doi.org/10.1016/j.jconhyd.2003.10.001
19. da Silva, J. C., do Amaral Vargas, E., Jr., & Sracek, O. (2009). Modeling multiphase reactive transport in a waste rock pile with convective oxygen supply. *Vadose Zone Journal, 8*, 1038–1050. https://doi.org/10.2136/vzj2008.0156
20. Coelho, P. C. S., Teixeira, J. P. F., & Gonçalves, O. N. B. S. M. (2011). In J. O. Nriagu (Ed.), *Encyclopedia of environmental health* (pp. 788–802). Elsevier.
21. Favas, P. J. C., Martino, L. E., & Prasad, M. N. V. (2018). In M. N. V. Prasad, P. J. de Campos Favas, & S. K. Maiti (Eds.), *Bio-Geotechnologies for mine site rehabilitation* (pp. 3–31). Elsevier.
22. Arnold, C. (2014). Once upon a mine: The legacy of uranium on the Navajo Nation. *Environmental Health Perspectives, 122*, A44–A49. https://doi.org/10.1289/ehp.122-A44
23. Kaksonen, A. H., Lakaniemi, A.-M., & Tuovinen, O. H. (2020). Acid and ferric sulfate bioleaching of uranium ores: A review #. *Journal of Cleaner Production, 264*, 121586. https://doi.org/10.1016/j.jclepro.2020.121586
24. Rogiers, T., et al. (2022). Molecular mechanisms underlying bacterial uranium resistance. *Frontiers in Microbiology, 13*, 822197. https://doi.org/10.3389/fmicb.2022.822197

25. Shen, Y., Zheng, X., Wang, X., & Wang, T. (2018). The biomineralization process of uranium(VI) by Saccharomyces cerevisiae, transformation from amorphous U(VI) to crystalline chernikovite. *Applied Microbiology and Biotechnology, 102*, 4217–4229. https://doi.org/10.1007/s00253-018-8918-4
26. Mkandawire, M., Vogel, K., Taubert, B., & Dudel, E. G. (2007). Phosphate regulates uranium (VI) toxicity to *Lemna gibba* L. G3. *Environmental Toxicology, 22*, 9–16.
27. Wufuer, R., et al. (2017). In S. Sariaslani & G. M. Gadd (Eds.), *Advances in applied microbiology* (Vol. 101, pp. 137–168). Academic Press.
28. Dobchuk, B., Nichol, C., Wilson, G. W., & Aubertin, M. (2013). Evaluation of a single-layer desulphurized tailings cover. *Canadian Geotechnical Journal, 50*, 777–792. https://doi.org/10.1139/cgj-2012-0119
29. Hall, S., & Coleman, M. (2012). Critical analysis of world uranium resources. In *U.S. Geological Survey scientific investigations report 2012-5239* (p. 56).
30. IAEA. (2004). In International Atomic Energy Agency (Ed.), *IAEA-TECDOC-1403* (Vol. 309). International Atomic Energy Agency.
31. Landa, E. (1980). *Isolation of uranium mill tailings and their component radionuclides from the biosphere; some earth science perspectives. Report no. 814*. Geological Survey.
32. Abdelouas, A., Lutze, W., & Nuttall, E. (1998). Chemical reactions of uranium in ground water at a mill tailings site. *Journal of Contaminant Hydrology, 34*, 343–361. https://doi.org/10.1016/S0169-7722(98)00097-7
33. Krupka, K., & Serne, R. (2002). *Geochemical factors affecting the behavior of antimony, Cobalt, Europium, Technetium, and Uranium in Vadose sediments*. Pacific Northwest National Laboratory (PNNL). https://doi.org/10.2172/15004491
34. Dong, W., & Brooks, S. C. (2006). Determination of the formation constants of ternary complexes of uranyl and carbonate with alkaline earth metals (Mg2+, Ca2+, Sr2+, and Ba2+) using anion exchange method. *Environmental Science & Technology, 40*, 4689–4695. https://doi.org/10.1021/es0606327
35. Kanzari, A., et al. (2017). Impact of incipient weathering on uranium migration in granitic waste rock piles from former U mines (Limousin, France). *Journal of Geochemical Exploration, 183*, 114–126. https://doi.org/10.1016/j.gexplo.2017.08.010
36. Skierszkan, E. K., et al. (2020). Geochemical controls on uranium release from neutral-pH rock drainage produced by weathering of granite, gneiss, and schist. *Minerals, 10*, 1104.
37. Han, T., et al. (2022). Immobilization of uranium during the deposition of carbonated hydroxyapatite. *Journal Taiwan Institute Chemical Engineers, 134*, 104331. https://doi.org/10.1016/j.jtice.2022.104331
38. Lahrouch, F., et al. (2021). Uranium retention on iron oxyhydroxides in post-mining environmental conditions. *Chemosphere, 264*, 128473. https://doi.org/10.1016/j.chemosphere.2020.128473
39. Stewart, B. D., Cismasu, A. C., Williams, K. H., Peyton, B. M., & Nico, P. S. (2015). Reactivity of uranium and ferrous iron with natural iron Oxyhydroxides. *Environmental Science & Technology, 49*, 10357–10365. https://doi.org/10.1021/acs.est.5b02645
40. Marshall, T. A., et al. (2014). Incorporation of uranium into hematite during crystallization from Ferrihydrite. *Environmental Science & Technology, 48*, 3724–3731. https://doi.org/10.1021/es500212a
41. Allard, T., Ildefonse, P., Beaucaire, C., & Calas, G. (1999). Structural chemistry of uranium associated with Si, Al, Fe gels in a granitic uranium mine. *Chemical Geology, 158*, 81–103.
42. Boland, D. D., Collins, R. N., Payne, T. E., & Waite, T. D. (2011). Effect of amorphous Fe(III) oxide transformation on the Fe(II)-mediated reduction of U(VI). *Environmental Science & Technology, 45*, 1327–1333. https://doi.org/10.1021/es101848a
43. Davis, J. A., Meece, D. E., Kohler, M., & Curtis, G. P. (2004). Approaches to surface complexation modeling of uranium(VI) adsorption on aquifer sediments. *Geochimica et Cosmochimica Acta, 68*, 3621–3641.

44. Waite, T. D., Davis, J. A., Payne, T. E., Waychunas, G. A., & Xu, N. (1994). Uranium(VI) adsorption to ferrihydrite: Application of a surface complexation model. *Geochimica et Cosmochimica Acta, 58*, 5465–5478.
45. Furcas, F. E., et al. (2023). Transformation of 2-line Ferrihydrite to goethite at alkaline pH. *Environmental Science & Technology, 57*, 16097–16108. https://doi.org/10.1021/acs.est.3c05260
46. Liu, H., Wei, Y., & Sun, Y. (2005). The formation of hematite from ferrihydrite using Fe(II) as a catalyst. *Journal of Molecular Catalysis A: Chemical, 226*, 135–140. https://doi.org/10.1016/j.molcata.2004.09.019
47. Abd El-Naby, H. H., & Dawood, Y. H. (2008). Natural attenuation of uranium and formation of autunite at the expense of apatite within an oxidizing environment, south Eastern Desert of Egypt. *Applied Geochemistry, 23*, 3741–3755. https://doi.org/10.1016/j.apgeochem.2008.09.011
48. Boekhout, F., et al. (2015). Uranium migration and retention during weathering of a granitic waste rock pile. *Applied Geochemistry, 58*, 123–135. https://doi.org/10.1016/j.apgeochem.2015.02.012
49. Miao, Z., Akyol, H. N., McMillan, A. L., & Brusseau, M. L. (2013). Transport and fate of ammonium and its impact on uranium and other trace elements at a former uranium mill tailing site. *Applied Geochemistry, 38*, 24–32. https://doi.org/10.1016/j.apgeochem.2013.08.002

Chapter 4
Uranium Behaviour in Abandoned Underground Mines

Abstract This chapter highlights uranium behaviour in abandoned underground mines, focusing on the mechanisms that govern its mobility, transformation, and attenuation from the time flooding begins to when the mine water is considered mature. By combining empirical field data, controlled laboratory studies, and advanced reactive transport modelling, this chapter aims to clarify the long-term fate of uranium in these flooded mines.

Keywords Mine flooding · Hydrogeochemical evolution · Metal mobility · Redox shifts · Alkaline drainage · Water-rock interaction · Ground stability · Gas release · Site remediation

4.1 Introduction

Until recently, underground uranium mining (UUM) was the primary method for extracting uranium ore from the earth. As a result, most abandoned uranium mines are found underground. The underground mining operations predominantly focused on uraninite-rich orebodies in granitic, sedimentary, or metamorphic rocks, using conventional drilling and blasting techniques to access deep-seated uranium deposits [1, 2].

During active operations, the mines are kept artificially dry through a dewatering process that removes water, allowing workings deep below the water table. When the mine is closed, the cessation of dewatering operations results in a progressive flooding of the mine voids, unless drainage adits are installed. Thus, a flooded mine is a direct and automatic consequence of the mine closure process [3, 4]. This transition from unsaturated to saturated conditions marks a critical inflection point in the geochemical and hydrogeological evolutions of closed mines [5–7].

The geochemical and hydrogeological evolutions can profoundly influence the mobility and fate of uranium within the flooded underground uranium mine system. For instance, uranium predominantly exists as the highly soluble hexavalent uranyl ion (UO_2^{2+}) in acidic and oxidizing conditions, which readily complexes with

M. Mkandawire, *Fate and Effects of Uranium in Abandoned Uranium Mines*,
https://doi.org/10.1007/978-3-032-10447-2_4

carbonate, phosphate, and sulfate ligands [8–11]. These complexes can enhance uranium transport through the mine water, potentially leading to widespread contamination of surrounding aquifers and surface water bodies. Concurrently, secondary mineral precipitation, such as the formation of uranyl phosphates or silicates, may act as transient sinks, modulating uranium concentrations during different stages of mine water evolution [12–15].

With time, the mine water matures, characterized by the stabilization of redox conditions, pH, and ionic strength. There is often a notable transition in the speciation and mobility of uranium. Reductive processes, driven by microbial activity or by abiotic reactions involving ferrous iron and organic matter, convert uranyl to less soluble tetravalent uranium species (U^{4+}). This transformation enhances immobilization through sorption or co-precipitation with iron oxyhydroxides [16, 17].

The cessation of mine dewatering also induces significant changes in subsurface hydrogeology and geophysics. Rising water tables alter hydraulic gradients, potentially mobilizing contaminants into adjacent strata or surface ecosystems. Geochemical gradients established during mining may be disrupted, leading to the re-equilibration of mineral phases and the reactivation of previously stable contaminant reservoirs.

4.2 Abandoned Underground Uranium Mines

4.2.1 Global Legacy

Abandoned underground uranium mines are prevalent in regions worldwide with a history of significant uranium extraction. Table 4.1 presents the approximated global distribution of abandoned underground uranium mines, estimated in 2023. In North America, the western United States, particularly New Mexico, Arizona, Utah, and Colorado, hosts hundreds of legacy mines. These sites have been linked to long-term environmental degradation and public health concerns due to ongoing

Table 4.1 Global distribution of abandoned underground uranium mines (approximate percentages)

Region	Estimated share (%)
North America	~35% (primarily the USA and Canada)
Eastern Europe	~15% (e.g., Czech Republic, Romania)
Central Asia	~20% (Kazakhstan, Uzbekistan, Kyrgyzstan, Tajikistan)
Africa	~10% (Namibia, Niger, Algeria)
East Asia	~10% (China, Mongolia)
South America and others	~10% (Argentina, Brazil, Australia, others)

Note: The figures are derived from the distribution of known uranium deposits and historical mining activity, particularly those that have ceased operations and remain unremediated contamination and insufficient cleanup efforts. In Canada, the Athabasca Basin in northern Saskatchewan remains a key area for uranium mining, with several underground operations now decommissioned [18, 19].

Australia, which possesses the world's largest uranium reserves, features legacy sites, such as Rum Jungle and Radium Hill, both of which have undergone various rehabilitation efforts. In Europe, countries like Germany, the Czech Republic, and Romania have extensive histories of uranium mining, with numerous abandoned sites still subject to environmental monitoring and remediation. In Africa, nations including Niger, Namibia, and South Africa host legacy uranium mines that continue to pose environmental challenges. Additionally, the former Soviet Union left behind a significant number of abandoned uranium mines in Kazakhstan, Uzbekistan, and Russia, many of which still represent environmental liabilities [20, 21].

4.2.2 Hallmarks of Abandoned UUMs

Abandoned underground uranium mines feature both subsurface and aboveground characteristics. Beneath the surface, extensive networks of tunnels, horizontal adits, and shafts can stretch for thousands of kilometres, flooding over time. On the surface, notable remnants include former access points and ventilation shafts. Additionally, evidence of an abandoned underground mine may consist of dilapidated or deteriorating pump houses, concrete pads and foundations, remnants of timber and steel supports, and scattered equipment, such as drill rods, fuel tanks, and piping [22].

Vertical shafts, horizontal adits, and collapsed entries can create direct hydraulic and pneumatic links between the surface and submerged underground tunnels. These connections significantly influence water recharge and outflow patterns. Under unstable ground conditions, such as in sinkholes and subsidence events, they can form preferential pathways that change the timing and direction of floodwaters entering the mine system. Moreover, open or poorly sealed shafts and drifts can enable communication between flooded voids, the atmosphere, and regional groundwater, affecting hydrostatic regimes and causing episodic exchanges of water and gas. This interaction also impacts the chemistry of the flooded mine pool.

The initial characteristics of a flooded mine depend on factors, such as the extent of oxygen exposure on mineral surfaces within mine tunnels and faults, the oxygenation level of the flooding water, and the geological background of the site. When flooding occurs mainly due to underground water influx, oxygen levels are close to zero. Under these conditions, most uranium, its decay products, and metals tend to remain adsorbed onto rock surfaces, colloids, and suspended particles or in mineral form [17, 23]. This is only true when (a) the water displaces air in the void spaces without mixing and (b) no significant mineral oxidation occurred prior to flooding.

Otherwise, the geochemistry of mine water is dynamic, with the mobility and lability of uranium and metals depending on the predominant coexisting minerals, which can lead to acid or alkaline drainage. Sulfide mineral dominance results in acid mine drainage, while other conditions may produce alkaline runoff. These environments cause shifts in redox potential as oxygen is consumed, leading to reducing conditions; variations in pH, temperature, and dissolved ligands influence processes, such as dissolution, adsorption, and precipitation; and complexation reactions modify the form and solubility of uranium.

Flooded mine voids present unique hazards compared to dry surface sites. Hidden shafts and submerged openings pose immediate physical risks of falls, while weakened pillars and roofs increase the risk of sudden collapses. Elevated levels of radon and other hazardous gases can accumulate in the atmosphere above flooded workings. Flooding often enhances the transport of dissolved and colloidal radionuclides, allowing their migration with groundwater or their discharge at springs and portals [24–26]. External gamma radiation near vent areas and contaminated surfaces remains a concern because submerged radioactive materials continue to emit penetrating radiation. The interaction among mine geometry, hydrodynamics, and lithologic heterogeneity can cause contaminant release to occur episodically, often triggered by water level changes and recharge events [27, 28].

Furthermore, ^{226}Ra produced in submerged materials continues to emit radon gas. When air spaces above the water connect, or hydraulic forces facilitate degassing, radon can migrate to the surface. Seasonal fluctuations and storm-induced recharge can disrupt these chemical states, altering solute levels and, in some cases, releasing contaminants.

4.3 Flooding of Underground Mines

Active underground mining often reaches depths of tens to hundreds, or even thousands, of metres below the natural water table. This process intersects aquifers that would rapidly flood tunnels and shafts without continuous pumping. Consequently, mines are continuously dewatered to keep working levels dry, ensuring safe access to orebodies while preventing hydraulic backpressure that could destabilize excavations. Dewatering systems range from deep-well pumps to peripheral drainage galleries, lowering the local water table around the mine voids. It also allows operators to manage inflows, maintain geotechnical stability, and safeguard both equipment and personnel. However, once active mining operations stop, dewatering efforts naturally cease, and the underground mine workings begin to flood. Figure 4.1 summarizes the flow of the stages in the mine flooding process.

4.3.1 Planned vs. Abandoned Mines

Once ore extraction and processing are complete, the need for dewatering ceases. There is no operational requirement to remove groundwater once no one is accessing the workings for production. In well-planned mine closures, engineers typically

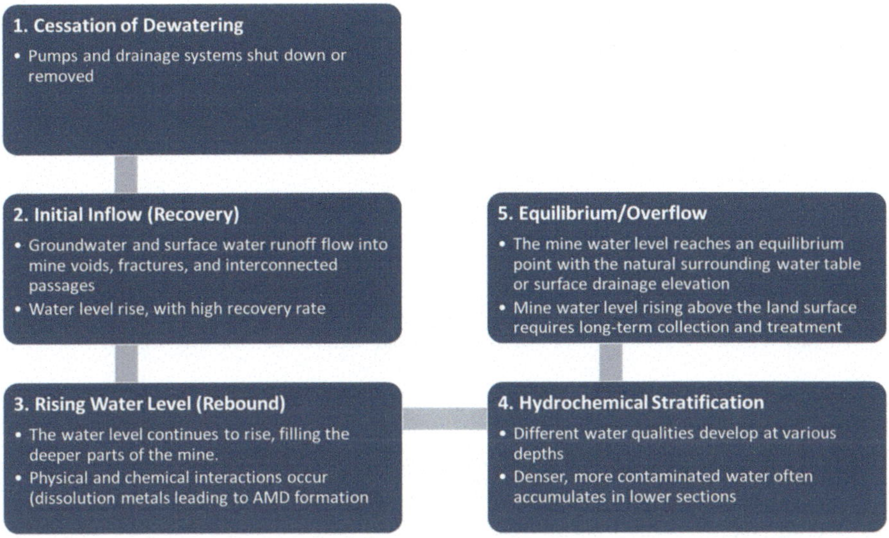

Fig. 4.1 Conceptualized flow of hydrological events/stages during the flooring process of underground mine flooding

devise a phased dewatering shutdown that allows water levels to rise in a controlled manner. Valves, bulkheads, or engineered spillways can regulate the rate of refilling, limit downstream discharge rates, and support post-closure monitoring of water quality as groundwater gradually returns to its original hydraulic gradient.

In contrast, abandoning a mine without proper closure planning results in uncontrolled flooding. Pumps are turned off, barriers corrode, and water fills the voids until pressure equilibrium is reached with the surrounding aquifers. This sudden flooding can erode portal seals, mobilize contaminants trapped in mine voids, and lead to surface sinkhole formation. Without the regulatory frameworks and engineering controls associated with a managed closure, abandoned mines pose long-term risks to groundwater quality, surface stability, and nearby water users.

4.3.2 Hydrogeochemical Behaviour of Uranium

The cessation of mining activities initiates a cascade of hydrogeochemical processes that continue to evolve long after the last ore is extracted. Once dewatering operations are halted, the underground workings begin to flood, marking a critical transition in the mine's geochemical regime. Oxygenated surface and groundwater infiltrate previously anoxic zones, triggering the oxidation of sulfide minerals, most notably pyrite (FeS_2), which are commonly associated with uranium ores. This reaction generates sulfuric acid, leading to the formation of acid mine drainage (AMD), a low pH solution enriched with dissolved metals and radionuclides. The AMD environment significantly alters the mobility and speciation of uranium, transforming it from relatively immobile tetravalent uranium (U^{4+}) into the highly soluble hexavalent uranyl ion (UO_2^{2+}).

The hydrogeochemical behaviour of uranium in these flooded systems is governed by a suite of reactions. Pyrite oxidation, a primary driver of AMD, proceeds via the reaction:

$$FeS_2 + 3.75O_2 + 3.5H_2O \rightarrow Fe(OH)_3 + 2SO_4^{2-} + 4H$$

This process not only acidifies the mine water, but also liberates sulfate and iron, which further participate in secondary reactions. Uraninite oxidation follows:

$$UO_2 + 0.5O_2 + 2H^+ \rightarrow UO_2^{2+} + H$$

The resulting uranyl ion readily forms aqueous complexes with ligands, such as carbonate, phosphate, and sulfate, enhancing its solubility and transport potential. These complexation reactions are represented as:

$$UO_2^{2+} + nL \rightarrow UO_2L_n^{2-n}$$

where L denotes the ligand species, under certain conditions, uranyl ions may adsorb onto iron oxyhydroxides formed during pyrite oxidation:

$$UO_2^{2+} + Fe(OH)_3 \rightarrow Fe(OH)_3 \cdot UO_2^{2+}$$

This adsorption can temporarily immobilize uranium, although changes in pH or redox conditions may reverse the process. In anoxic environments, microbial or abiotic reduction of uranyl to U^{4+} can occur:

$$UO_2^{2+} + 2e^- \rightarrow U^{4+} \rightarrow UO_2(s)$$

This transformation leads to the precipitation of uranium as insoluble UO_2, effectively removing it from the aqueous phase.

4.3.3 Potential Alkaline Mine Drainage Formation

In certain abandoned underground uranium workings, the legacy of carbonate-rich host rocks can give rise to alkaline drainage rather than the acid-dominated effluents more commonly associated with metal mines. When groundwater or surface water infiltrates tunnels cut through limestone or calcite-bearing strata, these minerals dissolve and consume free protons, raising pH. This neutralization process can offset acidity generated by the slow oxidation of uranium oxides or minor sulfide minerals, resulting in slightly basic waters that carry dissolved uranium and other elements in forms different from those found under acidic conditions.

Field studies of decommissioned uranium operations have recorded pH values ranging from mildly acidic to above neutral, with localized pockets of pH 8 or higher where calcite dissolution dominates. Under these alkaline conditions, uranium often remains in its hexavalent state, forming stable carbonate complexes that can migrate until changes in redox potential or pH trigger precipitation of secondary phases, such as autunite or schoepite. Because carbonate buffering can mask ongoing oxidative reactions, monitoring programs must track not only pH but also uranium speciation, carbonate concentration, and redox indicators to understand contaminant mobility.

Alkaline conditions increase aqueous uranium solubility, being predominantly hexavalent uranyl speciation (UO_2^{2-} and $UO_2(CO_3)_3^0$), and decrease sorption to mineral surfaces. Oxidation of reduced uranium minerals or sorbed U(IV) releases UO_2^{2+} + H_2O, while complexation with CO_3^{2-} shifts the partitioning strongly towards the aqueous phase. The aqueous phase can limit precipitation of low-solubility U(IV) phases, so elevated pH, high carbonate, and the presence of Ca^{2+} commonly promote long-range uranium transport rather than immobilization [29, 30].

Redox-driven reduction and secondary mineral reactions remain the primary pathways for immobilizing uranium if reducing conditions can be established or maintained. Microbial or chemical reduction converts U(VI) to sparingly soluble U(IV) (UO_2(s), uraninite) via electron transfer reactions ($UO_2^{2+} + 2e^- \rightarrow UO_2(s)$), and U can also be sequestered by sorption or co-precipitation with iron(III) oxides/hydroxides or by incorporation into secondary carbonate minerals where carbonate is limited.

Conversely, oxidative recharging, fluctuating Eh/pH, colloid-facilitated transport, and competition for sorption sites, especially under high ionic strength, can remobilize previously immobilized uranium; therefore, predicting plume behaviour requires integrating redox potential, pH, dissolved carbonate and calcium, mineralogy (notably Fe oxides and carbonates), and microbial activity in site-specific models. In hydrology, a plume is a body of one fluid moving through another surrounding fluid due to differences in properties, such as density, temperature, or contaminant concentration.

4.4 Underground Uranium Mine Flooding Process

4.4.1 Driving Factors

Flooding of underground uranium workings is mainly driven by the rebound of connate and fissure groundwater rather than direct surface inputs. Figure 4.2 illustrates the installation of dewatering wells that keep the underground mine dry during active operations. In a capped or dewatered mine void, when pumping ceases, the surrounding aquifer system re-equilibrates through interconnected fractures, brecciated ore zones, and intact strata. Due to these deep mines lying below the influence of ephemeral surficial flow, the principal source of refill water is meteoric recharge that has percolated into fractured bedrock well beyond the footprint of open pits or shafts. In abandoned operations, pumps are often switched off abruptly, causing hydraulic heads to recover rapidly and water to rise along high-permeability drifts and haulage ways, with little regard for geotechnical or geochemical staging.

As flooding progresses following unplanned closures, oxygenated water infiltrates sulfide-rich backfill and unremediated stopes, triggering pyrite oxidation and generating acid mine drainage. The drop in pH liberates U(VI) from mineral surfaces and complexes it with sulfate or carbonate ligands. Redox fronts form at the periphery of inundated galleries, and evolving Eh-pH conditions lead to transient trapping of uranium in iron hydroxide precipitates, which are then remobilized by reductive dissolution. Hydrodynamic dispersion smooths steep head gradients into adjacent rock, but localized upwelling can breach remaining dry voids, mixing

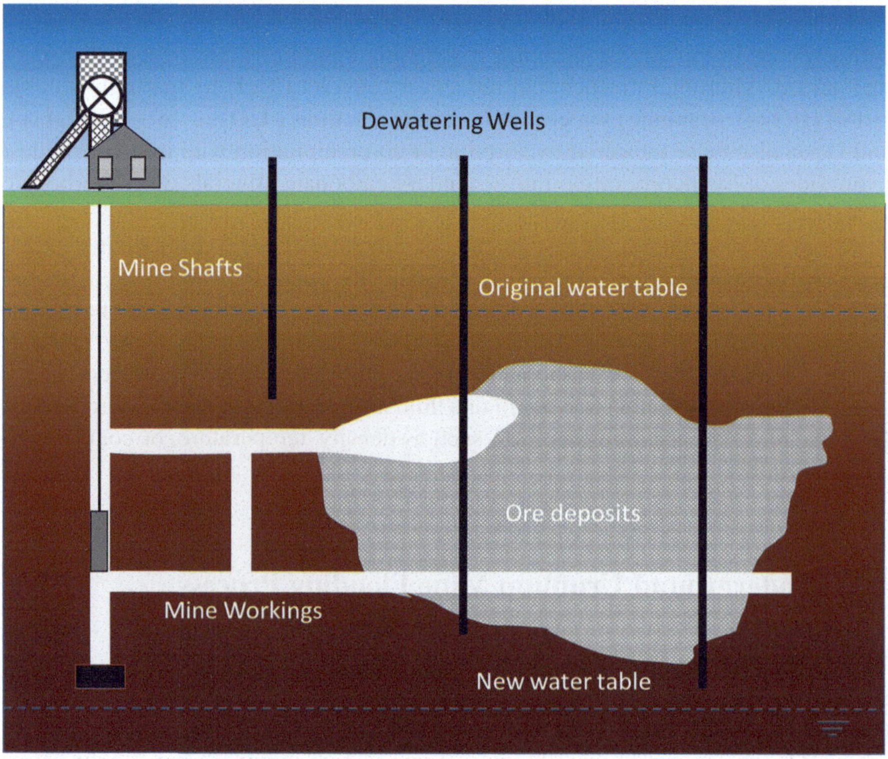

Fig. 4.2 Illustration of lowering the water table through the dewatering process during active mining

oxidized and reduced zones in an erratic, spatially heterogeneous pattern that defies simple prediction.

In contrast, a properly planned closure manages the inundation to suppress sulfide oxidation, control acid generation, and ensure occupant safety by permanently preventing access to dangerous workings and gas-filled voids. Early in the closure design, all potential water pathways that could concentrate flow or collapse under rising pressure are identified and eliminated; loose timbers are removed; loose sidewall blocks are filled; and unstable caverns are backfilled. Engineered cutoff walls or grout curtains, coupled with surface and subsurface diversion channels, guide recharge water along predetermined routes.

Key to this controlled approach is the establishment of hydrological and hydrogeochemical models that predict how mine waters will evolve as the void space fills. By forecasting pH, Eh, sulfate load, and uranium speciation over time, treatment systems can be redesigned to neutralize acidic inflows, precipitate or adsorb U(VI) in situ, and promote a stable reducing environment that favours the formation of insoluble uraninite or adsorption onto sulfide surfaces. Geophysical surveys, such as time-lapse electrical resistivity, track the advance of the water front and the development of reducing zones, enabling managers to adjust barriers or treatment thresholds on the fly.

Abandoned mines lacking these systematic controls flood haphazardly. Water exploits the path of least flow resistance, collapsed stopes, legacy sumps, and breccia pipes, leading to rapid hydrological connectivity between contaminated zones and the regional aquifer. Without pre-emptive hazard removal, targeted filling or diversion of infrastructure, and robust monitoring, the timing and geochemical evolution of the flood remain unpredictable. The result is dispersed uranium plumes with complex speciation and redox heterogeneity, complicating long-term containment and remediation strategies.

4.4.2 Hydrogeochemical Evolution

As mine water evolves during post-closure, it passes through distinct stages. The first phase is the flooding phase, marked by rapid oxidation of exposed minerals and the dissolution of soluble salts, resulting in peak acidity and dissolved metal concentrations. This is followed by a transitional phase during which redox gradients develop, and microbial communities begin to influence geochemical transformations. Secondary mineral formation may commence, partially attenuating contaminants. Eventually, the mine water reaches a mature state, characterized by stabilized pH, redox potential, and ionic strength. During this phase, uranium mobility may decrease due to sorption, co-precipitation, or reductive immobilization. However, persistent AMD conditions can prolong uranium transport, especially in zones where buffering capacity is low or microbial reduction is inhibited (Fig. 4.3).

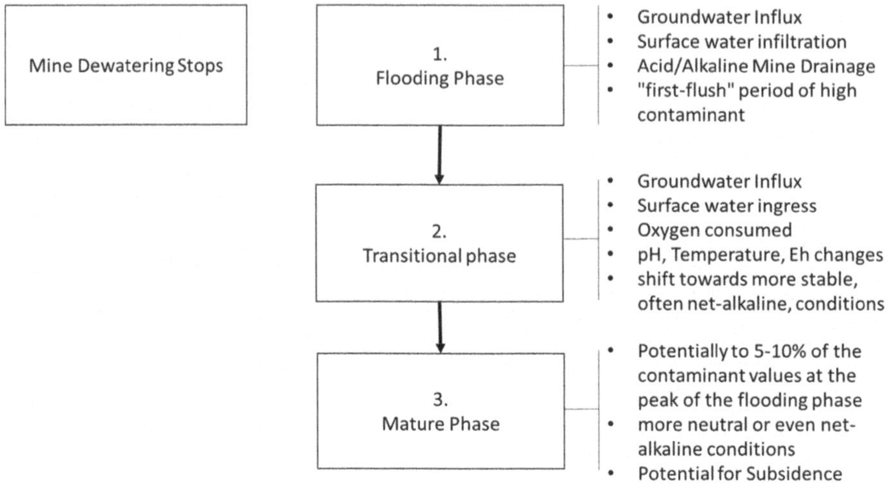

Fig. 4.3 Phases in hydrogeochemical evolution in a flooded underground mine, starting after the dewatering process stops

4.4.2.1 Mine Flooding Phase

The mine flooding phase is characterized by a sudden influx of oxygenated water into voids and fractured host rocks, resulting in significant changes in redox potential (Eh) and pH. In turn, the changing Eh and pH influence uranium speciation and mobility. In the initial oxic pulse, residual U(IV) minerals, particularly uraninite (UO_2), experience rapid oxidative dissolution, forming uranyl ions as shown in the reaction:

$$UO_2(s) + \frac{1}{2}O_2 + 2H^+ \rightarrow UO_2^{2+} + H_2O$$

The process yields soluble uranyl ion (UO_2^{2+}), which then partitions into a suite of aqueous complexes depending on the geochemistry of the milieu.

For instance, in mines with uranium deposited in association with elevated carbonate activity, the uranium dissolution in mine water equilibrates with carbonate host rocks or percolates through calcareous tailings, and uranyl exhibits a propensity to form highly stable ternary carbonate complexes. Such complexes significantly influence stability and geochemistry, as well as metal mobility, in flowing mine systems.

$$UO_2^{2+} + 3CO_3^{2-} \rightarrow UO_2(CO_3)_3^{4-}$$

To a lesser extent, hydroxyl or sulfate complexes (e.g., $UO_2(OH)^+$ and $UO_2SO_4(aq)$) enhance uranium's aqueous residence time and reduce its sorption affinity for iron and aluminium (oxyhydr)oxide surfaces, promoting long-range transport.

As microbial respiration and abiotic reactions with reduced mineral phases consume dissolved oxygen, the system drifts towards anoxic conditions. In this reducing regime, Fe^{2+} generated by reductive dissolution of ferric oxides and biogenic sulfide (S^{2-}) serve as electron donors for uranyl reduction:

$$UO_2^{2+} + 2e^- \rightarrow UO_2(s)$$

Moreover, in phosphate-rich microenvironments often created by the decomposition of organic matter or phosphate-bearing gangue, secondary U(IV) phosphates may precipitate, sequestering uranium in minimally soluble mineral phases. The reaction of the process is:

$$3UO_2^{2+} + 4PO_4^{3-} \rightarrow U_3(PO_4)_4(s)$$

The kinetics and extent of these redox transformations are modulated by pH, carbonate concentration, and the abundance of reactive iron or sulfur species. Acidic floodwaters accelerate UO_2 dissolution by increasing proton activity, whereas neutral to slightly alkaline conditions favour carbonate complexation and aqueous persistence of U(VI). Elevated bicarbonate levels shift the U(IV)/U(VI) stability boundary to more reducing potentials, effectively widening the window in which uranyl complexes remain mobile. Conversely, abundant Fe^{2+} or S^{2-} and active microbial consortia (e.g., *Geobacter* and *Desulfovibrio* spp.) expedite U(VI) reduction, promoting precipitation of UO_2 or U(IV) phosphates even at modestly circumneutral pH.

Hydrological parameters, particularly water residence time and flow pathways, govern contact between uranium-bearing solids, electron donors, and microbial catalysts. In low-flow zones and dead-end cavities, prolonged anoxia fosters complete U(VI) reduction and solid-phase accumulation. In contrast, in high-flux channels, the repeated alternation between oxic and anoxic pulses can create a "redox shuttle", repeatedly mobilizing and immobilizing uranium. Temperature further accelerates both dissolution and reduction kinetics, meaning seasonal or depth-dependent thermal gradients can imprint spatial heterogeneity on uranium distributions within flooded mine workings.

4.4.2.2 Transitional Phase

The transitional phase starts when dewatering stops and continues until the flooded workings reach a new hydrological steady state. This period is influenced by coupled hydrological, geochemical, and geomechanical processes that occur over timescales from months to decades. It is marked by strong interactions among advancing water, newly exposed uranium-bearing minerals, and the fracture and pore network of the host rock; these interactions determine the timing and composition of emerging mine waters and the developing subsurface flow system.

Hydrogeological Behaviour and Flow System Evolution

When pumps stop, the hydraulic gradient imposed by dewatering disappears, and the regional water table begins to recover. Water influx into mine voids occurs through a combination of:

1. Fracture flow driven by ambient hydraulic gradients and episodic recharge.
2. Matrix diffusion in low-permeability lithologies.
3. Rapid channelized flow in which open paleoflows, faults, or mine workings provide high-transmissivity pathways.

The rise of the water surface in the workings is controlled by mass balance between inflow (i.e., recharge from the surface, inflow from adjacent aquifers, seepage through fractured host rock, and any perched water bodies) and storage in the void space and compressible pore volume. Storage comprises cavernous void volume, fracture porosity, and matrix porosity; in fractured crystalline rocks, effective specific yield is commonly much lower than in porous sediments, which prolongs the recovery.

Transient groundwater flow during the transition is governed by the diffusivity (D) given by:

$$D = \frac{K}{S_s}$$

where K is the hydraulic conductivity, and S_s is the specific storage.

In fractured systems, apparent diffusivity is strongly scale-dependent because fracture connectivity and aperture distributions dominate K, and because matrix diffusion into low-permeability blocks imposes an effective dual-domain response. Early in flooding, flow is often non-Darcy within large open conduits; as water levels rise and hydraulic head gradients decline, flow regimes revert to laminar, gravity-dominated flow and discharge loci migrate to inherent low points or pre-existing seeps. Where mine voids intersect confining units or perched aquifers, new discharge points (adits, springs, and wetlands) may become active once the hydraulic head in workings exceeds local thresholds.

Hydraulic interactions with connected regional aquifers may cause a drawdown recovery in the surrounding groundwater system and establish new head gradients that can propagate away from the mine. The timescale for equilibration with regional groundwater depends on the mine void's transmissivity and effective radius of influence. It can be long when transmissivity is low or vertical hydraulic communication is limited.

Geochemical Pathways and Uranium Behaviour

The geochemistry of floodwater evolves through a sequence of interrelated processes: oxygenation/dissolved oxygen dynamics, redox buffering, metal-sulfide oxidation, complexation, sorption/desorption, precipitation/dissolution, and mixing/dilution. Uranium behaviour during the transitional phase is strongly controlled by redox potential (Eh), pH, ligand availability (especially, carbonate and sulfate), and surface reactions on iron oxides, clays, and organic matter.

Primary uranium in most ore deposits exists as reduced U(IV) phases (e.g., uraninite, UO_2) or is embedded within sulfide or silicate matrices. When exposed to oxygen-rich floodwaters, oxidative dissolution converts U(IV) to U(VI) in solution. The oxidation process of U(IV) solids can be summarized as a combination of electron transfer and proton or ligand exchange, where $UO_2(s)$ is transformed into uranyl species, for example:

$$UO_2(s) + \frac{1}{2}O_2 + 2H^+ \rightarrow UO_2^{2+} + H_2O$$

The rate of oxidative dissolution depends on oxygen availability, microbial catalysis (e.g., heterotrophic or chemolithotrophic bacteria that accelerate sulfide and iron oxidation and thereby maintain oxidative conditions), the surface area of finely divided uranium minerals, and the presence of catalytic oxidants, such as Mn(IV) oxides.

Once in solution, uranyl (UO_2^{2+}) speciation controls transport. In near-neutral to alkaline waters with appreciable dissolved inorganic carbon, uranyl forms stable carbonate complexes (e.g., $UO_2(CO_3)_2^{2-}$, $UO_2(CO_3)_3^{4-}$) that significantly increase uranium solubility and mobility. The formation constants for these carbonate complexes are high, so carbonate availability (from dissolved CO_2, carbonate minerals, or organic matter mineralization) is a primary determinant of long-range transport. Conversely, in low-carbonate and reducing environments, uranyl is more likely to be removed from solution via adsorption onto freshly formed Fe(III) oxyhydroxides, co-precipitation with secondary minerals, or reduction back to U(IV) and precipitation as sparingly soluble phases.

Sorption to mineral surfaces strongly influences aqueous uranium concentrations. Most minerals contain metal oxides, such as Fe oxides and Mn oxides, as well as phyllosilicates. The reaction can involve surface complexation, for example, on Fe-oxides.

$$FeOH + UO_2^{2+} \rightleftharpoons FeOUO_2^{+} + H^+$$

The surface charge, pH, and competing cations (Ca^{2+}, Mg^{2+}, and Pb^{2+}) affect the partitioning coefficient K_d. The reaction works the same with Mn-oxides and phyllosilicates, just by substituting Fe-oxide. High pH generally reduces sorptive capacity on metal oxide surfaces, while the presence of carbonate competes with uranyl for sorption sites and forms aqueous complexes that reduce sorptive removal. This is attributed to uranyl speciation being highly pH-dependent. At high (alkaline) pH, the negative charge of both the uranyl species (which transition to more negatively charged hydroxide and carbonate complexes) and the metal oxide surfaces increases. This mutual repulsion reduces the sorptive capacity.

Sulfate produced by sulfide oxidation indirectly enhances uranium mobility. Sulfate increases ionic strength and can stabilize certain aqueous complexes, and in some circumstances, the presence of sulfate and high acidity favours metal mobility. Conversely, under strongly reducing conditions where sulfate reduction is active (driven microbially by organic carbon or hydrogen from corrosion and water-rock interactions), sulfide production can precipitate some metals as sulfides. Still, uranium typically remains in solution until it encounters reducing microenvironments that enable U(VI) reduction to U(IV) and precipitation.

Redox heterogeneity is a common feature during transition. Oxic inflows from surface recharge and degassing near shafts create oxidizing fronts that move into previously reduced rock, forming a shifting reaction zone marked by iron oxidation, acid generation, and pulses of dissolved metals and sulfate. Behind this front, reducing conditions can re-establish, enabling processes, such as denitrification, iron reduction, and sulfate reduction. This vertical and lateral redox zoning results in complex uranium distribution patterns: higher concentrations in oxic zones where U(IV) oxidation and carbonate complex formation are active, and lower concentrations in reducing microniches where reduction immobilizes U (IV).

Matrix diffusion and sorption buffering produce prolonged tailing of uranium concentrations after the first flush. Dissolved uranium trapped in the fracture network and sorbed to mineral surfaces slowly exchanges with the rising bulk water, often producing a long-term low-to-moderate concentration discharge even after initial spikes subside. In many uranium mines, the "first flush" is characterized by a rapid spike in U, SO_4, Fe, and acidity, followed by an exponential or multi-rate decline controlled by desorption kinetics, secondary mineral formation (e.g., jarosite, schwertmannite under acidic conditions; uranyl phosphates or carbonates under neutral to alkaline conditions), and dilution.

Coupled Hydromechanical and Reactive Transport Processes

Reactive transport during transitional flooding couples advection, dispersion, diffusion, surface complexation, aqueous complex formation, redox reactions, precipitation/dissolution, and sorption-desorption kinetics. Advection dominates in conduits and well-connected fractures, while matrix diffusion and slow desorption control longer-term release from rock matrices. Precipitation of iron oxyhydroxides following sulfide oxidation can sequester uranium by co-precipitation or adsorption; however, these precipitates can later be reductively dissolved in anoxic zones, releasing bound uranium again. Changes in ionic strength and pH resulting from mixing and mineral dissolution/precipitation alter sorption capacities and mineral stability fields over time and space.

Geomechanical changes affect hydraulic properties and, therefore, reactive transport. Re-saturation increases pore pressures, reducing effective stress and potentially increasing fracture apertures or inducing shear that can either enhance permeability (through fracture dilation) or reduce it (through collapse and clogging by fines and secondary precipitates). Chemical alteration of supporting materials, acid attack on cementitious backfill, sulfate attack on concrete, or mineralogical weakening of pillars can further modify permeability structure and reactive surface area. This feedback produces non-linear and sometimes abrupt changes in flow paths and contaminant release rates.

Temporal Sequence and Emergent Behaviours

The sequence commonly observed in uranium mine transitions begins with an initial rapid inflow (first flush) that carries high concentrations of oxidation products, acid, sulfate, iron, and dissolved uranium. As oxygen is depleted in closed regions and secondary minerals precipitate, concentrations decline but do not necessarily return to background; instead, a prolonged period of moderated discharge may persist due to slow desorption and diffusion from rock pores. Over longer durations, post-flooding evolution depends on whether reducing conditions dominate (favouring immobilization of uranium as U(IV)) or oxidizing conditions persist (sustaining U(VI) mobility). Episodic recharge events, seasonal water table oscillations, or progressive chemical-front migration can cause pulses of remobilization. In fractured settings, preferential pathways may sustain downstream transport even as bulk conditions otherwise decline.

4.4.2.3 Mature State

In the mature state, flooded underground uranium mines are fundamentally defined by the chemical and physical pathways that determine uranium speciation, mobility, and long-term sequestration within the confined subsurface environment. The flooded workings form compartmentalized reservoirs where hydrodynamic constraints, redox stratification, and ligand availability govern whether uranium remains immobilized as reduced solids, is retained on mineral surfaces, or persists in aqueous transportable forms. Because the geometry of shafts, drifts, stopes, and fractures restricts mixing and fosters dual-domain exchange, uranium behaviour in these systems is controlled by coupled advective, diffusive, and reactive processes that operate over a wide range of timescales.

Hydraulically, the underground setting produces a residence time distribution that strongly influences uranium fate. Rapid advective flow occurs in open drifts, major fractures, and connected conduits, allowing relatively swift transfer of dissolved species to discharge points. In contrast, the surrounding rock matrix, backfill, and low-permeability blocks act as long-term storage domains where dissolved and sorbed uranium phases diffuse slowly in and out. This dual-domain mass transfer creates classic early-time pulses followed by protracted low-rate release: an initial "first flush" of uranium derived from immediate oxidative dissolution of exposed uraninite and other U(IV) phases in high surface area material is followed by long-tailed desorption and matrix diffusion that sustain measurable aqueous concentrations for decades to centuries.

Redox potential is the primary control on uranium speciation and, therefore, on mobility. Under reducing conditions, uranium is predominantly tetravalent and incorporated into sparingly soluble U(IV) phases such as uraninite or adsorbed into authigenic reduced phases; microbial and abiotic reduction of U(VI) to U(IV) is a principal immobilization mechanism where suitable electron donors (organic carbon, reduced iron, or sulfide) and microbial communities persist. Conversely, in oxic or sub-oxic compartments, common near shafts, recharge fractures, or during episodic oxygenated inflow, uranium is oxidized to hexavalent uranyl (UO_2^{2-} and $UO_2(CO_3)_3^{4-}$) stabilize U(VI) in solution and reduce its affinity for mineral surfaces, thereby increasing transport distances and persistence in groundwater.

Surface reactions and secondary mineral formation are the dominant short- to medium-term sinks for dissolved uranium in flooded underground environments. Sorption to iron oxyhydroxides, manganese oxides, phyllosilicates, and organic coatings occurs through surface complexation reactions that are strongly pH- and ionic strength-dependent; competitive adsorption by Ca^{2+} and dissolved carbonate can lower effective sorption capacity. Precipitation of uranyl minerals, such as carbonates, phosphates or mixed Fe-U phases, sequesters uranium in metastable solid forms. These secondary sinks, however, are not permanent: reductive dissolution of iron oxides under anoxic diagenesis or changes in pH and carbonate chemistry can remobilize previously sequestered uranium, producing episodic release events tied to internal biogeochemical cycling.

Matrix diffusion and desorption kinetics account for the characteristic long-term tailing in uranium release. Uranium that sorbed to fracture walls or that diffused into micro-porous grain boundaries exchanges back into the flow field at rates controlled by diffusion coefficients and kinetic desorption parameters. Because these processes are slow relative to advective flushing, they impart a "memory" to the system: aqueous concentrations decline non-linearly, often following multi-rate decay patterns rather than simple exponential decline. This behaviour is exacerbated in crystalline host rocks typical of many uranium deposits, where matrix porosity is low and diffusion pathways long, resulting in very prolonged low-level fluxes even when active sources are attenuated.

Redox heterogeneity and micro-scale hotspots govern localized immobilization and remobilization. Microbial hotspots, niches where organic matter or reduced minerals fuel sulfate and iron reduction, can create microniches with strong reducing potential that convert U(VI) to the insoluble U(IV) form. Adjacent oxic fractures can oxidize reduced solids, producing steep chemical gradients and shifting reaction fronts. The spatial juxtaposition of these microenvironments within confined underground compartments produces patchy uranium distributions: zones of high dissolved U downgradient from oxidizing fronts, juxtaposed with zones of precipitated

U in anoxic pockets. These fine-scale patterns scale up to influence bulk discharge chemistry via mixing at fracture intersections or during hydrologic perturbation.

Colloidal and particulate pathways also influence uranium transport in underground flooded mines. Colloidal iron oxyhydroxides and organic-rich particles can scavenge uranyl species and transport them as suspended phases; mechanical resuspension at discharge points or during inflow events can liberate particle-bound uranium into the water column, enhancing bioavailability at receptors. Furthermore, authigenic coatings and layered precipitates that form on fracture walls can act as reactive barriers, modifying flow-borne uranium concentrations until mechanical or chemical disruption removes or alters these coatings.

Interactions with co-contaminants and major ions modulate uranium chemistry. Sulfate generation from sulfide oxidation, for example, influences ionic strength and can promote the formation of soluble sulfate complexes with other metals while indirectly affecting uranium by altering sorbent surfaces. Phosphate, if present, can precipitate with uranyl to form low-solubility uranyl-phosphate minerals, providing a potentially robust sink under certain conditions; conversely, high carbonate concentrations undermine phosphate precipitation and favour uranyl-carbonate persistence. Radium isotopes, which often accompany uranium, have their own geochemical pathways tied to sulfate and barite dynamics and can be mobilized or sequestered independently or concurrently with uranium, depending on solution chemistry.

Temporal evolution of uranium concentration in discharges from flooded underground workings is therefore the outcome of competing and time-dependent processes: initial oxidative dissolution, sorption and secondary mineral trapping, slow matrix diffusion and desorption, episodic remobilization driven by hydrologic or redox perturbations, and particle-associated transport. Predicting this evolution requires mechanistic reactive transport models that account for fracture-dominated advection, dual-domain diffusion, kinetically limited redox reactions, and surface complexation. Field characterization to parameterize such models must quantify mineralogy of primary and secondary phases, sorption isotherms and kinetics, fracture network properties, matrix diffusivity, and the presence and activity of relevant microbial populations.

In sum, the mature underground flooded uranium mine is best understood as a chemically stratified, hydraulically constrained reactor in which uranium cycles among dissolved, sorbed, and solid phases according to redox state, ligand chemistry, and transport connectivity. Its long-term fate is governed less by simple removal or decay and more by the balance between persistent internal sources, slow diffusive release, metastable mineral sinks, and episodic processes that can remobilize sequestered uranium.

4.5 Other Effects of Mine Flooding

Flooding of underground workings produces a cascade of coupled hydrogeological, geomechanical, and geochemical responses that evolve over timescales from hours to decades. The initial ingress of water immediately changes the boundary conditions of the subterranean system, redistributing hydraulic gradients, reactivating sealed or formerly isolated flow paths, and initiating chemical reactions between pore fluids, dissolved gases, and exposed rock surfaces. For experienced practitioners, understanding these processes requires viewing the flooded mine not as a static void filled with water but as a dynamic, multiphase system in which pressure, chemistry, and stress interact and feed back on one another. The following subsections describe those interactions in detail and highlight key mechanisms that drive long-term behaviour and operational risks.

4.5.1 Modification of Mining-Induced Hydrogeology

When a mine floods, the first-order hydrogeological effect is the progressive recovery of groundwater to its pre-mining or new equilibrium piezometric conditions. The rate and pattern of recovery are controlled by the hydraulic properties of the host rocks, the connectivity of excavations, and the transient head gradients imposed by surface-water inputs or connected aquifers. In many settings, flooded workings act as large internal reservoirs; water levels within shafts and chambers equilibrate with regional potentiometric surfaces only after conduit flow and fracture network storage have been re-established. This recovery is commonly nonlinear: early rapid rises occur where high-permeability pathways or open shafts exist, whereas slower, diffusion-dominated equilibration governs flow into tight rock matrices or isolated voids.

Flooding alters drainage behaviour by converting systems designed for active dewatering into passive collectors. Drainage reversals and the formation of perched water tables are common, especially where engineered drainage relied on continuous pumping. As flows reorganize, new hydraulic short circuits may develop through fractures, faults, or mine-to-mine connections, altering both the direction and magnitude of subsurface fluxes. Where mines interconnect, intentionally through workings or inadvertently through fractures and legacy boreholes, the composite flood response becomes a function of the most significant and permeable linkages: pressure can propagate rapidly over long distances, causing synchronous level changes in otherwise separated compartments.

Floodwater quality follows predictable yet complex trends driven by dilution, redox evolution, and water-rock interactions. Initially, entering surface or shallow groundwater can dilute residual pore fluids and flush soluble salts and suspended fines. Over weeks to years, however, geochemical processes dominate. Oxidation of previously exposed sulfide minerals, acidification in poorly buffered systems, release of metals, and mobilization of metalloids or radionuclides can occur as oxygenated water reacts with fresh rock surfaces. In contrast, prolonged anoxia behind flood fronts may promote reductive dissolution, precipitation of secondary minerals, and sequestration or remobilization of contaminants depending on pH, Eh, and ligand availability. Temporal trends, therefore, often show an early spike in easily leached constituents followed by longer-term shifts linked to evolving redox and mineral surface equilibria.

Hydrogeochemical processes within flooded mines are sensitive to the prior hydrothermal and mining history. Reactive surfaces created by blasting and collapse present a high surface area for sorption and redox reactions. The presence of secondary mineral coatings, such as iron oxides, clays, or sulfates, moderates contaminant release by providing sinks that can be saturated or altered by changing chemistry. Additionally, the influx of organic-rich waters or sediments can drive heterotrophic microbial activity, accelerating reductive pathways that alter metal speciation and gas production. The interdependence of physical transport and chemical kinetics means that simple mixing models frequently fail to capture observed concentration-time patterns; instead, reactive transport frameworks that couple advective flow, diffusion into matrix porosity, and surface reaction kinetics better describe long-term water quality evolution.

4.5.2 Geomechanical Effects

The mechanical response of the rock mass to flooding is governed by changes in effective stress and pore-pressure distribution, coupled with time-dependent alterations in strength from chemical and hydromechanical processes. Rapid pressurization of previously drained voids can reduce effective stresses on pillars and hanging walls, promoting heave, dilation, or sudden yield in structures that were marginally stable under dewatered conditions. Conversely, as the water level rises and buoyant forces act on the collapse debris, localized support from block-to-block contact may be lost, enabling differential movement that manifests at the surface as subsidence.

Heaving and sinking at the mine scale are often spatially complex. In some cases, inflowing water lifts unconsolidated backfill or weakly cemented material, producing upward movement that can propagate to overlying strata. In other instances, progressive saturation weakens near-surface weathered horizons, causing downward compaction and gradual surface settlement. Where flooding interacts with preexisting geologic structures, stress redistribution can concentrate deformation along faults or shear zones. These preferential failure surfaces may then nucleate larger displacements or collapse events.

Flooding can also influence the timing and frequency of tectonic-like events within the mine. Variations in pore pressure that transiently load or unload faults and major discontinuities can trigger seismicity, ranging from microseismicity associated with rockburst risk to larger felt events in high-stress domains. Such induced seismicity is often episodic and linked to pressure-front propagation and pressure-stabilization periods. Monitoring trends in microseismic energy releases and correlating them with measured water level changes enables practitioners to infer active stress-transfer mechanisms and evaluate the likelihood of larger instability.

Subsidence is the long-term, cumulative expression of many of these processes. The magnitude and pattern of surface subsidence following flooding reflect the spatial distribution of voids, the competency of remaining pillars, and the compressibility of overburden units. In settings where extensive pillar failure or roof collapse occurs after saturation, subsidence can be abrupt and severe. In contrast, diffuse consolidation due to pore-pressure equilibration produces progressive, basin-like settlements. Predicting subsidence, therefore, demands integrated hydromechanical modelling that captures both immediate mechanical weakening and slower deformation driven by creep, dissolution, and chemical softening.

4.5.3 Gas Emanation During Flooding

The gas regime within a flooded mine changes substantially as water replaces air-filled voids and alters redox and adsorption environments. Immediately after flooding begins, trapped air is gradually displaced, concentrating gases in pockets and potentially driving these pockets through escape routes that sometimes vent at the surface. The chemistry of gases evolves with the hydrochemical conditions: oxygen is rapidly consumed at reactive surfaces, promoting anaerobic processes that generate reduced gases.

Radon and its short-lived progeny require special attention because flooding modifies both production and transport. Radon emanation from uranium-bearing host rocks depends on the availability of pore space and the diffusion pathways to the void. Flooding reduces gaseous diffusion but can increase radon release into water, from which radon can later partition back into gas phases as water degasses, potentially producing episodic pulses at vents and shafts. Additionally, changes in humidity, temperature, and pressure associated with water inflow influence daughter-product plate-out and airborne concentrations in partially flooded galleries, altering radiological exposure risks to personnel and complicating long-term environmental assessments.

Other gases, notably methane, can be mobilized or generated in flooded conditions, particularly where organic matter or hydrocarbon-bearing strata are present or where secondary uranium mineralization coincides with organic-rich horizons. Methane generation under reducing conditions can arise from microbial methanogenesis once oxygen is depleted. In systems containing coal, carbonaceous material, or hydrocarbon-bearing inclusions, the intrusion of water and subsequent anoxia can liberate sorbed methane or promote in situ gas generation. The presence of methane introduces an explosion risk in partially flooded atmospheres and alters buoyancy and pressure regimes, affecting ventilation and gas migration. Monitoring and managing gas emanation, therefore, requires not only detection of concentration levels but also an understanding of their source mechanisms, partitioning behaviour between water and gas phases, and potential for episodic release driven by pressure transients.

4.6 Remediation and Restoration Options

Remediation and restoration of abandoned underground uranium mines are vital parts of responsible environmental management but may not always be necessary. Remediation and restoration efforts should be undertaken when abandoned underground mines pose long-term risks due to remaining contamination. Thus, the efforts should aim to reduce the dangers posed, restore ecological health, and safeguard public health. The decision to undertake remediation depends on various site-specific factors, including contaminant levels, exposure pathways, regulatory requirements, and future land-use plans.

4.6.1 Circumstances Warranting Remediations

Remediation is warranted when contaminants such as uranium, radium, arsenic, or nitrate compounds pose a significant threat to human health or the surrounding environment. This is especially true when there is evidence of ongoing contaminant migration into groundwater, surface water, or ecologically sensitive areas. Regulatory frameworks may also mandate cleanup to meet environmental standards, particularly if the site is located near populated areas or within protected zones. Additionally, if the land is intended for future use, whether for redevelopment, conservation, or community access, restoration is necessary. In some cases, public concern or Indigenous land rights may further compel action to restore trust and ecological balance.

Conversely, remediation may not be warranted when contaminant concentrations are stable, below regulatory thresholds, and show no signs of migration. If natural attenuation processes are effectively reducing contaminant levels over time, and monitoring confirms a downward trend, active intervention may be unnecessary. Remote, inaccessible sites with no foreseeable exposure pathways to humans or sensitive ecosystems may also be candidates for non-intervention. Furthermore, if a cost-benefit analysis reveals that the financial burden of remediation far outweighs the environmental or health benefits, it may be prudent to forgo restoration. In some instances, remediation activities themselves could cause more ecological harm than leaving the site undisturbed, particularly in fragile or recovering habitats.

4.6.2 Some Applicable Remediation Techniques

When remediation is deemed necessary, several techniques are available, each suited to specific site conditions and contaminant profiles. Common remediation approaches and the scenarios in which they are recommended include the following:

(a) Pump and Treat Technique

This technique involves extracting groundwater, treating it aboveground, and re-injecting or discharging it. If contaminants are mobile and exceed safe levels, hydraulic control is recommended.

(b) Passive Barriers Technique

This technology involves using reactive materials (e.g., zero-valent iron) placed in the underground in the plum flow direction to intercept and neutralize contaminants. It is recommended for situations where groundwater flow is predictable; a long-term, low-maintenance solution is needed.

(c) Bioremediation

This involves using living organisms, most often microorganisms, to immobilize metallic contaminants or degrade organic pollutants in situ or ex situ. It is used where microorganisms can induce precipitation or absorption through a bioredox process or by organic or biodegradable contaminants present; conditions support microbial activity.

(d) Dimensioning Water Treatment Facilities

It involves designing scalable treatment systems for surface or groundwater. It is recommended for long-term water management where multiple contaminant types are present.

These techniques can be used alone or combined, depending on the site's complexity and desired results. A detailed site evaluation, including hydrogeological analysis, contaminant profiling, and stakeholder consultation, is crucial before choosing the right remediation approach. Ultimately, the aim is to make sure that any action is both environmentally effective and economically viable while also considering the social and cultural context of the affected area.

4.7 Concluding Remarks

This chapter outlines the typical behaviour of uranium in abandoned underground uranium mines, emphasizing the primary processes that influence its movement, form, and potential environmental release. The patterns described reflect what can generally be expected under common geological, climatic, and hydrological conditions. However, actual outcomes are intrinsically site-specific, as factors such as local rock chemistry, groundwater flow, mineralogy, and climate significantly affect the rates, pathways, and magnitudes of uranium migration. This synthesis aims to establish a conceptual baseline for interpreting post-closure observations and guiding monitoring priorities, rather than predicting the exact conditions at any particular mine. Notably, many environmental and radiological risks discussed can be significantly mitigated or avoided through careful planning for closure and long-term management, with an emphasis on groundwater control, containment of reactive materials, and engineered barriers. Therefore, readers should view these conclusions as a framework for assessment and decision-making, which can be further refined through site investigations and customized closure strategies.

References

1. Al Hashimi, A., Evans, G. J., & Cox, B. (1996). Aspects of the permanent storage of uranium tailings. *Water, Air, and Soil Pollution, 88*, 83–92.
2. Enderle, G. J., & Friedrich, K. (1999). Uranium mining in East Germany ("Wismut"): Health consequences, occupational medical care and workers' compensation. *International Archives of Occupational and Environmental Health, 72*, M042–M049.
3. Bowen, G. G., Dussek, C., & Hamilton, R. M. (1998). Pollution resulting from the abandonment and subsequent flooding of Wheal Jane Mine in Cornwall, UK. *Geological Society, London, Special Publications, 128*, 93–99. https://doi.org/10.1144/gsl.sp.1998.128.01.09
4. Pyle, G. G., Swanson, S. M., & Lehmkuhl, D. M. (2002). Toxicity of uranium mine receiving waters to early life stage fathead minnows (Pimephales promelas) in the laboratory. *Environmental Pollution, 116*, 243–255.
5. Roig, M. G., et al. (1995). Enzymically-enhanced extraction of uranium from biologically leached solutions. *International Biodeterioration and Biodegradation, 35*, 93–127.
6. Whitehead, P., & Neal, C. (2005). Bioremediation of acid mine drainage: The Wheal Jane Mine wetlands project. *Science of the Total Environment, 338*, 1.
7. Zhang, T., et al. (2023). A review: The formation, prevention, and remediation of acid mine drainage. *Environmental Science and Pollution Research International, 30*, 111871–111890. https://doi.org/10.1007/s11356-023-30220-5
8. Faizrakhmanov, F. F. (2024). Speciation of uranium in trace concentrations in aqueous solutions of complex chemical composition. *Radiochemistry, 66*, 268–273. https://doi.org/10.1134/S1066362224020164
9. Sessler, J. L., Melfi, P. J., & Pantos, G. D. (2006). Uranium complexes of multidentate N-donor ligands. *Coordination Chemistry Reviews, 250*, 816–843.
10. Fiedor, J. N., Bostick, W. D., Jarabek, R. J., & Farrell, J. (1998). Understanding the mechanism of uranium removal from groundwater by zero-valent iron using X-ray photoelectron spectroscopy. *Environmental Science and Technology, 32*, 1466–1473.
11. Meinrath, G. (1998). *Uranium Mining and Hydrogeology II. Proceedings* (Vol. 5, pp. 256–265). Sven von Loga.
12. Boekhout, F., et al. (2015). Uranium migration and retention during weathering of a granitic waste rock pile. *Applied Geochemistry, 58*, 123–135. https://doi.org/10.1016/j.apgeochem.2015.02.012
13. Jerden, J. J. L., & Sinha, A. K. (2006). Geochemical coupling of uranium and phosphorous in soils overlying an unmined uranium deposit: Coles Hill, Virginia. *Journal of Geochemical Exploration, 91*, 56–70.
14. Jerden, J. L., Sinha, A. K., & Zelazny, L. (2003). Natural immobilization of uranium by phosphate mineralization in an oxidizing saprolite-soil profile: Chemical weathering of the Coles Hill uranium deposit, Virginia. *Chemical Geology, 199*, 129–157.
15. Kanzari, A., et al. (2017). Impact of incipient weathering on uranium migration in granitic waste rock piles from former U mines (Limousin, France). *Journal of Geochemical Exploration, 183*, 114–126. https://doi.org/10.1016/j.gexplo.2017.08.010
16. Wolkersdorfer, C. (2008). *Water management at abandoned flooded underground mines*. Springer. ISBN: 978-3-540-77330-6.
17. Rapantova, N., Licbinska, M., Babka, O., Grmela, A., & Pospisil, P. (2013). Impact of uranium mines closure and abandonment on groundwater quality. *Environmental Science and Pollution Research International, 20*, 7590–7602. https://doi.org/10.1007/s11356-012-1340-z
18. IAEA. (2018). *World distribution of uranium deposits (UDEPO) 2016 edition*. International Atomic Energy Agency.
19. Duxbury, A. C., Mackenzie, F. T., & Byrne, R. H. (2024). In The Editors of Encyclopædia Britannica (Ed.), *Encyclopedia Britannica*. Britannica Education.
20. Association, W. N. (2025). Nuclear information World Nuclear Association.

21. Agency, N. E., & Agency, I. A. E. (2025). Uranium 2024: Resources, production and demand. Report No. NEA No. 7683. Nuclear Energy Agency, Organisation for Economic Co-Operation and Development.
22. Wolkersdorfer, C. (2008). *Water management at abandoned flooded underground mines: Fundamentals, tracer tests, modelling, water treatment* (p. 486). Springer.
23. Smedley, P. L., & Kinniburgh, D. G. (2023). Uranium in natural waters and the environment: Distribution, speciation and impact. *Applied Geochemistry, 148*, 105534. https://doi.org/10.1016/j.apgeochem.2022.105534
24. Deb, D., & Chakma, S. (2023). Colloid and colloid-facilitated contaminant transport in subsurface ecosystem, a concise review. *International Journal of Environmental Science and Technology, 20*, 6955–6988. https://doi.org/10.1007/s13762-022-04201-z
25. Mal'kovskii, V. I. (2011). Deposition of heterogeneous radiocolloid from groundwater on enclosing rocks. *Doklady Earth Sciences, 436*, 39–41. https://doi.org/10.1134/S1028334x11010077
26. Eyrolle, F., & Charmasson, S. (2001). Distribution of organic carbon, selected stable elements and artificial radionuclides among dissolved, colloidal and particulate phases in the Rhône River (France): Preliminary results. *Journal of Environmental Radioactivity, 55*, 145–155.
27. Colas, E., Kukla, P. A., Amann, F., & Back, S. (2025). Geological and mining factors influencing further use of abandoned coal mines—A multi-disciplinary workflow towards sustainable underground storage. *Journal of Energy Storage, 108*, 115101. https://doi.org/10.1016/j.est.2024.115101
28. Newman, C. P., Walton-Day, K., Runkel, R. L., & Wilkin, R. T. (2023). Mechanisms of water-rock interaction and implications for remediating flooded mine workings elucidated from environmental tracers, stable isotopes, and rare earth elements. *Applied Geochemistry, 157*, 1–17. https://doi.org/10.1016/j.apgeochem.2023.105769
29. Wei, X., Chen, H., Zhu, F., & Li, J. (2024). Microbial community structure in an uranium-rich acid mine drainage site: Implication for the biogeochemical release of uranium. *Frontiers in Microbiology, 15*, 1412599. https://doi.org/10.3389/fmicb.2024.1412599
30. Camarero, P. L., Moreira, C. A., Targa, D. A., Duz, B. G., & Pereira, H. G. (2022). Analysis of acid drainage flow zones in a rocky massif in a uranium mine from structural and geophysical diagnoses. *Mine Water and the Environment, 41*, 303–316. https://doi.org/10.1007/s10230-021-00827-6

Chapter 5
Uranium Dynamics in Pit Lakes

Abstract This chapter examines uranium dynamics in abandoned open-pit lakes from mine closure through natural maturation. Understanding uranium behaviour under these conditions is critical for evaluating long-term environmental risks and guiding remediation strategies. Although no documented case of an abandoned uranium pit lake exists, insights are drawn from laboratory experiments, microcosm trials, and advanced geochemical and hydrochemical modelling. These approaches simulate physicochemical processes, such as redox reactions, sorption-desorption equilibria, and complexation under fluctuating hydrological and mineralogical conditions. Integrated computer simulations predict uranium transport and speciation over decadal timescales, considering pH, redox potential, and pit-lake water chemistry. By bridging empirical data and theoretical modelling, this chapter provides a predictive framework for assessing uranium mobility and ecological implications in post-open-pit mining landscapes.

Keywords Open-pit mining · Mine flooding · Stratified waters · Redox gradients · Metal mobility · Sediment interactions · Colloidal transport · Water quality

5.1 Introduction

Understanding uranium behaviour in abandoned open-pit uranium mines is essential for assessing long-term environmental risks and informing future remediation strategies. Although no specific site has been formally documented as an abandoned open-pit uranium mine, this chapter explores uranium dynamics through a synthesis of laboratory experiments, trials, and advanced physical, geochemical, and hydrochemical modelling. These investigations are grounded in data from a broad spectrum of abandoned open-pit mines, not limited to uranium extraction, enabling informed extrapolation of uranium-specific processes.

Pit lakes, also known as pit lakes, are man-made bodies of water formed in former open-pit mines and exhibit steep walls, a high depth-to-surface-area ratio, and water quality that can range from alkaline to acidic, depending on the surrounding

M. Mkandawire, *Fate and Effects of Uranium in Abandoned Uranium Mines*,
https://doi.org/10.1007/978-3-032-10447-2_5

geology. They often have high levels of dissolved solids and can contain high levels of acid, sulfate, and dissolved metals, especially in the lower, anoxic layers. Pit lakes take a long time to stabilize and may require active rehabilitation to become ecologically viable [1–3].

This chapter delves into the physicochemical transformations of uranium under post-open-pit mining conditions where a mine is abandoned, and all processes are nature-driven, including oxidation-reduction reactions, sorption-desorption equilibria, and complexation with naturally occurring ligands. Simulated scenarios replicate the influence of fluctuating water tables, seasonal hydrology, and mineralogical evolution within pit walls and sediments. Microcosmos trials, controlled outdoor systems that mimic mine environments, provide valuable insights into uranium mobility, bioavailability, and retention mechanisms under quasi-natural conditions.

Computer simulations integrate these findings to model uranium transport and speciation over decadal timescales, accounting for variables, such as pH, redox potential, and the presence of competing ions. While the absence of a real-world case study imposes limitations, the convergence of empirical and theoretical approaches provides a robust framework for understanding uranium behaviour in such settings [4, 5]. Thus, this chapter also aims to bridge the gap between theoretical modelling and environmental application, offering a predictive lens through which the legacy of uranium mining can be assessed, even in the absence of direct site documentation.

5.2 Uranium Open-Pit Lakes

5.2.1 Types

There are several types of open-pit uranium mining, including conventional, strip, terrace, and highwall mining. Selecting an appropriate open-pit mining method for uranium extraction is a complex decision influenced by a confluence of geological, hydrological, environmental, and economic factors. Each technique, whether conventional, strip, terrace, or highwall mining, imparts distinct morphological and geochemical characteristics to the post-mining landscape, particularly the pit lakes that often form upon cessation of dewatering activities. These lakes are critical zones for uranium dynamics, as they serve as repositories for residual ore, weathered minerals, and process-affected waters.

The geometry and depth of the uranium orebody are primary determinants in the choice of mining method. Conventional open-pit mining is typically employed when the deposit is deep and vertically concentrated. This method involves the sequential removal of overburden and ore in benches, resulting in a deep, steep-walled excavation [6]. Upon mine closure, such pits often intersect regional groundwater systems, leading to the formation of deep lakes with pronounced thermal and chemical stratification. Stratification in open-pit lakes is caused by the formation of distinct water layers with different densities, driven primarily by temperature differences (thermal stratification) and variations in dissolved solids (chemical or salinity stratification). As illustrated in the comparison between a natural and an open-pit lake in Fig. 5.1, the steep, sheltered morphology of mine pits often enhances the stability of this layering by reducing wind mixing.

Stratification is a crucial physical and chemical characteristic of developing pit lakes, arising from the interplay of water density gradients, temperature variations, and solute concentrations. In contrast to natural lakes, pit lakes often exhibit extreme depth-to-surface-area ratios and steep walls, which enhance their tendency to stratify and resist mixing. This stratification considerably influences the distribution, speciation, and mobility of uranium and other contaminants, particularly in systems affected by acid mine drainage (AMD).

The exposure of uranium-bearing minerals to atmospheric oxygen and infiltrating water during and after mining promotes oxidative dissolution, particularly the transformation of relatively immobile U(IV) to the more soluble U(VI) species. Consequently, these lakes often exhibit elevated dissolved uranium concentrations, especially in the epilimnion, where oxic conditions prevail [7, 8].

In contrast, strip mining is favoured when uranium deposits are shallow and laterally extensive. This method involves removing long, narrow strips of overburden to access the ore, resulting in broad, shallow pits. The lakes formed in such settings tend to have large surface areas and shallow depths, which influence their thermal behaviour and mixing regimes. The shallow nature of these lakes facilitates greater interaction between the water column and atmospheric gases, enhancing photochemical and biogeochemical processes that can mobilize uranium through complexation with dissolved organic matter or colloidal iron phases. Moreover, the

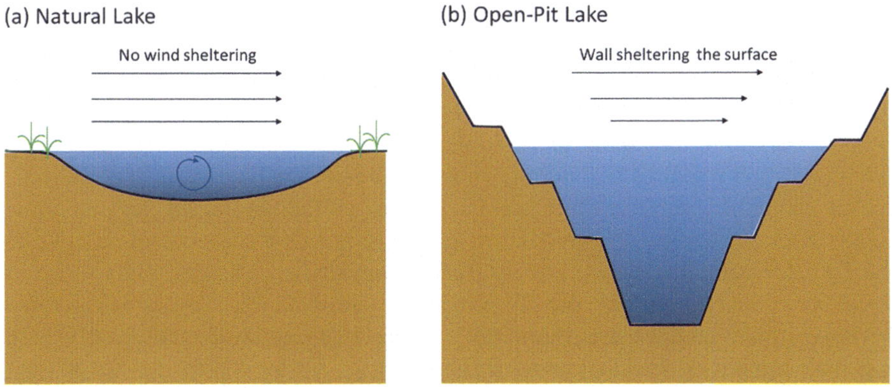

Fig. 5.1 Comparison of morphologies of a (**a**) natural and (**b**) open-pit lake, showing that the sheltered morphology of mine pits and depth, which often enhances the stability and reduces wind mixing. **Note**: Some natural lakes, especially deeper ones, commonly exhibit thermal stratification. Famous examples of stratified lakes include deep, ancient rift lakes like Baikal, Tanganyika, and Malawi

relatively high surface-area-to-volume ratio increases the susceptibility of these lakes to evaporation, potentially concentrating dissolved uranium and other solutes.

Terrace mining is employed on moderate-relief terrain, where the orebody is accessed via a series of stepped benches constructed along the slope. The resulting pit lakes are often irregular in shape and depth, leading to spatial heterogeneity in water chemistry and sediment interactions. These lakes may exhibit localized zones of anoxia, particularly in deeper pockets, which can foster reductive conditions conducive to the precipitation of uranium as U(IV) minerals, such as uraninite. However, the dynamic nature of redox conditions in these lakes, driven by seasonal turnover and variable inflows, can periodically remobilize uranium, complicating predictions of long-term contaminant behaviour.

Highwall mining, though less commonly applied in uranium extraction, is occasionally used to recover ore from exposed seams in the highwalls of previously mined pits. This method results in narrow, elongated voids that may fill with water post-mining, forming confined lakes with limited hydrological connectivity and low flow-through. The restricted volume and minimal mixing can lead to persistent reducing conditions, stabilizing uranium in its tetravalent form. Nevertheless, these systems are vulnerable to episodic oxygenation events, such as heavy rainfall or groundwater intrusion, that can disrupt redox equilibria and trigger the release of uranium into the aqueous phase [9].

5.2.2 Pit-Lake Formation Process

The genesis of a pit lake begins at the cessation of open-pit mining operations, marking a pivotal transition from active resource extraction to passive hydrological transformation. During mining, extensive dewatering systems are employed to keep the pit dry and operational, often lowering the water table significantly below its natural level. Once mining activities conclude and dewatering ceases, the engineered void is left to interact with its surrounding hydrogeological environment, initiating the formation of a lake [6, 10, 11].

With the pumps turned off, groundwater begins to rebound, gradually infiltrating the pit through fractures and porous strata in the surrounding rock. This influx is complemented by surface water inputs such as rainfall, snowmelt, and, in some cases, diverted streams or rivers. However, these are typically constrained by the mine's catchment area, defined by its artificial boundaries. The rate and volume of water accumulation are governed by the local water balance, a dynamic interplay between inflow sources and evaporative losses that can vary significantly with regional climate and seasonal patterns (Fig. 5.2).

As the pit fills, the emerging lake begins to exhibit distinct physical and chemical characteristics. The steep geometry of the mine walls and the high depth-to-surface-area ratio promote thermal and chemical stratification, often resulting in a stable, layered water column. The lower strata, isolated from atmospheric mixing, tend to become anoxic, fostering the accumulation of dissolved metals and sulfates,

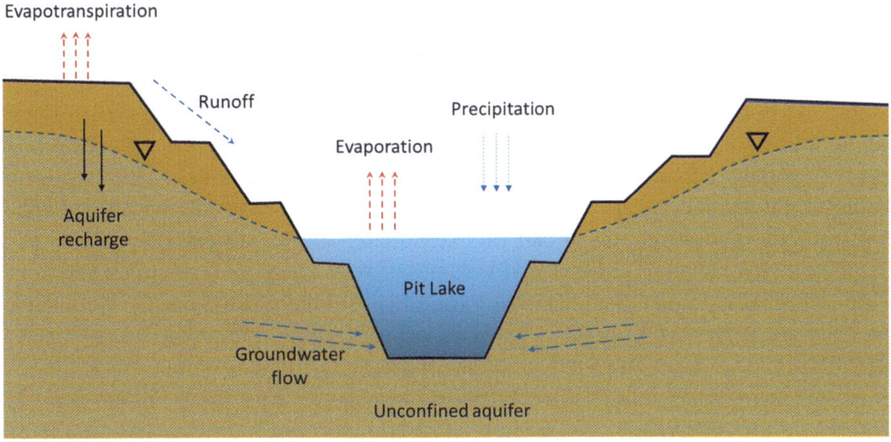

Fig. 5.2 Rebounding of the water table after cessation of dewatering leading to the formation of an open-pit lake. Water surface runoff, direct precipitation from the atmosphere, and ground fill the open pit

particularly in pits where sulfide minerals were exposed during excavation. These geochemical processes can lead to acid generation, producing water with low pH and elevated concentrations of contaminants, such as arsenic, copper, and zinc [12, 13].

The evolution of water quality in the pit lake is a gradual, complex process. Initially, the lake may exhibit extreme conditions, ranging from highly acidic to alkaline, depending on the mineralogy of the pit walls and the surrounding geology. Over time, natural attenuation processes, such as dilution, precipitation, and microbial activity, may help stabilize water chemistry. However, in many cases, passive recovery is insufficient, and active rehabilitation strategies are required. These may include chemical amendments, aeration, or bioremediation to mitigate acidity and metal concentrations, thereby enhancing the lake's ecological viability.

As it matures, a pit lake may take decades to millennia to stabilize, depending on its size, depth, and environmental context. In some instances, with sustained management, these lakes can transition into functional aquatic ecosystems, supporting biodiversity and even recreational use. Yet, the path to such outcomes is neither linear nor guaranteed, underscoring the importance of integrated hydrogeological, geochemical, and ecological assessments in post-mining land-use planning.

5.2.3 *Characteristics of Open-Pit Mine Lakes*

Open-pit mine lakes are artificial water bodies that form in the voids left by open-pit mining operations. These lakes are defined by unique physical, chemical, and biological characteristics that distinguish them from natural lakes.

Physically, pit lakes are characterized by their steep, engineered walls and a notably high depth-to-surface-area ratio. Unlike natural lakes that typically have broad catchments and gradual shorelines, pit lakes are confined within the sharply contoured boundaries of the former mine, resulting in limited surface inflows. Interactions with groundwater systems heavily influence their hydrology. They may receive groundwater inflows or, conversely, act as recharge zones. The long-term stability of the pit lakes hinges on a delicate balance between groundwater exchange and surface evaporation, which can vary seasonally and regionally [14, 15].

Chemically, pit lakes exhibit a wide range of water quality, from fresh and non-toxic to highly saline and acidic. This variability is mainly governed by local geology and the types of minerals exposed during mining. Many pit lakes contain elevated levels of total dissolved solids, with conductivity and solute concentrations often increasing with depth. When sulfide-bearing rocks are exposed to oxygen and water, acid mine drainage can occur, leading to high concentrations of sulfate and dissolved metals, such as arsenic, copper, manganese, and zinc. These conditions are exacerbated in the lower layers of the lake, which are frequently anoxic, oxygen-deprived, and chemically stratified, preventing vertical mixing and trapping contaminants in the deeper layers.

Biologically, newly formed pit lakes tend to have low biodiversity. The absence of organic matter and the harsh chemical conditions limit the establishment of aquatic life, notably higher trophic levels, such as fish and aquatic plants. Over time, however, these lakes may undergo a slow process of ecological succession. Natural stabilization can take decades to millennia, depending on the lake's size, chemistry, and surrounding environment. With targeted rehabilitation efforts, such as nutrient amendments, aeration, or the use of constructed wetlands, pit lakes can be transformed into ecologically viable habitats. In some cases, they have even been repurposed for recreational use, supporting fishing, boating, and wildlife conservation.

5.3 Hydrogeochemical Evolution of Open-Pit Lakes

5.3.1 Hydrological Evolution

The hydrological evolution of open-pit lakes is a dynamic and multifaceted process that unfolds in distinct stages following the cessation of mining activities. There are three arbitrary sequences of transformation stages of pit-lake formation as outlined in Fig. 5.3. This evolution is intricately linked to the broader hydrogeochemical transformations that define the post-mining landscape. Upon the termination of dewatering operations, typically maintained during active mining to keep the pit dry, the regional groundwater table begins to rebound. This initiates the first phase of lake formation, characterized by the gradual inundation of the pit void.

Initially, the pit functions as a terminal sink, a hydrologically closed system where water inflows, primarily from groundwater seepage, direct precipitation, and surface runoff, accumulate without significant outflow. In arid and semi-arid

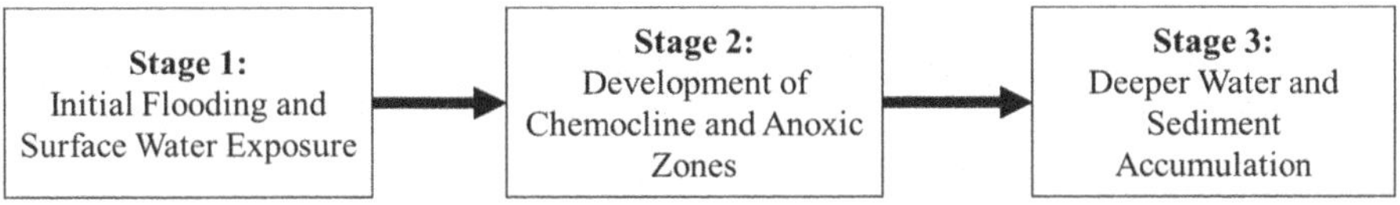

Fig. 5.3 The arbitrary sequence of transformation stages involved in pit-lake formation upon cessation of mine dewatering

climates, evaporation dominates the water balance, leading to a net water loss and a progressive concentration of dissolved solutes. This results in elevated salinity and the potential for acidic conditions, particularly in pits intersecting sulfide-rich lithologies. The oxidation of sulfide minerals such as pyrite (FeS_2) generates sulfuric acid, which lowers pH and mobilizes metals into the aqueous phase, a process described by the reaction:

$$FeS_2 + 3.75O_2 + 3.5H_2O \rightarrow Fe(OH)_3 + 2SO_4^{2-} + 4H^+$$

This geochemical evolution is compounded by the lack of flushing mechanisms in terminal sinks, allowing contaminants to accumulate over time.

As the lake continues to fill and the regional water table re-establishes, the system may transition into a throughflow regime. This occurs when the hydraulic head within the lake exceeds that of the surrounding aquifers, reversing the direction of groundwater flow. The lake then begins to discharge water into adjacent groundwater systems, effectively becoming a hydrologically open system. This transition is governed by Darcy's law, which describes the volumetric flow rate Q of groundwater through a porous medium:

$$Q = -K \cdot A \cdot \frac{dh}{dl}$$

where K is the hydraulic conductivity of the aquifer, A is the cross-sectional area perpendicular to flow, and $\frac{dh}{dl}$ is the hydraulic gradient. Establishing throughflow conditions enhances the potential for dilution and attenuation of contaminants, thereby influencing the long-term water quality of the pit lake.

The overall water balance of an open-pit lake can be expressed as:

$$\Delta S = P + R + G_{in} - E - G_{out}$$

where ΔS is the change in lake storage, P is precipitation, R is surface runoff, G_{in} and G_{out} are groundwater inflow and outflow, respectively, and E is evaporation. The relative magnitudes of these components are influenced by climatic conditions, pit geometry, and the hydrogeological properties of the surrounding strata [7, 9, 16].

The hydrological trajectory of a pit lake is thus not only a function of physical processes but also a driver of hydrogeochemical evolution. The interplay between water fluxes and geochemical reactions, such as mineral dissolution, precipitation, and redox transformations, creates a complex and evolving aquatic system. Understanding these processes is critical for predicting long-term water quality and informing effective closure and remediation strategies [17].

5.3.2 Geochemical Evolution

During the filling stage, the abrupt transition from a drained excavation to a hydrologically active basin exposes large areas of freshly fractured rock and mine waste to water and oxygen, establishing the primary water-rock reaction pathways that control early chemistry [18]. The first pulses of groundwater and surface water contacting reactive sulfide-bearing surfaces produce rapid increases in conductivity and dissolved metal concentrations as oxidative weathering commences; these early fluxes set the initial inventory of dissolved constituents and particulate loads that will be redistributed or attenuated later in the lake's life. The dominant oxidative reaction for pyrite can be written as a net stoichiometric expression that links oxygen consumption to acidity generation and sulfate release:

$$FeS_2 + 3.75O_2 + 3.5H_2O \rightarrow Fe(OH)_3 + 2SO_4^{2-} + 4H^+$$

Chalcopyrite and other metal sulfides undergo analogous oxidation, releasing Cu, Zn, As, and other metals to the nascent water column and increasing the acid load that controls subsequent speciation and solubility [19].

As the lake deepens and physical gradients develop, thermal and density stratification commonly produce a persistent chemocline that separates an oxic epilimnion from deeper, chemically distinct waters. The deep, narrow geometry and limited fetch of many pits favour stable layering and reduced wind-driven mixing, so that the chemocline becomes the locus of steep redox and pH gradients that control speciation. Within and across the chemocline, redox couples such as O_2 / NO_3^-, Fe^{3+}/Fe^{2+}, and SO_4^{2-} / H_2S define sequential electron acceptor zones; the position and sharpness of the chemocline are therefore a function of mixing regime, inflow chemistry, and internal oxygen demand. These gradients also concentrate particulate iron and manganese oxides at the interface, which act as transient scavengers for trace metals until reductive dissolution releases associated contaminants to deeper waters [19–21].

In the anoxic bottom waters and sediments, redox-sensitive transformations dominate element mobility. Microbial and abiotic reduction of ferric iron increases dissolved ferrous iron according to:

$$Fe^{3+} + e^- \rightarrow Fe^{2+}$$

while microbial sulfate reduction converts sulfate to hydrogen sulfide:

$$SO_4^{2-} + 8H^+ + 8e^- \rightarrow H_2S + 4H_2O$$

The produced H_2S can precipitate dissolved metals as insoluble metal sulfides:

$$M^{2+} + H_2S \rightarrow MS(s) + 2H^+$$

This sequence can immobilize metals in sediments, but it can also release acidity and, under some conditions, remobilize elements if redox conditions oscillate; the net outcome depends on organic carbon availability, sulfate supply, and sediment geochemistry [20]. The balance among evaporative concentration, carbonate buffering, and ongoing subaqueous weathering controls salinity, buffering, and long-term stabilization. Carbonate neutralization follows:

$$CaCO_3 + 2H^+ \rightarrow Ca^{2+} + CO_2 + H_2O$$

But carbonate reserves are often limited in sulfide-dominated pits, so that pH recovery may be slow without external alkalinity inputs. Over decadal timescales, internal cycling determines whether a pit lake evolves towards neutral, metal-attenuating conditions or remains acidic and metal-rich [22]. The internal cycling involves precipitation of secondary minerals, adsorption onto oxides, and burial in sediments.

5.3.3 Water Quality Evolution

During the initial filling stage, freshly exposed sulfide minerals on pit walls oxidize rapidly, producing classic acid mine drainage (AMD) chemistry: for pyrite oxidation:

$$FeS_2 + 3.5O_2 + H_2O \rightarrow Fe^{2+} + 2SO_4^{2-} + 2H^+$$

followed by ferrous iron oxidation

$$4Fe^{2+} + O_2 + 4H^+ \rightarrow 4Fe^{3+} + 2H_2O$$

and hydrolysis/precipitation of ferric iron:

$$Fe^{3+} + 3H_2O \rightarrow Fe(OH)_3(s) + 3H^+$$

These coupled reactions lower pH and mobilize metals (Fe, Al, Zn, and Cu) and sulfate, establishing a low-pH, high-metal, high-sulfate water column that controls early transport and toxicity. In the transitional stage, dilution by surface and groundwater inflows, carbonate buffering, and mineral precipitation begin to moderate acidity; carbonate complexation and neutralization reactions, such as:

$$CO_3^{2-} + H^+ \rightarrow HCO_3^-$$

and the mass balance effect:

$$C = C_0 \frac{V_0}{V}$$

for conservative solutes, illustrate how dilution and volume change alter concentrations. Redox gradients develop as oxygen is consumed, enabling microbial processes, notably heterotrophic respiration and early sulfate reduction. So, the sulfate reduction reaction:

$$SO_4^{2-} + 2CH_2O \rightarrow H_2S + 2HCO_3^-$$

and dissimilatory iron reduction:

$$Fe(OH)_3 + e^- + 3H^+ \rightarrow Fe^{2+} + 3H_2O$$

These two reactions become important, promoting metal remobilization or sequestration depending on local conditions. In the mature stage, physical stratification and, in arid settings, evaporative concentration drive salinity increases and permanent density layering; evaporative concentration of dissolved salts follows the inverse-volume relation above and can produce terminal-sink hypersaline brines that concentrate conservative and non-conservative constituents. Elevated nutrient loads stimulate phytoplankton and algal blooms that change optical properties and oxygen dynamics, while microbial community shifts towards anaerobic metabolisms (sulfate reduction and methanogenesis), reconfiguring metal cycling and redox-sensitive partitioning. Throughout these stages, small shifts in pH, redox potential, or hydrologic balance can flip dominant controls between aqueous complexation, adsorption to Fe/Al oxides, and precipitation/co-precipitation, so predictive models must couple hydrology, thermodynamic speciation, and biogeochemical kinetics to capture the evolving water quality of pit lakes.

5.4 Uranium Geochemical Behaviour in Pit Lakes

The geochemical behaviour of uranium in pit lakes is governed by a sequence of transformations that reflect the evolving physical and chemical conditions during lake formation. From initial exposure to oxygenated water to eventual sequestration in sediments, uranium undergoes complex speciation changes that influence its mobility, bioavailability, and long-term environmental impact.

5.4.1 Initial Flooding and Surface Water Exposure

Upon cessation of mining and dewatering, the pit begins to fill with groundwater and surface water. During this early phase, uranium is primarily mobilized through leaching from exposed rock surfaces, particularly from lithologies, such as granites, pegmatites, and uraniferous shales. In the oxygen-rich, acidic conditions typical of initial flooding, uranium predominantly exists in its hexavalent form as the highly soluble uranyl ion (UO_2^{2+}). This species readily forms aqueous complexes with sulfate and carbonate ions, enhancing its mobility in the water column.

Groundwater inflows may also contribute dissolved uranium, depending on the regional hydrogeochemistry and the presence of uranium-bearing minerals in the aquifer matrix. The combination of oxidative dissolution and complexation leads to elevated uranium concentrations in the surface layers of the nascent lake, posing potential risks for downstream transport and ecological exposure.

5.4.2 Development of Chemocline and Anoxic Zones

As the lake deepens and stratifies, a chemocline often develops, delineating the transition from oxygenated surface waters to anoxic bottom layers. A chemocline is a sharp vertical layer in a body of water where the water's chemical properties change drastically, separating an upper layer from a lower one. This transition zone is characterized by a steep gradient, often separating oxygen-rich surface water from deep, anoxic water with higher concentrations of dissolved solids and gases. This stratification marks a critical turning point in uranium geochemistry. In the anoxic zone, microbial activity, particularly from sulfate-reducing and iron-reducing bacteria, creates reducing conditions that favour the transformation of uranium from its mobile hexavalent state (U(VI)) to the less soluble tetravalent form (U(IV)).

Reduced uranium tends to precipitate as insoluble minerals such as uraninite (UO_2) or coffinite ($U^{IV}SiO_4$), effectively immobilizing it within the water column or onto particulate surfaces. The availability of reductants influences this process, the presence of nucleation sites, and the kinetics of microbial metabolism. In some cases, uranium may also associate with organic matter or colloidal particles, such as iron and aluminium oxyhydroxides, forming stable complexes that further limit its mobility.

5.4.3 Deeper Water and Sediment Accumulation

Over time, reduced or colloidal uranium settles to the lakebed, where it is incorporated into accumulating sediments. In these deeper layers, diagenetic processes may form polymeric and nanocrystalline uranium-silica species, particularly in silica-rich environments. These forms are often highly stable and resistant to remobilization, suggesting a pathway for long-term sequestration.

The speciation of uranium continues to evolve in the sediment, influenced by redox gradients, microbial activity, and mineral transformations. Stable-isotopic signatures in sediment cores have revealed consistent uranium accumulation patterns over millennial timescales, indicating that, once immobilized, uranium can remain effectively sequestered under favourable geochemical conditions.

In a nutshell, uranium in pit lakes transitions from a highly mobile, oxidized state during initial flooding to a largely immobile, reduced or colloid-bound form in deeper, anoxic zones and sediments. This transformation is mediated by both abiotic and biotic processes, with microbial reduction playing a central role in uranium immobilization.

5.4.4 Uranium Redox Transformations

The behaviour of uranium in pit lakes is significantly influenced by its redox chemistry, which affects its solubility, mobility, and environmental fate. In aqueous environments, uranium primarily exists in two oxidation states: hexavalent uranium (U(VI)) and tetravalent uranium (U(IV)). The transition between these states is strongly affected by factors such as redox potential (Eh), pH, and the presence of complexing agents and microbial communities.

5.4.4.1 Oxidation and Dissolution in Oxygenated Conditions

In oxygenated early pit-lake waters, uranium is rapidly converted to the soluble hexavalent uranyl ion, which forms strong aqueous complexes with carbonate and sulfate, thereby controlling its mobility.

During flooding of a mine void, reduced uranium minerals such as uraninite ($UO_2(s)$) are exposed to oxygenated surface water and groundwater and undergo oxidative dissolution. A commonly used net reaction for this process is:

$$UO_2(s) + \frac{1}{2}O_2 + 2H^+ \rightarrow UO_2^{2+} + H_2O$$

It converts relatively insoluble U(IV) to the soluble uranyl ion, UO_2^{2+}. Oxidation can be abiotic (e.g., molecular O_2 and Mn/Fe oxides) or microbially mediated. The produced UO_2^{2+} is the central reactive species for subsequent complexation and transport in oxygenated pit-lake waters [23].

Once in the hexavalent state, uranyl strongly coordinates oxyanions. In carbonate-bearing, neutral to alkaline waters, the dominant speciation shifts to uranyl-carbonate complexes. Stepwise complexation produces species, such as $UO_2CO_3(aq)$, $UO_2(CO_3)_2^{2-}$, and $UO_2(CO_3)_3^{4-}$; in the presence of calcium, mixed Ca-uranyl-carbonate complexes (e.g., $Ca_2UO_2(CO_3)_3^0$) are important because they further stabilize dissolved uranium and reduce adsorption to mineral surfaces. The relative abundance of these species increases strongly with pH and dissolved inorganic carbon, and their thermodynamic stability controls uranium persistence in the water column [24].

In acidic, sulfate-rich waters, a common condition in early stages of pit lakes affected by acid mine drainage, uranyl preferentially forms uranyl-sulfate complexes. Typical aqueous equilibria include $UO_2^{2+} + SO_4^{2-} \rightleftharpoons UO_2SO_4(aq)$ and $UO_2^{2+} + 2SO_4^{2-} \rightleftharpoons UO_2(SO_4)_2^{2-}$. Spectroscopic and solution studies show these complexes dominate U(VI) speciation at low pH and elevated sulfate concentrations, increasing uranium solubility relative to simple hydrolysed forms [25, 26].

Complexation alters the balance between aqueous transport and removal processes. Complexed uranyl species are less likely to sorb to iron oxides and clays and less likely to precipitate as hydroxides or secondary uranyl minerals, so complexation increases the aqueous residence time and lateral mobility of uranium. Hydrolysis and precipitation reactions that remove U(VI) from solution include, for example, $UO_2^{2+} + 2OH^- \rightarrow UO_2(OH)_2(s)$ or formation of hydrated uranyl oxides (e.g., schoepite-like phases), but these pathways are suppressed when strong carbonate or sulfate complexes dominate. The interplay among redox state, pH, ligand concentrations, and competing sorption sites therefore determines whether uranium remains in the water column or is immobilized in sediments [27]. Overall, oxygenation drives U(IV) → U(VI) conversion and complexation with carbonate or sulfate, and these complexes are the primary chemical controls on uranium mobility in early pit-lake environments.

5.4.4.2 Reduction and Precipitation in Anoxic Zones

As a lake stratifies and a sharp chemocline forms, the deeper waters become anoxic because physical mixing is limited and microbial respiration consumes available oxygen; under these reducing conditions, the dominant dissolved form of uranium, the uranyl ion UO_2^{2+} (U(VI)), becomes thermodynamically susceptible to reduction to U(IV), a change that greatly lowers uranium solubility and controls its fate in the water column and sediments [28].

The core redox step is the two-electron reduction of the uranyl ion to tetravalent uranium, which can be written in its simplest half-reaction form as:

$$UO_2^{2+} + 2e^- \rightarrow UO_2(s)$$

In natural anoxic waters, this electron transfer is commonly coupled to the oxidation of organic electron donors (e.g., acetate) or to microbial respiration. A representative overall stoichiometry for acetate-driven reduction is often written as:

$$CH_3COO^- + 4UO_2^{2+} + 4H_2O \rightarrow 2HCO_3^- + 4UO_2(s) + 9H^+$$

The reaction illustrates how organic carbon oxidation supplies the electrons that drive U(VI) → U(IV) precipitation [28, 29].

Microbial mediation is central: dissimilatory metal-reducing bacteria such as *Geobacter* and *Shewanella* spp. can use uranyl as a terminal electron acceptor during anaerobic respiration, transferring electrons from intracellular metabolism to extracellular U(VI) via multiheme c-type cytochromes, conductive pili or "nanowires", and periplasmic electron carriers; these biological pathways accelerate reduction rates and localize U(IV) formation at cell surfaces or in the immediate extracellular matrix [30, 31]. Experimental and spectroscopic studies show that microbial reduction can yield both nanoparticulate and crystalline forms of U(IV), and that the structural form of the product (e.g., poorly ordered mononuclear U(IV) versus crystalline uraninite) depends on microbial species, ligands present, and geochemical conditions [28, 29].

Once reduced, uranium precipitates as uraninite (nominally UO_2), a crystalline, low-solubility phase, and, in silica-rich porewaters, as coffinite, commonly represented as $UsiO_4$ or more precisely as $U^{IV}SiO_4$ with variable hydroxyl substitution; formation of coffinite reflects the availability of dissolved silica and the tendency of U(IV) to complex with silicate ligands, producing a solid phase that is stable under persistent reducing conditions and that accumulates in sediments where it effectively immobilizes uranium.

The long-term stability of these U(IV) phases is conditional: exposure to oxidants such as oxygen, nitrate, or manganese oxides can reoxidize U(IV) back to soluble U(VI), remobilizing uranium into the water column. Thus, the persistence of immobilization depends on sustained anoxia, continued organic-electron supply or sorptive protection within sediments, and the mineralogical form of the precipitate, factors that together determine whether uranium remains sequestered as uraninite, coffinite, or more labile U(IV) species [28].

5.4.4.3 Colloidal and Organic Associations

Below the chemocline, colloidal iron and aluminium oxyhydroxides, clay minerals, and natural organic matter (NOM) provide abundant reactive surfaces that bind the uranyl ion UO_2^{2+}, thereby lowering dissolved uranium concentrations and mobility even when full reduction to U(IV) does not occur. Surface complexation on hydrous ferric oxides can be represented by a generalized reaction in which a surface hydroxyl site (≡FeOH) binds uranyl, releasing a proton:

$$\equiv \text{FeOH} + \text{UO}_2^{2+} \rightleftharpoons \equiv \text{FeOUO}_2^{+} + \text{H}^{+}$$

This type of inner-sphere complexation and outer-sphere electrostatic adsorption has been demonstrated across ferric oxyhydroxide phases and controls uranyl partitioning between colloids and solution. The strength and reversibility of these associations depend on pH, ionic strength, competing cations, and the crystallinity of the oxide colloid, with poorly crystalline phases (e.g., ferrihydrite) often exhibiting high sorption capacities [32, 33].

Natural organic matter interacts with uranyl via complexation to carboxylate and phenolic functional groups, forming soluble or colloid-bound U-NOM complexes that can retard or facilitate transport, depending on colloid stability. A simplified stoichiometry for complexation with a humic ligand (HA) is:

$$\text{UO}_2^{2+} + \text{HA} \rightleftharpoons \text{UO}_2 - \text{HA}^{+}$$

There is competitive partitioning between NOM and mineral surfaces, with NOM sometimes displacing uranyl from oxide surfaces or partitioning uranyl onto mobile organic colloids. The NOM can substantially alter uranium speciation and mobility under circumneutral to acidic conditions [34].

In silica-rich lakes, diagenetic reactions in sediments promote the formation of polymeric and nanocrystalline uranium-silica species, which are distinct from simple surface complexes and represent condensed, often poorly crystalline solids with low remobilization potential. These phases can form by association of uranyl with dissolved silica, followed by condensation or by reductive transformation coupled to silica incorporation; a conceptual redox-coupled precipitation can be written as:

$$\text{UO}_2^{2+} + 2\text{e}^{-} + \text{SiO}_2(\text{aq}) \rightarrow \text{U}^{\text{IV}} - \text{SiO}_2(\text{s})$$

However, the stoichiometry and structure are variable and often polymeric or nanocrystalline. Diagenetic stabilization of such U-Si phases and their burial in sediments create long-term sinks for uranium, provided reducing conditions and low porewater oxidant fluxes persist [35]. Diagenetic processes are the physical, chemical, and biological changes sediments undergo after deposition but before metamorphism, transforming loose material into solid sedimentary rock (lithification) through compaction, cementation, dissolution, recrystallization, and mineral replacement, driven by burial, pressure, temperature, and water-rock interactions. These changes reduce porosity, expel water, alter mineral composition, and create features like fossils, concretions, and petroleum.

Overall, adsorption to mineral colloids, complexation with NOM, and formation of uranium-silica polymers act together to immobilize uranium in anoxic sedimentary environments, with the balance among these pathways controlled by pH, redox state, silica availability, and the abundance and character of colloids and organic ligands.

5.4.4.4 Uranium Speciation Dynamics

Uranium speciation in pit lakes evolves with depth and redox conditions, transitioning from mobile uranyl complexes in surface waters to stable, immobile forms in deeper layers and sediments. These transformations are central to understanding uranium mobility and long-term environmental risk.

The speciation of uranium in pit lakes is a dynamic process governed by the lake's vertical stratification and redox gradients, which evolve over time as the lake matures. In the epilimnion, the oxygen-rich surface layer, uranium predominantly exists in the uranyl ion. This form is highly soluble and forms stable aqueous complexes, particularly with carbonate ions, such as $UO^{2(CO^{3})_{2}^{2-}}$ and $UO_2(CO_3)_3^{4-}$. These complexes are favoured at neutral to alkaline pH and in the presence of dissolved inorganic carbon, thereby enhancing uranium's mobility in surface waters [36–38].

As water depth increases into the hypolimnion, the lake's lower layer becomes increasingly anoxic due to limited mixing and microbial oxygen consumption. Under these reducing conditions, uranium undergoes a redox transformation from U(VI) to U(IV), a process often mediated by dissimilatory metal-reducing bacteria. The reduction reaction can be represented as:

$$UO_2^{2+} + 2e^- \rightarrow UO_2(s)$$

The resulting U(IV) species, primarily in the form of uraninite ($UO_2(s)$), is sparingly soluble and tends to associate with colloidal particles, such as iron sulfides or organic matter. This transformation significantly reduces uranium's mobility, effectively immobilizing it within the hypolimnion or facilitating its downward transport to the sediments [39, 40].

In the sediment compartment, uranium is further stabilized through incorporation into mineral phases or by sorption onto organic and inorganic matrices. It may precipitate as phosphate minerals, such as autunite ($Ca(UO_2)_2(PO_4)_2 \cdot 10–12H_2O$), or bind to humic substances and clay minerals. These processes are influenced by sediment geochemistry, including pH, redox potential, and the presence of ligands, such as phosphate, carbonate, and sulfide.

The long-term immobilization of uranium in pit lakes depends primarily on the persistence of reducing conditions. Disturbances such as seasonal turnover, groundwater inflows, or anthropogenic activities can reintroduce oxygen into the hypolimnion or sediments, triggering the oxidative dissolution of U(IV) back to its mobile U(VI) form. This reoxidation can be described by the reaction:

$$UO_2(s) + 0.5O_2 + 2H^+ \rightarrow UO_2^{2+} + H_2O$$

Such reoxidation events can remobilize previously sequestered uranium, posing renewed environmental risks. Therefore, understanding the redox-driven speciation dynamics of uranium is essential for predicting its long-term behaviour in pit lakes, assessing ecological and human health risks, and designing effective remediation and monitoring strategies.

5.4.5 Uranium Complexation

Uranium complexation involves the formation of stable chemical species when uranium ions interact with ligands in water. In pit lakes, this process is key to understanding uranium's movement, availability to living organisms, and potential for long-term storage. The mechanisms of complexation are influenced by uranium's oxidation state, the lake's chemistry, and the presence of organic and inorganic ligands.

5.4.5.1 Complexation in Oxidizing Conditions

Under oxic conditions, the linear dioxo uranyl cation UO_2^{2+} is the thermodynamically stable oxidation state, and its speciation is governed by ligand availability and pH. In neutral to alkaline pit lakes where bicarbonate is abundant, uranyl-carbonate complexes dominate and greatly enhance uranium solubility. Stepwise complexation produces species, such as $UO_2CO_3^0$, $UO_2(CO_3)_2^{2-}$, and the tris-carbonate $UO_2(CO_3)_3^{4-}$. Formation of these anionic complexes increases uranium mobility by preventing adsorption to mineral surfaces and by stabilizing uranium in solution, with stability constants and temperature dependence that control the relative abundance of mono-, di-, and tri-carbonate species. Thermodynamic and spectroscopic studies quantify these equilibria and show that carbonate complex stability is a primary control on uranium transport in carbonate-hosted and bicarbonate-rich pit lakes [24, 41].

In acidic, sulfate-rich waters typical of acid mine drainage, uranyl forms strong sulfate complexes that also increase solubility and facilitate downstream transport. Complexes such as $UO_2SO_4^0$ and $UO_2(SO_4)_2^{2-}$ form by ligand exchange with water and protonated surface sites, and their prevalence rises with increasing SO_4^{2-} concentration and decreasing pH; these species reduce the tendency of uranyl to sorb to iron oxyhydroxides and clays, thereby enhancing aqueous uranium concentrations in oxidized pit-lake surface waters influenced by sulfide oxidation [25].

Less abundant but geochemically meaningful are uranyl complexes with phosphate and fluoride, which can become significant where phosphate is released from weathering or fertilizers and where fluorite-bearing lithologies supply fluoride. Phosphate complexation yields species such as $UO_2PO_4^-$ and can promote either aqueous complexation or precipitation of uranyl-phosphate minerals, depending on saturation, while fluoride forms species like UO_2F^+. These ligands alter the charge and coordination environment of uranyl, thereby affecting its sorption behaviour and mobility [42, 43].

Mechanistically, complex formation proceeds by ligand substitution at the uranyl equatorial plane, often accompanied by deprotonation of ligands (e.g., bicarbonate to carbonate) and by changes in ionic strength and pH that shift equilibrium constants. Representative reactions include $UO_2^{2+} + CO_3^{2-} \rightleftharpoons UO_2CO_3^0$ and $UO_2^{2+} + SO_4^{2-} \rightleftharpoons UO_2SO_4^0$, with further stepwise additions forming di- and tri-ligand complexes; the net effect in oxidizing pit lakes is that ligand-driven complexation, rather than reduction, primarily controls uranium solubility and transport until conditions change to favour adsorption, precipitation, or reduction.

5.4.5.2 Complexation in Reducing Conditions

Under reducing conditions, the central redox step is the two-electron conversion of the uranyl ion to tetravalent uranium, commonly written as $UO_2^{2+} + 2e^- \rightarrow U^{4+}$ or, when precipitating directly as uraninite, $UO_2^{2+} + 2e^- \rightarrow UO_2(s)$. Microbial respiration and abiotic reductants supply the electrons that drive this transformation; microbial pathways both accelerate reduction and influence the mineral form of the product, so that biologically mediated reduction often controls whether crystalline uraninite or more disordered U(IV) phases form [31, 44].

Once reduced, U(IV) undergoes hydrolysis and precipitation. Tetravalent uranium hydrolyses strongly and can precipitate as crystalline uraninite (i.e., nominally UO_2) or as amorphous hydroxide/oxyhydroxide phases. A simple hydrolysis/precipitation representation is $U^{4+} + 4\,OH^- \rightarrow U(OH)_4(s)$, while the formation of uraninite from aqueous uranyl via reduction and condensation is represented by $UO_2^{2+} + 2e^- \rightarrow UO_2(s)$. The mineralogy (i.e., crystalline vs. amorphous) and particle size are controlled by reduction rate, ligand availability, and the presence of nucleation surfaces, and these factors determine the long-term stability and solubility of the precipitates [28].

In organic-rich anoxic zones, organic ligand complexation strongly influences uranium partitioning. Humic and fulvic acids bind both residual U(VI) and U(IV) through carboxylate, phenolic, and other functional groups, producing colloidal or nanoparticulate organo-uranium complexes that can either remain suspended or settle to sediments. Conceptually, uranyl complexation with a humic ligand (HA) can be written as:

$$UO_2^{2+} + HA \rightleftharpoons UO_2 - HA^+$$

The U(IV) may form stronger inner-sphere complexes or be incorporated into organic matrices as:

$$U^{4+} + 2HA^- \rightleftharpoons U(HA)_2\,(s/colloid)$$

These organo-complexes can protect U(IV) from rapid reoxidation or, conversely, enhance transport if colloids remain mobile; the net effect depends on colloid stability, redox gradients, and sedimentation rates [45].

Colloidal associations and co-precipitation with mineral colloids are significant in transitional redox zones where conditions fluctuate. Uranium can sorb to or be occluded within iron and manganese oxyhydroxide colloids, aluminium hydroxides, and clays; for example, uranyl sorption to ferric surface sites is often written as:

$$\equiv \mathrm{FeOH} + \mathrm{UO_2^{2+}} \rightleftharpoons \equiv \mathrm{FeOUO_2^{+}} + \mathrm{H^{+}}$$

During reduction, Fe(III) colloids may themselves be transformed, redistributing sorbed U and producing non-uraninite U(IV) products or biomass-associated U(IV) that differ in solubility and reactivity from crystalline uraninite. These dynamic sorption-desorption and transformation cycles control whether uranium remains immobilized in sediments or becomes remobilized during redox oscillations [46].

Together, hydrolysis/precipitation, organic complexation, and colloidal interactions form a coupled set of processes that determine uranium fate in reducing pit-lake zones; the relative importance of each pathway is set by electron donor supply, organic matter quality, colloid abundance and stability, and the persistence of reducing conditions.

5.4.5.3 Complexation Influence on Uranium Speciation Dynamics

Uranium speciation in pit lakes is highly dynamic: complexation reactions with carbonate, sulfate, phosphate, fluoride, mineral surfaces, and natural organic matter shift with pH, redox potential, ionic strength, and ligand supply. These shifts control uranium mobility, bioavailability, remediation difficulty, and long-term sequestration.

The speciation trajectory begins with the dominant aqueous form under oxic conditions, the uranyl ion $\mathrm{UO_2^{2+}}$, which undergoes rapid ligand exchange in response to changing water chemistry. In neutral to alkaline waters, stepwise carbonate complexation stabilizes the uranyl ion as $\mathrm{UO_2CO_3^0}$, $\mathrm{UO_2(CO_3)_2^{2-}}$, and $\mathrm{UO_2(CO_3)_3^{4-}}$ via reactions such as:

$$\mathrm{UO_2^{2+}} + \mathrm{CO_3^{2-}} \rightleftharpoons \mathrm{UO_2CO_3^0}$$

and subsequent additions of $\mathrm{CO_3^{2-}}$; the high stability of these anionic complexes under bicarbonate-rich conditions promotes aqueous transport and reduces sorption to mineral surfaces. In acidic, sulfate-rich waters derived from sulfide oxidation, uranyl forms neutral and anionic sulfate complexes, e.g., $\mathrm{UO_2SO_4^0}$ and $\mathrm{UO_2(SO_4)_2^{2-}}$, through ligand substitution at the uranyl equatorial plane ($\mathrm{UO_2^{2+}} + \mathrm{SO_4^{2-}} \rightleftharpoons \mathrm{UO_2SO_4^0}$), increasing solubility and downstream mobility.

Ionic strength and pH modulate complex stability and surface interactions: increasing ionic strength screens electrostatic attraction and can favour inner-sphere complexation or desorption, while pH controls ligand protonation states (e.g., bicarbonate ⇄ carbonate) and surface charge on oxides, thereby shifting partitioning between dissolved complexes and adsorbed phases. Surface complexation on iron oxyhydroxides is commonly represented as:

$$\equiv \text{FeOH} + \text{UO}_2^{2+} \rightleftharpoons \equiv \text{FeOUO}_2^{+} + \text{H}^{+}$$

Additionally, competitive binding by carbonate or natural organic matter (NOM) can displace sorbed uranyl from the solid phase into solution or onto mobile colloids [47].

Organic complexation alters both mobility and bioavailability. Humic and fulvic acids bind uranyl through carboxylate and phenolic groups (i.e., $\text{UO}_2^{2+} + \text{HA} \rightleftharpoons \text{UO}_2 - \text{HA}^{+}$), producing soluble or colloidal organo-uranium complexes that may be more bioavailable to plankton and benthic organisms or, if colloids aggregate, may enhance sedimentation and sequestration. The balance between stabilization in suspension and removal to sediments depends on colloid stability, redox gradients, and microbial processing of NOM [24, 48].

These dynamic speciation pathways create remediation challenges because stable aqueous complexes (notably carbonate and sulfate species) resist simple precipitation or adsorption treatments. Effective remediation, therefore, often requires altering bulk geochemistry (e.g., lowering pH, removing bicarbonate, or inducing reducing conditions) to shift equilibria towards insoluble U(IV) phases such as uraninite ($UO_2(s)$) via $\text{UO}_2^{2+} + 2e^{-} \rightarrow \text{UO}_2(s)$ or to promote strong sorption/immobilization on engineered sorbents.

Long-term sequestration in sediments occurs when uranium is incorporated into stable mineral phases (uraninite, coffinite, and U-Si polymers) or becomes tightly bound to refractory organic matter; however, oxygen intrusion or other oxidants can re-oxidize U(IV) to U(VI), remobilizing uranium as soluble complexes, so the permanence of sequestration is contingent on sustained reducing conditions, mineralogy, and organic protection. Predicting fate and designing monitoring or remediation, therefore, requires integrated aqueous speciation modelling, mineralogical characterization, and microbial ecology to capture the coupled chemical and biological controls on uranium complexation and transport.

5.4.6 *Role of Organic Matter*

Organic matter accumulates in open-pit lakes as terrestrial plant debris, algal biomass, and microbial detritus settle into sheltered basins. Limited outflow, slow decomposition under stratified or anoxic conditions, and continual inputs from runoff and shoreline erosion promote buildup and sedimentation, forming organic-rich layers that fuel diagenesis and influence water chemistry.

Organic matter plays a multifaceted and critical role in the geochemical behaviour of uranium in pit lakes, particularly in influencing its mobility, speciation, and long-term sequestration. As pit lakes evolve from freshly flooded, oligotrophic systems to more biologically active environments, the accumulation and transformation of organic matter, both particulate and dissolved, become increasingly crucial in uranium dynamics.

5.4.6.1 Complexation with Dissolved Organic Matter (DOM)

Dissolved organic matter (DOM), especially humic and fulvic acids, controls uranium behaviour in pit lakes by forming strong, often multidentate complexes with the uranyl ion UO_2^{2+} ; these complexes can both enhance aqueous mobility and, under different conditions, promote aggregation and sedimentary sequestration.

Dissolved organic matter derived from plant decay and microbial activity presents a heterogeneous array of functional groups (i.e., carboxylates, phenolics, hydroxyls, and occasionally phosphoryl moieties) that act as electron-donating ligands to the uranyl equatorial plane. Laboratory and field studies show that uranyl-humate complexes are thermodynamically stable across a broad pH range and can form inner-sphere chelates as well as outer-sphere electrostatic associations; experimental biphasic systems demonstrate that humic flocs can scavenge and carry the bulk of dissolved uranyl out of solution during aggregation events [49].

Mechanistically, complexation proceeds by ligand substitution at the uranyl equatorial coordination sites, often accompanied by deprotonation of DOM functional groups. A simplified representation of uranyl binding to a humic ligand (HA) is $UO_2^{2+} + HA \rightleftharpoons UO_2 - HA^+$, but the accurate stoichiometry is variable and can include 1:1 chelates, 1:2 complexes, and polymeric assemblies. Ion-exchange and spectroscopic studies quantify a range of conditional stability constants that depend on molecular weight fraction, aromaticity, and pH; fulvic acids typically form weaker, more soluble complexes than higher-molecular-weight humic fractions, which influences whether uranium remains dissolved or partitions to colloids and particles. There are multiple preferential binding motifs, including chelation via carboxylate groups and coordination through hydroxo or phenolic oxygens, which explain why DOM can form both compact chelates and extended, nanoparticulate organo-uranium assemblies [50, 51].

The net effect of DOM complexation is context-dependent. In many pit-lake surface waters, uranyl-DOM complexes increase apparent solubility, allowing uranium to persist in solution under conditions that would otherwise favour adsorption or mineral precipitation; carbonate or sulfate ligands can compete with DOM, shifting speciation towards mixed-ligand complexes. Conversely, when DOM aggregates into flocs or associates with mineral colloids, large negatively charged uranyl-organic complexes can be removed from the water column by flocculation and sedimentation, effectively transferring uranium to the benthos. Ionic strength, pH, and redox state modulate these pathways: higher ionic strength and the presence of divalent cations promote DOM coagulation and particle settling, while reducing conditions can transform U(VI) to U(IV) and alter DOM binding modes. These coupled chemical and physical processes explain why DOM both complicates remediation, by stabilizing uranium in solution, and can also facilitate long-term sequestration when organo-mineral aggregates are buried in sediments [52].

5.4.6.2 Adsorption onto Particulate Organic Matter

Particulate organic matter (POM) scavenges dissolved uranium through coordinated adsorption and ligand exchange at functional groups, converting dissolved uranyl into particle-associated forms that either settle to sediments or remain mobile as organo-mineral colloids depending on geochemical and biological controls.

Particulate organic matter is detrital plant fragments, algal biomass, microbial cells, and extracellular polymeric substances. They present abundant carboxyl, phenolic, hydroxyl, and phosphoryl sites that bind the uranyl ion UO_2^{2+} by inner-sphere chelation and outer-sphere electrostatic association. Inner-sphere binding can be represented conceptually by $RCOO^- + UO_2^{2+} \rightleftharpoons RCOO - UO_2^+$, where adjacent functional groups form multidentate chelates that markedly increase affinity relative to simple electrostatic adsorption. These molecular-scale interactions control initial partitioning from solution to particles and are central to observed high affinities of uranium for organic-rich phases in natural sediments and wetlands [53].

Environmental conditions modulate adsorption strength and POM behaviour. pH governs ligand protonation (reducing binding at low pH when RCOOH predominates), ionic strength compresses electrical double layers and alters outer-sphere interactions, and divalent cations (Ca^{2+} and Mg^{2+}) or Fe^{3+} can form ternary bridges or compete for sites, changing apparent sorption pathways. In many systems, increasing salinity reduces diffuse-layer adsorption but promotes coagulation of POM and associated uranium, shifting the balance from dispersed colloids to rapidly settling flocs [54, 55].

Biological activity and particle surface architecture strongly influence uranium fate. Microbial colonization and extracellular polymeric substances create high-surface-area, chemically heterogeneous microenvironments that concentrate uranyl and favour the formation of nanoparticulate organo-uranium complexes. These biologically mediated microenvironments can both stabilize U(VI) against abiotic transformation and provide nucleation sites for reductive precipitation when anoxia develops, linking adsorption to subsequent redox-driven immobilization processes [38].

Physical aggregation and flocculation determine whether POM-bound uranium is transported or buried. Coagulation driven by cation bridging, pH shifts, or biological exudates converts colloidal organo-uranium complexes into larger aggregates that settle, transferring uranium to the benthic zone where diagenetic processes can further transform it. Once in sediments, POM-bound uranium may be reduced microbially to U(IV) and precipitate as uraninite or be incorporated into refractory organic matrices, producing long-term sinks; conversely, oxidation or changes in porewater chemistry (e.g., increased bicarbonate) can remobilize uranium back into soluble uranyl complexes [56].

5.4.6.3 Microbial-Mediated Reductive Binding

Microbial communities in anoxic pit-lake zones enzymatically reduce soluble U(VI) to insoluble U(IV), driving precipitation of uraninite and formation of organo-uranium phases that nucleate on cell surfaces, within biofilms, or become entrapped in extracellular polymeric substances (EPS), thereby stabilizing uranium against remobilization.

In anoxic layers where electron acceptors such as oxygen and nitrate are depleted, dissimilatory metal-reducing bacteria (i.e., *Geobacter* and *Shewanella* spp.) couple oxidation of organic electron donors to the reduction of uranyl UO_2^{2+} . Electrons derived from the oxidation of substrates (e.g., acetate) are transferred to U(VI) and drive the two-electron reduction to tetravalent uranium, which can be written as $UO_2^{2+} + 2e^- \rightarrow U^{4+}$ or, when precipitating directly, $UO_2^{2+} + 2e^- \rightarrow UO_2(s)$. A representative overall stoichiometry coupling acetate oxidation to U(VI) reduction is:

$$CH_3COO^- + 4UO_2^{2+} + 4H_2O \rightarrow 2HCO_3^- + 4UO_2(s) + 9H^+$$

It illustrates how organic carbon oxidation supplies electrons for reductive immobilization.

Microbial reduction proceeds by both direct and indirect electron transfer mechanisms. Direct extracellular electron transfer involves outer-membrane multiheme c-type cytochromes and conductive pili or nanowires that deliver electrons from the cell interior to extracellular U(VI) species, enabling reduction at or near the cell surface. Indirect pathways include enzymatically produced reductants (e.g., biogenic Fe(II) or sulfide) that abiotically reduce U(VI) in the vicinity of cells. These mechanisms concentrate reduced uranium at microbial interfaces and promote nucleation of uranium minerals on cell envelopes and within biofilms, where local supersaturation and templating by organic functional groups favour formation of nanoparticulate or crystalline U(IV) phases.

Following reduction, biogenic mineral formation commonly yields uraninite (UO_2) as discrete nanoparticles or poorly ordered mononuclear U(IV) species; the mineral form depends on reduction rate, ligand environment, and microbial physiology. Rapid enzymatic reduction often produces poorly crystalline or mononuclear U(IV) that may later recrystallize to uraninite, whereas slower processes or high organic ligand concentrations favour non-uraninite U(IV) forms. Hydrolysis of U(IV) can also produce amorphous hydroxide phases via $U^{4+} + 4OH^- \rightarrow U(OH)_4(s)$, particularly where local pH and ligand conditions permit.

Organic encapsulation further stabilizes reduced uranium. Extracellular polymeric substances (EPS) and biofilm matrices provide a dense network of carboxyl, hydroxyl, and phosphoryl groups that can physically entrap U(IV) nanoparticles and form strong inner-sphere complexes, reducing exposure to oxidants and limiting remobilization. EPS-mediated encapsulation also promotes aggregation and burial of uranium-bearing biomass, linking microbial activity to sedimentary sequestration. Conversely, disturbance or oxygen intrusion can expose these biogenic U(IV) phases to oxidants, re-oxidizing U(IV) to soluble U(VI) and reversing immobilization.

Collectively, microbial reduction, surface-mediated nucleation, and EPS encapsulation form an integrated pathway by which organic-rich, anoxic pit-lake zones immobilize uranium; understanding the balance among these processes is essential for predicting long-term stability and for designing bioremediation strategies that harness or protect biogenic U(IV) sinks.

5.4.6.4 Sediment Sequestration

Sediment sequestration in pit lakes plays a critical role in the long-term immobilization of uranium, primarily through interactions with organic matter, mineral phases, and diagenetic processes that stabilize uranium in immobile forms.

As pit lakes mature, organic-rich sediments accumulate at the bottom of the lake, forming a geochemically reactive zone that serves as a long-term sink for uranium. These sediments are often composed of detrital material, microbial biomass, and decaying organic matter, creating a reducing environment conducive to uranium immobilization. One of the primary mechanisms by which uranium is sequestered in these sediments is through complexation and co-precipitation with organic ligands and mineral phases. Under anoxic conditions, uranium in its hexavalent form (UO_2^{2+}) is reduced to the tetravalent state (U(IV)), which is far less soluble. The reduction can be microbially mediated, following the reaction:

$$UO_2^{2+} + 2e^- \rightarrow UO_2(s)$$

The resulting U(IV) species, typically in the form of uraninite ($UO_2(s)$), can adsorb onto or co-precipitate with iron sulfide minerals such as pyrite (FeS_2) or bind directly to humic substances and other organic macromolecules. These interactions significantly reduce uranium mobility by incorporating it into solid phases that are stable under reducing conditions [57].

In silica-rich environments, another important pathway for uranium immobilization involves the formation of stable uranium-organic polymers. These polymers can incorporate into nanocrystalline uranium-silica phases, which are structurally robust and resistant to dissolution. Recent studies have demonstrated that such uranium-silica associations form through diagenetic processes, in which uranium initially complexed with organic matter becomes progressively incorporated into amorphous or poorly crystalline silica matrices over time. This process not only enhances the physical stability of uranium-bearing phases but also reduces their susceptibility to remobilization under fluctuating geochemical conditions [58].

Furthermore, diagenetic stabilization plays a crucial role in the long-term retention of uranium in sediments. As organic matter undergoes microbial degradation and humification, uranium bound to these organic substrates becomes increasingly recalcitrant. This transformation involves the condensation of organic molecules and the formation of strong covalent or coordination bonds with uranium, effectively locking it within the sediment matrix. Even if redox conditions shift towards more oxidizing conditions, the tightly bound uranium is less likely to be reoxidized and released into the aqueous phase, thereby remaining immobilized.

These sedimentary processes underscore the importance of organic matter in mediating uranium behaviour in pit lakes. The interplay between microbial activity, mineral precipitation, and organic complexation creates a multifaceted barrier to uranium mobility. However, the long-term effectiveness of this natural attenuation strategy depends on the stability of sediment redox conditions. Any perturbation that introduces oxygen or alters the geochemical equilibrium, such as sediment resuspension, groundwater inflow, or seasonal turnover, could compromise the integrity of these sequestration mechanisms and lead to uranium remobilization.

5.4.7 Role of Humic Substances

In pit lakes, humic substances play a significant role in uranium's geochemical behaviour. Humic substances are complex organic molecules derived from the decomposition of plant and microbial matter. They include humic acids, fulvic acids, and humin and are rich in functional groups, such as carboxyl, phenolic, and hydroxyl groups. These groups provide multiple reactive sites capable of forming stable complexes with metal ions, including uranium. Table 5.1 provides a brief overview of humic substances.

5.4.7.1 Uranium-Humic Substance Interaction

Under oxidizing conditions, the predominant uranyl ion (UO_2^{2+}), a linear dioxo cation, readily interacts with humic substances, forming coordination complexes through direct bonding with oxygen donor atoms. The resulting uranyl-humic complexes can vary in structure and stability depending on factors such as pH, ionic strength, and the specific composition of the humic material.

At circumneutral to slightly alkaline pH, the carboxyl and phenolic groups on humic molecules are deprotonated, enhancing their ability to bind uranyl ions. These interactions can lead to the formation of inner-sphere complexes, in which uranium is tightly bound to the organic matrix, or outer-sphere complexes, which involve weaker electrostatic interactions. The stability of these complexes often keeps uranium in solution longer than it would otherwise be, potentially increasing its mobility within the lake system.

However, the presence of humic substances can also facilitate uranium immobilization. In deeper, anoxic zones of the lake, microbial activity may reduce uranyl to the less soluble tetravalent uranium (U(IV)). This reduced form of uranium has a strong affinity for organic matter and can precipitate or become tightly bound to humic substances. In some cases, uranium may be encapsulated within extracellular

Table 5.1 A quick summary of types of humic substances

Type	Structure	Key features
Humic acid	*Empirical formula:*	• Dark brown in colour
	$C_xH_yO_zN_w$ (typical range: $C_{20-60}H_{10-30}O_{10-40}N_{0-5}$)	• Soluble in alkaline; insoluble at pH < 2
	Molecular formula:	• High molecular weight; aromatic; many carboxyl and phenolic groups
	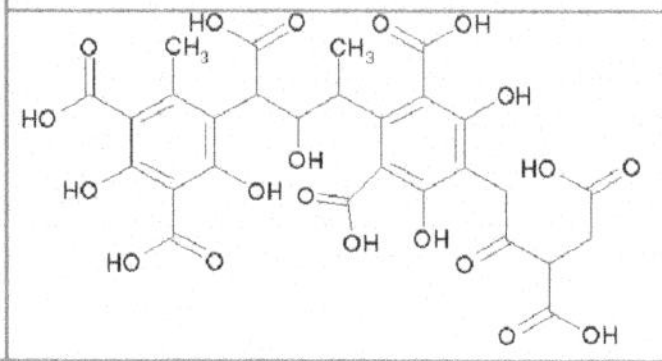	• Forms humates in alkaline solutions
Fulvic acid	***Empirical formula:***	• Yellow in colour
	$\mathbf{C_xH_yO_zN_w}$ (lower C and higher O per C than humic acid)	• Soluble in both acid and alkaline
	Molecular formula:	• Lower molecular weight; more oxygenated and acidic functional groups
	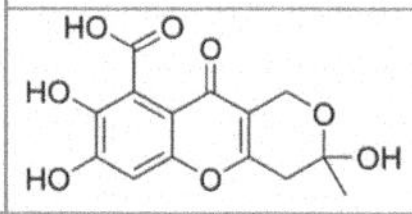	• Higher mobility in water
Humin	***Empirical formula:***	• Black or dark brown in colour
	• Heterogeneous macromolecular network; no single formula	• Insoluble in water at all pH levels
	The molecular structure of a humin fragment:	• Highly recalcitrant
	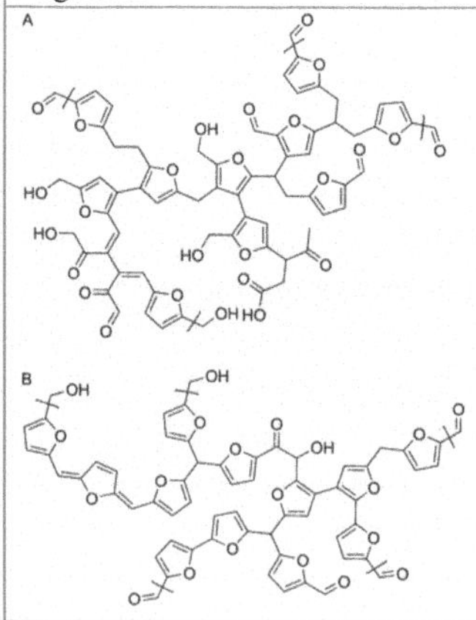	• Strongly bound to mineral matrix; contributes to stable soil/sediment organic matter

(continued)

Table 5.1 (continued)

Type	Structure	Key features
Humic-like substances (aquatic)	Complex mixtures approximated by empirical elemental ratios (C:N:O)	• Variable solubility; often dissolved organic matter
		• Derived from aquatic and terrestrial sources
		• Influence colour and light attenuation in water

Note: Humic substances are a heterogeneous group (i.e., humic acid, fulvic acid, and humin) with no single fixed molecular formula; representative empirical formulas or compositional ranges are used where possible

polymeric substances (EPS) produced by microbial biofilms, further stabilizing it against remobilization.

Over time, uranium bound to humic substances may be transported to the lakebed and incorporated into sediments. In these environments, diagenetic processes can form stable uranium-organic complexes or even nanocrystalline uranium-silica phases, particularly in silica-rich systems. These sedimentary associations are often highly resistant to dissolution, providing a long-term sink for uranium and reducing its environmental risk.

Thus, humic substances act as both facilitators and mitigators of uranium mobility. While they can enhance uranium solubility and transport in surface waters, they also contribute to its sequestration in anoxic zones and sediments. The dual role of humic substances underscores the importance of understanding organic matter dynamics in pit lake systems, particularly when assessing uranium fate and designing remediation strategies.

5.4.7.2 Humic Substances for Uranium Immobilization

The composition and behaviour of humic substances in pit lakes are influenced by complex interactions among geological, hydrological, and biological factors. Humic substances, which include humic acids, fulvic acids, and humin, are not uniform. Their molecular structures and functional group densities vary significantly depending on the source material and environmental conditions. This variability has direct implications for uranium complexation and immobilization.

In young pit lakes, especially those formed in arid or semi-arid regions or in pits with minimal organic input, humic substances are typically scarce and poorly developed. The organic matter present is often derived from sparse vegetation or microbial colonization, resulting in low-molecular-weight fulvic acids with fewer reactive sites. These lakes tend to have limited capacity for uranium binding, and uranyl ions remain relatively mobile, especially in the presence of competing ligands like sulfate or carbonate.

In contrast, older or more biologically productive pit lakes, particularly those in temperate or boreal regions, accumulate more complex, higher-molecular-weight humic substances over time. These lakes benefit from greater terrestrial input, leaf litter, root exudates, and microbial biomass, which enrich the water and sediments with carboxylic and phenolic groups capable of forming strong inner-sphere complexes with uranium. In such systems, uranium is more likely to be sequestered through adsorption, complexation, and co-precipitation with organic matrices.

The geological context also influences the composition of humic substances. Pit lakes formed in coal-bearing strata or organic-rich shales may inherit pre-existing humic materials from the host rock, which can immediately interact with uranium during the early flooding phase. These naturally occurring humic substances may already possess a high affinity for uranyl ions, contributing to early-stage immobilization.

5.4.7.3 Nexus of Microbial-Humic Substance Interactions

Microorganisms play a central role in the biogeochemical cycling of uranium in pit lakes, particularly through their interactions with humic substances. These interactions are multifaceted, involving both direct and indirect mechanisms that influence uranium speciation, mobility, and long-term sequestration. Due to their high content of carboxyl, phenolic, and hydroxyl functional groups, humic substances serve as both chemical ligands and substrates for microbial metabolism. In oxygenated surface waters, these substances can bind uranyl ions (UO_2^{2+}) through coordination with their functional groups, forming stable aqueous complexes. While this complexation can enhance uranium solubility and mobility, it also reduces its bioavailability and toxicity by sequestering it in non-reactive forms.

As pit lakes stratify and develop anoxic zones, microbial communities adapted to low-oxygen conditions become increasingly active. Among these, dissimilatory metal-reducing bacteria (DMRB), such as *Geobacter* and *Shewanella* species, play a pivotal role in uranium redox transformations. These microbes can use uranyl ions as terminal electron acceptors during anaerobic respiration, reducing U(VI) to the less soluble U(IV). The reduced uranium precipitates as uraninite (UO_2) or binds to organic matrices, including humic substances.

Humic substances themselves can act as electron shuttles, facilitating microbial reduction of uranium even when direct contact between microbes and uranium is limited. In this capacity, humic molecules accept electrons from microbial metabolism and transfer them to uranyl ions, promoting their reduction. This indirect pathway enhances the efficiency of uranium immobilization, especially in environments where physical constraints limit microbe-metal interactions.

Moreover, microbial biofilms, aggregates of microorganisms embedded in extracellular polymeric substances (EPS), often incorporate humic-like materials. These biofilms provide microenvironments with steep redox gradients and high concentrations of organic ligands, ideal for uranium sequestration. Within these biofilms, uranium can be reduced, adsorbed, or co-precipitated with biogenic minerals, forming stable associations that resist remobilization.

In sediments, microbial degradation of organic matter continues to influence uranium behaviour. As humic substances break down or transform, they may release bound uranium or alter its speciation. However, under sustained reducing conditions, uranium tends to remain immobilized, either as mineral precipitates or as part of stable organic complexes. Diagenetic processes may further stabilize uranium by forming nanocrystalline uranium-silica or uranium-organic polymers.

Microorganisms and humic substances form a synergistic system that governs uranium fate in pit lakes. Microbes catalyse redox transformations, while humic substances provide both chemical binding sites and electron transfer capabilities. Together, they contribute to the attenuation and long-term sequestration of uranium, making their interaction a critical focus for understanding and managing uranium-contaminated aquatic environments.

5.4.7.4 Microbial Electron Shuttling in Uranium Immobilization

In the anoxic zones of pit lakes, where oxygen is depleted and redox conditions are low, microorganisms play a vital role in transforming uranium from its mobile hexavalent form (U(VIU)) to its immobile tetravalent form (U(IV)). One of the key mechanisms by which microbes facilitate this transformation is through electron shuttling, a process that allows for indirect reduction of uranium even when microbes are not in direct contact with uranium ions.

Electron shuttling involves the transfer of electrons from microbial metabolism to extracellular electron acceptors via intermediate compounds. In the context of uranium geochemistry, this means that microbes can reduce uranyl ions (UO_2^{2+}) to insoluble uranium species without physically interacting with them. Instead, they rely on electron shuttles, molecules capable of accepting electrons from microbial cells and donating them to uranium in the surrounding environment.

Humic substances, particularly humic and fulvic acids, are among the most effective natural electron shuttles. These complex organic molecules contain redox-active functional groups, such as quinones, which can reversibly accept and donate electrons. In anoxic pit-lake conditions, microbes such as *Geobacter* and *Shewanella* species oxidize organic substrates (e.g., acetate or lactate) and transfer electrons to humic substances. The reduced humic molecules then diffuse through the water column or sediment pore spaces and reduce uranyl ions to U(IV), which precipitates as uraninite (UO_2) or binds to organic and mineral surfaces.

This indirect reduction pathway is particularly important in environments where uranium is not readily accessible to microbial cells due to physical separation or low concentrations. Electron shuttling extends the spatial range of microbial influence, allowing for broader zones of uranium immobilization. Moreover, it enhances the efficiency of bioremediation strategies by leveraging naturally occurring or supplemented humic substances to facilitate uranium reduction.

Beyond humic substances, other compounds, such as iron oxides and manganese oxides, can also serve as electron shuttles. However, humic substances are particularly relevant in pit lakes due to their abundance in organic-rich sediments and their dual roles as electron shuttles and uranium-binding agents.

Thus, microbial electron shuttling represents a powerful mechanism for uranium immobilization in pit lakes. By coupling microbial metabolism with the redox activity of humic substances, this process enables uranium to be transformed into stable, insoluble forms, even in complex, spatially heterogeneous environments. Understanding and harnessing this mechanism is essential for predicting uranium behaviour and designing effective biogeochemical remediation strategies in post-mining aquatic systems.

5.5 Maturing Process of Uranium and AMD in Pit Lakes

The evolution of uranium and acid mine drainage (AMD) in pit lakes is a complex, multiphase process that unfolds over years to decades. As pit lakes transition from freshly flooded excavations to chemically and biologically stratified systems, uranium undergoes a series of transformations, driven by redox gradients, mineral interactions, microbial activity, and organic matter dynamics, that determine its long-term fate. Simultaneously, AMD constituents, such as sulfate, iron, and metals, evolve in response to similar processes, often interacting with uranium geochemistry in synergistic or antagonistic ways. The maturation process of uranium and AMD in a pit mine involves several distinct processes, including, inter alia, sedimentation and immobilization of uranium, fixation of uranium in sediments, long-term sequestration, and remobilization.

5.5.1 Uranium Sedimentation and Immobilization

Uranium in maturing pit lakes transitions from a mobile hexavalent form in oxygenated waters to largely immobilized tetravalent phases in anoxic sediments; this immobilization is driven by redox stratification, microbial reduction, and co-precipitation with iron and sulfur minerals.

In the early stages of pit lake development, uranium exists predominantly as soluble U(VI) in the form of the uranyl ion (UO_2^{2+}), which is highly mobile in oxygenated, acidic waters. As the lake deepens and thermal and chemical stratification establishes a chemocline, the water column separates into an oxygenated epilimnion and an anoxic hypolimnion; this persistent redox gradient is the primary physical control that sets the stage for uranium removal from the water column and transfer into sediments.

Within the anoxic bottom waters and at the sediment-water interface, redox-driven reduction of U(VI) to U(IV) is the central geochemical step that decreases uranium solubility and promotes sedimentation. Dissimilatory metal-reducing bacteria, notably members of the *Geobacter* and *Shewanella* genera, enzymatically reduce uranyl to tetravalent uranium, often coupling uranium reduction to the oxidation of organic carbon or other electron donors. Microbial biofilms and aggregates enhance local reducing microenvironments and can concentrate uranium at the interface where precipitation initiates [31, 59].

The mineralogical form of immobilized uranium varies with kinetics, electron donor availability, and the presence of complexing ligands. Crystalline uraninite (UO_2) is a common product of rapid, localized reduction, but non-crystalline or mononuclear U(IV) species also form under many environmental conditions; these amorphous phases can be intimately associated with biomass and extracellular polymeric substances, which both nucleate and stabilize reduced uranium but may also render it more susceptible to reoxidation if conditions change. Binding to organic matter and sorption onto mineral surfaces, therefore, contribute to initial accumulation and the textural character of uranium-bearing sediments [60].

Co-precipitation and sorption with iron and sulfur minerals are major pathways that enhance long-term immobilization. During neutralization of acid mine drainage and under sulfidic conditions, iron oxides and iron sulfides form and can scavenge U(VI) or incorporate U(IV) into mixed phases, producing composite particles that settle rapidly from the water column. Experimental and field studies show that iron-mediated coagulation and occlusion within Fe-oxide coatings can sequester uranyl, while electrochemical and ligand-assisted processes can accelerate interception and precipitation of uranium from solution [61, 62].

Iron sulfides such as mackinawite and pyrite are particularly important in sulfidic sediments: abiotic reductive immobilization by freshly precipitated mackinawite can convert U(VI) to less soluble forms and incorporate uranium into sulfide matrices, producing mixed-phase precipitates that are often more stable under sustained reducing conditions than pure uranium minerals alone. The combined action of redox stratification, microbial reduction, organic complexation, and co-precipitation with Fe-S and Fe-oxide phases therefore governs the transition from dissolved uranium to sedimentary, immobilized reservoirs in maturing pit lakes [61–63].

5.5.2 Sediment Fixation and Long-Term Sequestration

Once uranium sediments, diagenetic processes drive further stabilization, transforming initially precipitated U(IV) phases into more recalcitrant forms. In silica-rich environments, uranium can reorganize into nanocrystalline uranium-silica polymers that are highly insoluble and resistant to remobilization, becoming physically incorporated into the sediment fabric and effectively isolating uranium from the overlying water column. These diagenetic transformations occur over decades to millennia. They can convert poorly ordered, noncrystalline U(IV) assemblages into more polymerized or mineral-bound states that are less reactive to transient redox perturbations [64].

The distinction between crystalline uraninite and noncrystalline U(IV) is important for long-term behaviour: noncrystalline U(IV) often forms initially and can be more labile, but diagenetic consolidation, silica polymerization, and incorporation into the sediment matrix progressively reduce its solubility and mobility, altering the potential for future remobilization. These transformations are mediated by local geochemistry, organic matter diagenesis, and the availability of silica and other ligands that promote polymer formation and physical encapsulation.

Sorption onto mineral surfaces is a parallel and complementary pathway for long-term fixation. Fresh and diagenetically altered iron oxyhydroxides, clays, and authigenic minerals provide abundant sorption sites; pH, redox potential, and competing cations control sorption capacity and speciation. Under oxic to suboxic conditions, uranyl sorption to iron oxyhydroxides and subsequent burial can be a dominant removal mechanism, while reductive dissolution of iron phases at depth can either release or re-sequester uranium depending on local redox dynamics [35].

In persistently anoxic sediments, inner-sphere complexation and incorporation into iron mineral lattices enhance stability. Reduced uranium species can form strong inner-sphere bonds with mineral surface functional groups or become structurally incorporated into mixed Fe-U phases, reducing their susceptibility to desorption and oxidation. Mechanistic studies show that incorporation pathways and the oxidation state distribution within Fe(II)/Fe(III) (oxyhydr)oxides influence how tightly uranium is bound and how resistant those associations are to subsequent redox shifts.

Organic matter also plays a dual role. It serves as an electron donor that drives reduction and as a sorbent that binds U(IV) and U(VI). Sorption to natural organic matter can strongly control uranium mobility, often producing highly stable associations in anoxic sediments. Still, these associations remain conditional on the persistence of reducing conditions and the chemical integrity of the organic matrix [65].

Over time, the sediment column functions as a geochemical sink, accumulating uranium and co-contaminants through precipitation, sorption, and diagenetic encapsulation. The long-term sequestration of uranium therefore reflects a balance between immobilizing processes (i.e., silica polymerization, mineral sorption, and incorporation into Fe-S and Fe-oxide phases) and the potential for remobilization if redox conditions, pH, or sediment structure are altered [54].

5.6 Uranium Remobilization

Despite the apparent stability of uranium in sediments, various processes can remobilize it. The primary driver of uranium mobilization in pit lakes is redox perturbation, which reintroduces oxygen into previously anoxic zones. The redox perturbations are often triggered by seasonal turnover, groundwater intrusion, or human activities. Under these conditions, the oxidation of U(IV) to U(VI) can resolubilize uranium, releasing it back into the water column [66]. Additionally, microbial remineralization of organic matter can modify local redox conditions, subsequently releasing bound uranium. As microbes degrade organic ligands, uranium previously complexed with humic substances may be released, particularly if this degradation shifts the redox balance towards more oxidizing conditions [66, 67]. Changes in pH and ionic strength can further disrupt sorption equilibria. For instance, acidification resulting from ongoing acid mine drainage (AMD) or external inputs can desorb uranium from mineral surfaces, enhancing its mobility. Similarly, fluctuations in carbonate or sulfate concentrations can alter uranium speciation, favouring more soluble complexes.

5.6.1 Natural pH and Ionic Strength Changes

The geochemical stability of uranium in pit lakes is strongly influenced by pH and ionic strength variations, which can disrupt sorption equilibria and alter uranium speciation. Acidification is one of the most critical drivers of uranium mobilization. In systems affected by ongoing acid mine drainage (AMD) or external acidic inputs, the pH of the water column and sediments can drop significantly, weakening the electrostatic interactions that bind uranium to mineral surfaces. Under acidic conditions, protonation of surface hydroxyl groups on minerals such as iron oxides or aluminosilicates reduces their capacity to adsorb uranyl ions (UO_2^{2+}). This desorption process can be represented as:

$$\equiv \text{FeOH} + \text{UO}_2^{2+} \rightleftharpoons \equiv \text{FeOUO}_2^{+} + \text{H}^{+}$$

As pH decreases, the equilibrium shifts towards the right, releasing UO_2^{2+} into solution. Acidification also promotes the dissolution of secondary minerals that previously sequestered uranium, further enhancing its mobility.

In addition to pH, fluctuations in ionic strength, which are particularly due to changes in carbonate and sulfate concentrations, play a decisive role in uranium speciation. At circumneutral to alkaline pH, carbonate ions (CO_3^{2-}) form highly stable aqueous complexes with uranyl, such as:

$$\text{UO}_2^{2+} + 3\text{CO}_3^{2-} \rightarrow \left[\text{UO}_2\left(\text{CO}_3\right)_3\right]^{4-}$$

These complexes are significantly more soluble than free uranyl ions, increasing uranium mobility in carbonate-rich waters. Similarly, elevated sulfate concentrations can lead to the formation of uranyl-sulfate complexes:

$$\text{UO}_2^{2+} + \text{SO}_4^{2-} \rightarrow \left[\text{UO}_2\text{SO}_4\right]^{0}$$

Such speciation changes reduce uranium's tendency to adsorb onto mineral surfaces, thereby favouring its persistence in the aqueous phase. These processes are particularly relevant in pit lakes where seasonal evaporation, groundwater inflow, or anthropogenic inputs alter ionic strength and pH over time.

The combined effect of acidification and complexation reactions underscores the dynamic nature of uranium mobility in pit lakes. Even in sediments where uranium had previously been immobilized, shifts in geochemical conditions can destabilize these phases, leading to renewed contamination risks to surrounding ecosystems and water resources.

5.6.2 Microbial Remobilization of Uranium

While microbial communities are instrumental in immobilizing uranium during the early stages of pit lake development, primarily through reductive precipitation and complexation, their role in uranium remobilization becomes increasingly significant as the lake matures and environmental conditions fluctuate. Microorganisms can both stabilize and destabilize uranium, depending on the prevailing redox conditions, availability of organic substrates, and the structure of microbial consortia.

5.6.2.1 Microbial Redox Cycling and Uranium Reoxidation

Dissimilatory metal-reducing bacteria (DMRB), such as *Geobacter* and *Shewanella* species, play a pivotal role in immobilizing uranium within anoxic pit-lake sediments. Under oxygen-deprived conditions, these bacteria use soluble uranyl ions (UO_2^{2+}) as terminal electron acceptors, coupling their reduction to the oxidation of organic substrates or hydrogen. The canonical half-reaction is:

$$UO_2^{2+} + 2e^- + 4H^+ \rightarrow UO_2(s) + 2H_2O$$

Here, uranium is converted from its soluble hexavalent form (U(VI)) to insoluble tetravalent uranium, most commonly precipitated as uraninite (UO_2). In biofilms of *Geobacter sulfurreducens*, conductive pili and outer-membrane cytochromes, such as OmcZ, enhance this reduction, enabling sustained immobilization even at high uranium concentrations. Similarly, *Shewanella oneidensis* expresses c-type cytochromes that facilitate extracellular electron transfer, precipitating U(IV) on the cell envelope [59, 68–71].

However, this immobilization is only secure under strictly reducing conditions. Any perturbation, whether through seasonal overturn, oxygen intrusion, or bioturbation, can reintroduce oxidants into the sediments, including oxygen and nitrate. At this point, facultative or aerobic bacteria such as *Pseudomonas* and chemolithoautotrophs such as *Thiobacillus denitrificans* may catalyse the reoxidation of U(IV) to U(VI), effectively reversing the immobilization process. The simplified reaction under aerobic conditions is:

$$UO_2(s) + \frac{1}{2}O_2 + 2H^+ \rightarrow UO_2^{2+} + H_2O$$

Under nitrate-rich, anaerobic conditions, *Thiobacillus* can employ UO_2 as an electron donor, oxidizing it to U(VI) even in the absence of oxygen:

$$5UO_2(s) + NO_3^- + 10H^+ \rightarrow 5UO_2^{2+} + \frac{1}{2}N_2 + 5H_2O$$

This nitrate-dependent U(IV) oxidation significantly enhances uranium dissolution, producing aqueous U(VI) and mononuclear uranyl species [72–74].

These redox transitions are especially concerning in pit lakes with shallow chemoclines or a history of frequent overturn events. In such systems, oxygenated surface water or groundwater-nitrates mixing can penetrate deeper sediments, activating diverse microbial communities. Aerobic *Pseudomonas* may directly oxidize uranium, while denitrifiers like *Thiobacillus* generate reactive nitrogen intermediates (e.g., NO^{2-}) that chemically oxidize UO_2, amplifying mobilization [72, 73, 75, 76].

Thus, the microbial-mediated reoxidation of uranium involves a tightly coupled interplay of microbial metabolism and redox dynamics. What begins as a reductive immobilization process can inadvertently shift to oxidative remobilization when the geochemical environment changes. In pit-lake management, maintaining long-term reducing conditions and mitigating oxygen or nitrate intrusion are therefore essential to prevent uranium resurgence into the water column.

5.6.2.2 Organic Matter Decomposition

Microbial communities in pit-lake sediments drive a cascade of organic matter transformations that directly affect uranium mobility. Heterotrophic bacteria and fungi enzymatically depolymerize complex organic matrices into smaller molecules that serve as electron donors for respiration. The organic matrices include humic and fulvic substances, lignin-derived polymers, and particulate organic carbon. During this degradation, organic ligands that had complexed U(VI) as the uranyl ion, UO_2^{2+} are altered or mineralized, releasing bound uranyl back into porewaters. The net effect depends on the balance between ligand breakdown, microbial uptake of released carbon, and secondary reactions that may re-sequester uranium [77].

Biological mechanisms that promote ligand-mediated uranium release include both direct enzymatic cleavage of ligand functional groups and indirect redox-driven pathways. Extracellular enzymes (e.g., oxidases, peroxidases, and hydrolases) attack aromatic and aliphatic components of humic substances, reducing the density of carboxylate ($-COO^-$) and phenolic (–OH) binding sites that coordinate uranyl through inner-sphere complexation. Simultaneously, microbial respiration alters local redox conditions: aerobic degradation maintains U as hexavalent uranyl (U(VI)), whereas anaerobic processes (e.g., sulfate reduction and iron reduction) can produce reducing conditions that transform U(VI) to less soluble U(IV) phases or, conversely, mobilize noncrystalline U(IV) via complexation with newly produced ligands [31].

Changes in the structure of organic ligands also shift uranium speciation. Humic acids bind uranyl through multiple functional groups to form stable humate-uranyl complexes, but progressive oxidation or cleavage of these groups lowers complex stability constants and increases the fraction of free UO_2^{2+} or formation of smaller, more mobile organo-uranyl complexes. In addition, microbial metabolites such as low-molecular-weight organic acids (e.g., acetate and oxalate) and siderophores can compete with humic binding sites, forming soluble complexes like $UO_2\left(C_2O_4\right)_n^{2-2n}$ or mixed organo-uranyl species that enhance aqueous mobility [48].

Sediment microenvironments control the rates and outcomes of these processes. Labile organic matter hotspots, zones with high particulate organic carbon and active respiration, promote rapid ligand turnover and transient increases in dissolved uranium. Conversely, in zones where microbial reduction of U(VI) to U(IV) is sustained, uranium may precipitate as uraninite (UO_2(s)) or associate with reduced iron sulfides. Yet, these reduced phases can be remobilized if organic degradation produces complexing ligands or if oxidants are introduced, for example, through seasonal mixing or bioturbation [78].

At the molecular level, experimental and spectroscopic studies show that loss or modification of carboxyl and phenolic groups on humic macromolecules measurably weakens uranyl sorption, and that anthropogenic or biogenic ligands can mobilize uranium from both U(IV) and U(VI) solid phases. The interplay of enzymatic degradation, redox transitions, and competitive complexation therefore governs whether organic matter decomposition leads to net immobilization or remobilization of uranium in pit-lake sediments [79–81].

5.6.3 Microbial Influence on Co-precipitation and Nanoparticle Dynamics

Microorganisms also influence the formation and dissolution of mineral phases that co-precipitate with uranium. Sulfate-reducing bacteria (SRB), for example, promote the formation of iron sulfides and other secondary minerals that can incorporate uranium through co-precipitation. However, these minerals are sensitive to redox changes; their dissolution under oxidizing conditions can release associated uranium.

In addition, microbial activity can affect the formation and stability of uranium-bearing nanoparticles. Biogenic uraninite often forms as nanocrystals within biofilms or is associated with extracellular polymeric substances (EPS). These nanoparticles may remain suspended in the water column or settle into sediments. However, their small size and high surface area make them susceptible to oxidative dissolution, especially if microbial communities shift towards aerobic metabolism.

5.6.4 Uranium Mobilization Via Chelation and Bioligands

Microbial communities in organically rich pit-lake sediments generate a suite of low-molecular-weight organic ligands, phosphorylated metabolites, and specialized chelators such as siderophores that alter uranium partitioning by forming soluble complexes with the uranyl ion, UO_2^{2+}. Organic ligands include carboxylic acids (e.g., citrate, oxalate, and acetate), phosphorylated metabolites, and specialized chelators, such as siderophores. Common microbial-produced ligands, their sources, and functions are summarized in Table 5.2. During fermentation and heterotrophic metabolism, microbes release these metabolites into porewaters, where they can compete with mineral surfaces and humic substances for uranyl, driving ligand-exchange reactions that displace sediment-bound uranium into solution. The net effect is controlled by ligand concentration, complex stability, and environmental parameters such as pH and ionic strength, with mildly acidic to neutral pH favouring the formation of many organo-uranyl complexes and thus enhanced mobility [82].

Carboxylate ligands produced biologically form well-defined coordination complexes with uranyl. For example, citrate and oxalate coordinate UO_2^{2+} through multiple oxygen donors to yield mono- and polynuclear species; generalized formulations include $UO_2\left(C_2O_4\right)_n^{2-2n}$ for oxalate complexes and a range of mono- to oligomeric uranyl-citrate species that can remain highly soluble and mobile in porewaters. Spectroscopic and structural studies demonstrate that uranyl-citrate complexes can be dimeric or trimeric and that their formation constants are sufficient to outcompete weaker sorption sites on clays and iron oxides under many natural conditions [83, 84].

Siderophores and pyoverdine act as hard Lewis-base donors and bind actinides with high affinity. Siderophores are microbial iron-scavenging molecules, such as desferrioxamine B. Pyoverdine is an organic compound that contains hydroxamate, catecholate, or carboxylate functional groups. These ligands can form stable complexes with U(VI), preventing precipitation and inhibiting sorption to mineral surfaces; experimental work shows that siderophores can complex uranyl, thereby maintaining uranium in solution across a range of pH and redox conditions, effectively increasing its aqueous residence time and transport potential. The ability of siderophores to stabilize U(VI) is crucial in microenvironments where iron limitation drives high siderophore production [85].

Mechanistically, microbial chelation mobilizes uranium via ligand-exchange and complexation reactions, lowering free uranyl activity and reducing sorption affinity. Complexation can also solubilize uranium from reduced U(IV) phases by forming organo-uranyl complexes or by promoting oxidative dissolution when coupled with microbial or chemical oxidants. Transport is further facilitated when organo-uranyl complexes associate with mobile colloids or dissolved organic matter, enabling long-range dispersal beyond the immediate sediment interface. The balance between microbial chelation-driven mobilization and concurrent immobilization processes (e.g., microbial U(VI) reduction and sorption to newly formed iron sulfides)

Table 5.2 The most relevant microbial ligand to uranium chemistry

Type	Examples	Remark
Organic acids	• Citrate • Oxalate • Acetate • Lactate	• Commonly produced during fermentation and respiration • Contain carboxyl groups that can coordinate with uranyl ions, forming soluble complexes
Siderophores	Major chemical classes (chelating moieties): • Catecholates (e.g., enterobactin, yersiniabactin) • Hydroxamates (e.g., ferrichrome, desferrioxamine, ferrioxamine B) • Carboxylates/hydroxamate-carboxylates (e.g., rhizoferrin, aerobactin) • Mixed-ligand (e.g., pyoverdine, azotobactin)	• High-affinity iron-chelating compounds secreted by bacteria and fungi under iron-limited conditions • Primarily designed to bind Fe^{3+}, many siderophores can also form complexes with uranyl ions due to similar coordination chemistry
Exopolysaccharides and extracellular polymeric substances (EPS)	• A mixture of functional groups (e.g., hydroxyl, carboxyl, amine)	• Capable of binding uranium. Biofilms produce these substances, which can trap uranium within microbial matrices
Phosphorylated metabolites	• Sugar phosphates (e.g., glucose 6-phosphate (G6P) and fructose 6-phosphate (F6P)) • Nucleoside monophosphates (e.g., uridine monophosphate (UMP) and thymidine monophosphate (TMP)) • Phosphorylated amino acids	• It can bind uranium ions via coordinate covalent interactions involving their phosphate groups • Multiple phosphate groups provide numerous binding sites

Note: Microorganisms generate various ligands as metabolic byproducts or as specialized compounds for nutrient acquisition

determines whether ligand production leads to net uranium release in a given pit-lake setting [83–85].

5.6.5 Chelation and Uranium Mobility in Pit Lakes

Chelation plays a pivotal role in modulating uranium (U) mobility within pit lakes, particularly through the biogeochemical activities of microbial communities during the three distinct stages of pit lake formation: flooding, stratification, and stabilization. These stages are characterized by evolving redox conditions and organic matter dynamics, which collectively influence uranium speciation and transport.

During the initial flooding stage, the influx of surface and groundwater into the decommissioned open pit introduces dissolved oxygen and organic matter, initiating a cascade of redox reactions. Uranium, primarily present as insoluble U(IV) species such as uraninite ($UO_2(s)$), may remain immobilized under reducing conditions. However, the presence of organic-rich sediments fosters microbial activity, particularly by heterotrophic bacteria that degrade organic matter. This degradation releases low-molecular-weight organic ligands, including carboxylic acids and siderophores, which can chelate uranium. For instance, citrate ($C_6H_5O_7^{3-}$), a common microbial metabolite, forms stable complexes with U(VI) as:

$$UO_2^{2+} + C_6H_5O73- \rightarrow UO_2\left(C_6H_5O_7\right)^-$$

This complexation enhances uranium solubility, facilitating its release from sediments into the aqueous phase.

As the lake progresses into the stratification stage, a meromictic structure often develops, characterized by a stable chemocline separating oxygenated surface waters from anoxic bottom layers. In the anoxic hypolimnion, microbial sulfate reduction and methanogenesis dominate, leading to the formation of sulfide minerals and the potential precipitation of uranium as uraninite or coffinite ($USiO_4$). However, periodic mixing events or seasonal overturn can introduce oxygen into previously anoxic zones, triggering redox transitions. These transitions stimulate shifts in microbial metabolism from anaerobic to aerobic pathways, intensifying the production of chelating agents such as siderophores, high-affinity iron-binding compounds that can also bind U(VI). For example, desferrioxamine B (DFOB), a well-known siderophore, forms strong complexes with uranyl ions:

$$UO_2^{2+} + (DFOB)^{3-} \rightarrow \left[(UO_2)(DFOB)\right]^-$$

Such complexes are highly soluble and can remobilize uranium that has been sequestered in sediments or adsorbed onto mineral surfaces.

In the final stabilization stage, the lake reaches a quasi-equilibrium state, but uranium mobility remains dynamic due to ongoing microbial processes and seasonal redox fluctuations. Humic substances, prevalent in organically enriched sediments, can both immobilize and mobilize uranium depending on their redox state and the availability of functional groups. Under oxidizing conditions, U(IV) may oxidize to U(VI), which readily forms soluble complexes with humic ligands:

$$UO_2^{2+} + (\text{Humic Acid})^{n-} \rightarrow \left[(UO_2)(\text{Humic Acid})^{2-n}\right]^-$$

These complexes can persist in the water column, contributing to long-range transport of uranium within and beyond the pit lake system [76, 86].

5.6.6 Uranium-Ligand Stability in Pit Lakes

The stability of uranium-ligand complexes plays a pivotal role in controlling uranium's mobility, bioavailability, and long-term behaviour in pit lakes. Uranium in these systems predominantly exists as the hexavalent uranyl ion, UO_2^{2+}, a hard Lewis acid that forms strong inner-sphere complexes with ligands containing oxygen donor atoms, such as carboxylate, phosphate, and hydroxyl groups. The strength and persistence of these complexes are governed by thermodynamic stability constants, which quantify the equilibrium between the complexed and free species. They are influenced by ligand denticity, donor atom type, pH, redox conditions, and the presence of competing ions [24].

Under alkaline conditions, carbonate ligands dominate uranium speciation due to their high affinity for uranyl. The formation of highly stable complexes such as $UO_2(CO_3)_3^{4-}$ and $UO_2(CO_3)_2^{2-}$ is favoured, significantly enhancing uranium solubility and mobility. These complexes are particularly prevalent in carbonate-rich pit lakes. The formation reaction for the tricarbonato complex can be represented as:

$$UO_2^{2+} + 3CO_3^{2-} \rightleftharpoons UO_2(CO_3)_3^{4-}$$

In contrast, in acidic environments influenced by acid mine drainage, sulfate complexes such as $UO_2SO_4^0$ are more common, though they are less stable than carbonate complexes. Organic ligands like citrate, oxalate, and humic substances also form stable complexes with uranyl. Citrate, for example, can act as a bidentate or tridentate ligand, forming complexes that remain stable across a broad pH range. Humic substances, due to their heterogeneous composition and high density of functional groups, create a variety of uranyl complexes whose stabilities depend on molecular weight and aromaticity.

The environmental conditions in pit lakes critically influence the stability of uranium-ligand complexes. pH is a primary control, as it affects both uranium speciation and ligand ionization. At low pH, ligands are protonated and less effective at binding uranyl, while at higher pH, deprotonation enhances their binding capacity. The general protonation equilibrium of a ligand (HL) is:

$$HL \rightleftharpoons H^+ + L^-$$

Only the deprotonated form, L^-, binds effectively to UO_2^{2+}, forming complexes such as:

$$UO_2^{2+} + L^- \rightleftharpoons UO_{2L^+}$$

The conditional stability of these complexes is thus pH-dependent, influenced by both the ligand's pK_a and the hydrolysis equilibria of uranyl. Hydrolysis reactions, which compete with ligand binding, include:

$$UO_2^{2+} + H_2O \rightleftharpoons UO_2OH^+ + H^+$$

$$UO_2OH^+ + H_2O \rightleftharpoons UO_2(OH)_2(aq) + H^+$$

At higher pH, these hydroxo species can polymerize or precipitate as $UO_2(OH)_2(s)$, effectively removing uranium from solution. Even small pH shifts (1–2 units) can significantly alter the balance between soluble complexes and solid phases.

Redox potential is another critical factor. Under reducing conditions, U(VI) is reduced to U(IV), which forms insoluble species such as $UO_2(s)$, thereby decreasing uranium mobility. Ligands that stabilize U(VI) may not bind U(IV) effectively, altering the speciation landscape.

Competing cations such as Ca^{2+}, Mg^{2+}, Fe^{3+}, and Al^{3+} can also influence uranium-ligand stability by either forming their own complexes with available ligands or by competing directly with uranyl for binding sites. This competition reduces the availability of free ligands for uranium complexation.

Surface interactions further modulate uranium mobility. At pH values above 4, freshly precipitated Fe/Al hydroxides can adsorb uranyl or facilitate co-precipitation. These interactions often involve inner-sphere complexation or the formation of ternary surface-ligand-uranyl species, particularly when surfaces are deprotonated.

In pit lakes impacted by acid mine drainage, low pH conditions initially favour the presence of mobile, weakly complexed uranyl species. As the system evolves, through natural attenuation, carbonate buffering, microbial sulfate reduction, or engineered liming, pH increases, promoting the formation of stable uranium-ligand complexes and reducing uranium mobility. However, seasonal or hydrological fluctuations can destabilize these complexes. A decrease in pH may lead to ligand protonation and uranium remobilization, while an increase may enhance complexation but also favour the formation of mobile uranyl-carbonate species.

Accurate modelling of uranium behaviour in pit lakes thus requires comprehensive thermodynamic data, including pH-dependent stability constants for uranyl hydrolysis, carbonate complexation, ligand protonation, and surface complexation reactions. These parameters are essential for predicting uranium speciation and mobility under site-specific geochemical conditions [27, 39, 86].

5.7 Limnological Influences on Uranium Dynamics

Open-pit lakes exhibit unique limnological processes that critically influence the geochemical behaviour and fate of uranium. These processes, particularly thermal and chemical stratification, redox gradients, pH variability, and microbial activity, interact in complex ways to control uranium mobility, speciation, and long-term sequestration. Figure 5.4 is a conceptual model of general processes that determine the geochemical dynamics of a pit lake.

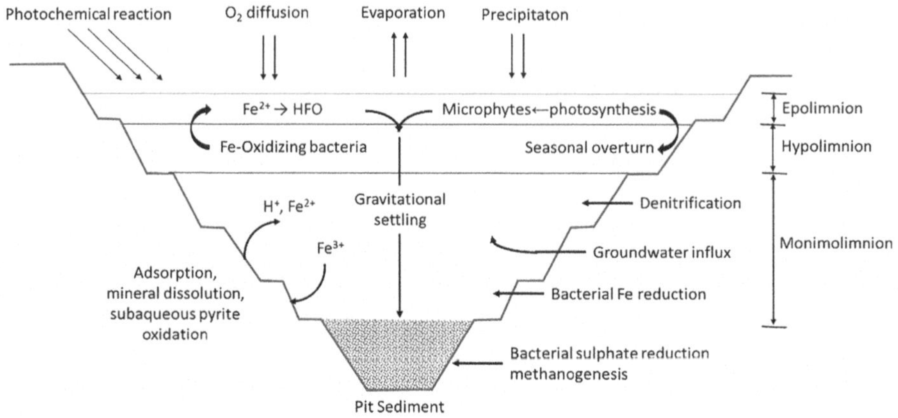

Fig. 5.4 A conceptual model of the processes occurring in a stratified open-pit lake that influence the geochemical dynamics within the lake

5.7.1 Limnological Process and Uranium

When mining ceases, and open pits fill with water, they often evolve into meromictic lakes, systems characterized by persistent stratification. This stratification arises due to the steep-sided morphology and depth of the pits, which inhibit wind-driven mixing. As a result, the water column separates into an upper, seasonally mixed oxic layer (the mixolimnion) and a deeper, anoxic monimolimnion, divided by a sharp chemocline. This vertical separation creates distinct geochemical compartments, isolating contaminants such as uranium in the deeper layers, where reducing conditions prevail and exchange with surface waters is minimal [11].

The redox potential and pH within these stratified layers are the primary chemical controls on uranium speciation. In the oxygen-rich mixolimnion, uranium predominantly exists as soluble hexavalent U(VI), often complexed with carbonate or sulfate ligands that enhance its mobility. In contrast, the anoxic monimolimnion favours the formation of tetravalent U(IV) species, such as uraninite (UO_2), which are far less soluble and tend to precipitate from solution. The chemocline often hosts steep redox and pH gradients, making it a dynamic zone where even minor vertical mixing can trigger rapid changes in uranium solubility, either precipitating it at redox boundaries or remobilizing it into the water column [86–88].

Microbial activity further modulates uranium dynamics. In the anoxic layers, certain bacteria can enzymatically reduce U(VI) to U(IV), effectively immobilizing uranium in sediments. These microbial processes are tightly coupled with the cycling of other elements, such as iron and sulfur, which also influence redox conditions. For instance, microbial sulfate reduction can generate sulfide, which reacts with iron to form iron sulfides, indirectly affecting uranium mobility by altering the availability of sorption sites or redox buffering capacity [87, 89].

Organic matter and colloids also play a significant role. Dissolved organic carbon can bind uranium, facilitating its transport across stratified layers or enhancing its retention in sediments. In oxic zones, iron and aluminium oxyhydroxides scavenge U(VI) from the water column, but under reducing conditions, these phases may dissolve, releasing previously sequestered uranium back into solution. Biological productivity in the photic zone, such as phytoplankton blooms, can increase pH and oxygen levels, promoting U(VI) stability and mobility. At the same time, biosorption by algae, fungi, and other biota can incorporate uranium into biomass, linking geochemical cycling to ecological pathways and potentially contributing to long-term sedimentary storage [90].

However, the stability of these stratified systems is not guaranteed. Seasonal turnover, extreme weather events, or anthropogenic disturbances can disrupt stratification, introducing oxygen into previously anoxic layers. Such events can oxidize U(IV) back to U(VI), remobilizing it and potentially leading to its migration into surface waters. This coupling of physical mixing with chemical and biological processes underscores the sensitivity of uranium dynamics to limnological changes. A delicate interplay of stratification, redox chemistry, pH, and microbial ecology governs the fate of uranium in open-pit lakes.

5.7.2 Uranium Dynamics in Stratified Pit Lakes

As a pit lake evolves, solar heating and limited vertical exchange establish a three-layer thermal structure: a warm, mixed epilimnion, a sharp metalimnion (thermocline), and a cold, dense hypolimnion; this physical stratification imposes strong gradients in dissolved oxygen, redox potential, and density that control uranium speciation and transport. In the surface epilimnion, continuous gas exchange and wind-driven mixing maintain oxic conditions, so uranium is stabilized in its hexavalent state, U(VI), most commonly as the uranyl ion UO_2^{2+} and as aqueous complexes with carbonate or sulfate, depending on pH and ligand availability. Under alkaline conditions, carbonate complexes (e.g., $UO_2(CO_3)_2^{2-}$) increase uranium solubility and mobility, whereas in more acidic or sulfate-rich waters, uranyl-sulfate complexes can predominate; these soluble species remain suspended in the water column and are subject to advective and diffusive transport within the epilimnion and across the metalimnion [86].

The metalimnion and the chemocline that commonly coincides with it act as a reactive interface where temperature-driven density gradients limit vertical mixing and create steep chemical gradients. Here, oxygen concentrations fall rapidly with depth and redox potential declines, producing microenvironments in which partial reduction of U(VI) can occur. The chemocline concentrates organic matter and electron donors delivered from the productive epilimnion or from settling particulates; these substrates fuel heterotrophic microbial communities capable of enzymatic or indirect reduction of uranyl to tetravalent uranium, U(IV). The metalimnion, therefore, functions both as a zone of transformation, where soluble U(VI) is converted to less soluble forms, and as a barrier that slows the downward flux of dissolved uranium, promoting localized precipitation or sorption at the interface [64, 86].

Beneath the chemocline, the hypolimnion commonly becomes anoxic and reducing, particularly in meromictic or strongly stratified pit lakes with limited turnover. Under these conditions, microbial reduction proceeds more completely, and U(IV) precipitates as uraninite (UO_2) or forms noncrystalline reduced U phases that sorb to organic matter and iron sulfide or oxide surfaces. These reduced phases have far lower aqueous solubility than U(VI), driving net removal of uranium from the dissolved pool and promoting accumulation in the sediments through particle settling and diagenetic incorporation. The long-term stability of sedimentary uranium depends on the mineralogy and crystallinity of the reduced phases; poorly crystalline U(IV) can be more labile and subject to reoxidation if oxygenated waters or oxidants are reintroduced, whereas crystalline uraninite is comparatively stable.

Microbial mediation is central to these redox transitions: dissimilatory metal-reducing bacteria and sulfate-reducing consortia use organic carbon or hydrogen as electron donors and couple their metabolism to U(VI) reduction, often in concert with iron and sulfur cycling, which further influences uranium partitioning. In pit-lake contexts, field studies document vertical segregation of uranium species consistent with these processes, with dissolved U(VI) dominating oxic layers and particulate or sediment-associated U(IV) concentrated below the chemocline. The interplay of thermal stratification, redox gradients, ligand chemistry, and microbial activity therefore governs uranium mobility, transformation, and long-term sequestration in stratified pit lakes [64, 86, 91].

5.7.3 Uranium in Mixis: Disruption and Redistribution

Mixis episodically converts sequestered, reduced uranium into mobile oxidized forms and redistributes particulate and colloidal uranium through the water column, with consequences driven by physical overturn, redox chemistry, and rapid shifts in microbial ecology.

The physical process of mixis, whether the twice-annual overturn of dimictic systems or more frequent disturbances in mono- or polymictic lakes, erases the thermal and density gradients that maintain a stable chemocline, allowing surface waters to penetrate to depth and promoting rapid exchange of dissolved gases and solutes. As the water column homogenizes, oxygen is entrained into formerly anoxic hypolimnetic waters, raising redox potential and altering the thermodynamic favourability of uranium species; reactive transport models show that such hydrodynamic forcing can rapidly change concentration gradients and drive transient pulses of dissolved contaminants as equilibrium and kinetic controls shift. The depth to which oxygenated water intrudes and the rate of overturn determine the spatial extent and magnitude of these chemical perturbations, with complete turnover producing the most extensive reoxidation of reduced phases [92].

Chemically, the principal consequence of reoxygenation is the oxidation of tetravalent uranium (U(IV)) to hexavalent uranium (U(VI)), a transformation that increases aqueous solubility and mobility. Solid-phase U(IV) phases formed under reducing conditions, whether crystalline uraninite or poorly crystalline U(IV) associated with organic matter and sulfides, are susceptible to oxidative dissolution when exposed to oxygen or other oxidants mobilized during mixis. This oxidation liberates uranium into the dissolved pool as uranyl species and their complexes, enabling upward and lateral transport by advective currents and diffusion. Field and laboratory studies of aquifer and sediment systems demonstrate that the return of oxic conditions can produce measurable remobilization of uranium from sediments, with the magnitude controlled by the speciation and mineralogical form of the reduced uranium and by the availability of complexing ligands and competing sorbents [93].

Mixis also redistributes colloidal and nanoparticulate uranium, which can be detached from sediments or formed in situ during redox transitions. Physical resuspension during overturn and shear-driven turbulence can mobilize particle-bound uranium, while changes in ionic strength, pH, and redox state alter colloid stability and aggregation behaviour. These processes change transport pathways because colloids can remain suspended longer than larger particles and can traverse the water column, increasing exposure risk to biota and complicating remediation or monitoring efforts. Observations from mining-impacted lakes indicate that oxygenation regimes strongly influence microbial activity and the net release of uranium from sediments; mildly oxygenated conditions may minimize release relative to fully oxic or fully anoxic extremes, underscoring the non-linear response of uranium mobility to mixing events [76].

Biologically, mixis reshapes microbial communities by favouring aerobic metabolisms and oxidants that outcompete anaerobic reducers. The loss of sustained reducing niches diminishes the activity of dissimilatory metal-reducing bacteria and sulfate reducers that previously maintained U(IV) sequestration. At the same time, aerobic heterotrophs and oxidizing chemolithotrophs can accelerate organic matter degradation and produce ligands that complex U(VI). These ecological shifts not only promote reoxidation and solubilization but also alter the production and consumption of electron donors and acceptors, thereby controlling secondary mineral formation and sorption capacity. The combined physical, chemical, and biological consequences of mixis therefore produce episodic pulses of dissolved uranium and particulate transport that must be anticipated in monitoring and management of stratified pit lakes.

5.7.4 Stratification and Mixis Implications for Uranium Fate

Stratification and episodic mixing create a seasonally dynamic uranium cycle in pit lakes, where physical overturns can convert long-sequestered, reduced uranium into mobile, oxidized forms and redistribute particulate and colloidal uranium throughout the water column. Stable thermal stratification establishes distinct chemical regimes that control where uranium is stored and how readily it can be remobilized. In the epilimnion, persistent oxygenation and higher temperatures favour U(VI) as soluble uranyl complexes, which remain available for transport and biological uptake. The metalimnion and chemocline concentrate gradients in oxygen, redox potential, and organic substrates; these gradients create reactive interfaces where partial reduction, sorption, and particle-associated sequestration of uranium occur. In the anoxic hypolimnion and underlying sediments, microbial respiration and diagenetic reactions drive redox potentials low enough to reduce U(VI) to U(IV), promoting precipitation as uraninite or association with organic matter and sulfide minerals. This vertical partitioning leads to a net accumulation of uranium in deep waters and sediments during prolonged stratification. However, the chemical forms and mineral associations that constitute this sequestration vary in crystallinity and stability, which, in turn, control susceptibility to later remobilization [86, 88].

When mixis occurs, whether seasonal dimictic overturn or episodic wind-driven mixing, the physical homogenization of the water column entrains oxygen into formerly reducing zones and disrupts the chemocline. Reoxygenation oxidizes U(IV) to U(VI), increasing its aqueous solubility and producing pulses of dissolved uranium that can be transported into surface waters and laterally exported. The extent of remobilization depends on the mineralogical form of reduced uranium, the rate and depth of oxygen penetration, and the availability of complexing ligands such as carbonate and sulfate that stabilize uranyl species. Turbulence and shear during overturn also resuspend sediments and detach colloidal or nanoparticulate uranium, altering particle-size distributions and prolonging suspension times; these processes change transport pathways and exposure risks because colloids can traverse the water column more readily than larger aggregates [86, 94].

Mixis also triggers rapid shifts in microbial community composition and metabolic pathways that feed back on uranium chemistry. Aerobic microbes proliferate as oxygen becomes available, outcompeting anaerobic metal- and sulfate-reducing bacteria that previously maintained U(IV) sequestration. Changes in organic matter degradation rates and ligand production further influence uranium complexation and sorption dynamics, sometimes accelerating the dissolution of reduced phases or stabilizing U(VI) in solution. Because these biological responses can be rapid and spatially heterogeneous, the chemical outcome of a mixing event is often non-linear and temporally variable, producing transient concentration spikes in surface waters and episodic releases from sediments.

These coupled physical, chemical, and biological dynamics complicate monitoring and remediation. Surface uranium concentrations can fluctuate seasonally, and sediment-bound uranium is not necessarily permanently fixed, so management strategies should incorporate temporal sampling across stratification cycles and consider interventions that reduce the frequency or intensity of redox perturbations or that stabilize reduced uranium phases through mineralogical or organic amendments. Long-term risk assessment must therefore account for the potential for episodic remobilization driven by mixis and the variable stability of reduced uranium forms in pit-lake sediments [64].

5.8 Benthic Processes Influencing Uranium Behaviour

Physical benthic processes, such as sedimentation, resuspension, and bioturbation, modulate the exposure of uranium-bearing phases to changing redox and chemical conditions. Resuspension events or burrowing organisms can reoxidize buried U(IV) and reintroduce U(VI) to the water column, while progressive burial under reducing conditions promotes long-term sequestration. The interplay of these chemical, microbial, and physical benthic mechanisms ultimately governs whether uranium in flooded open pits is retained in sediments as stable U(IV) minerals or remains a mobile contaminant in the lake system.

Benthic sediments and the microbial and geochemical reactions that occur at the sediment-water interface are primary controls on uranium mobility in flooded open-pit mines; redox-driven mineral transformations, sorption/desorption on iron and aluminium phases, sulfate reduction and sulfide precipitation, organic matter interactions, and bioturbation together determine whether uranium is sequestered as insoluble U(IV) minerals or remains mobile as U(VI) complexes.

At the sediment-water interface, redox gradients develop sharply because oxygen is consumed rapidly by organic matter oxidation, while deeper porewaters remain reducing. Under oxic conditions, uranium exists mainly as the hexavalent uranyl ion, UO_2^{2+}, which forms stable aqueous carbonate complexes such as $UO_2(CO_3)_2^{2-}$ and $UO_2(CO_3)_3^{4-}$ in alkaline waters, promoting mobility. When sediments become reducing, microbial and abiotic electron transfer reduce U(VI) to tetravalent U(IV) according to the half-reaction $UO_2^{2+} + 2e^- \rightarrow UO_2(s)$, producing sparingly soluble uraninite or other U(IV) phases and thereby immobilizing uranium in the benthos. The depth and persistence of the reducing zone in pit-lake sediments therefore control the long-term sink strength for uranium [86].

Iron cycling in benthic sediments exerts a dominant sorptive control. In oxic microzones, Fe(III) oxyhydroxides (e.g., ferrihydrite and goethite) scavenge U(VI) via surface complexation and co-precipitation, removing dissolved uranium from solution and transferring it to particulate phases. Under reducing conditions, microbial or chemical reductive dissolution of Fe(III) minerals releases Fe^{2+} and any previously sorbed U back into porewaters, coupling iron diagenesis to uranium remobilization. The reductive dissolution can be represented conceptually as $Fe(OH)_3 + e^- \rightarrow Fe^{2+} + OH^-$, with concomitant release of sorbed UO_2^{2+}.

Sulfate reduction in organic-rich benthic layers produces hydrogen sulfide (H_2S) via microbial sulfate reduction:

$$SO_4^{2-} + 8e^- + 10H^+ \rightarrow H_2S + 4H_2O$$

Sulfide can precipitate with metal cations or form U-S phases. While uraninite remains a common U(IV) product, sulfide-rich conditions may favour the formation of U-bearing sulfide or thiolate complexes and promote co-precipitation with metal sulfides, altering the mineralogy and stability of immobilized uranium. The relative rates of Fe and sulfate reduction, controlled by organic carbon availability and microbial community composition, therefore determine whether U is fixed as uraninite or associated with sulfide minerals.

Organic matter and dissolved organic carbon in sediments influence uranium through complexation and electron shuttling. Humic substances and reduced organic ligands can complex U(VI), enhancing transport in porewaters, or serve as electron donors for U(VI)-reducing bacteria that catalyse U(IV) precipitation. Microbial communities, including known U(VI)-reducing genera, mediate direct enzymatic reduction and indirect controls via iron and sulfur cycling, making microbial ecology a central benthic mechanism for uranium fate.

Microbial ligand production also contributes to the remobilization of uranium from sediments. In organically enriched sediments, uranium may be bound to humic substances, mineral surfaces, or precipitated as uranium minerals. However, microbial degradation of organic matter can release ligands that re-complex uranium, increasing its solubility and mobility. This process is especially active during redox transitions, where microbial communities shift from anaerobic to aerobic metabolism. The production of organic acids and siderophores during these shifts can disrupt existing uranium associations, leading to resolubilization and potential migration into the water column.

5.9 Concluding Remarks and Perspective

This chapter has presented a general synthesis of the processes governing uranium dynamics in open-pit lakes, recognizing that the scenarios described are conceptual frameworks rather than prescriptive templates. Open-pit lakes form when mining excavations fill with water, and their steep morphology, depth, and hydrological setting predispose them to develop persistent stratification, distinct redox zones, and complex interactions among water, sediments, and biota. From initial pit inundation through the establishment of mixolimnion and monimolimnion layers, physical, chemical, and biological processes evolve together to create the conditions that ultimately control uranium speciation, mobility, and long-term storage.

Thermal and chemical stratification is a primary limnological control: the separation of an oxygenated surface layer from a deeper, often anoxic layer produces sharp chemoclines that isolate geochemical compartments and concentrate redox-sensitive elements. Hydrodynamic forcing, seasonal turnover, storm events, groundwater exchange, and episodic mixing periodically couple these compartments, with profound consequences for uranium. Redox potential and pH are the dominant geochemical levers that determine whether uranium exists as mobile hexavalent complexes or as reduced, insoluble tetravalent phases; carbonate, sulfate, iron, and organic ligands modulate these equilibria and influence sorption, precipitation, and colloidal transport. Sediment-water interactions, including adsorption to iron and aluminium oxyhydroxides and incorporation into organic and mineral matrices, provide pathways for sequestration, while reductive dissolution and reoxidation events can remobilize uranium that has been previously immobilized.

Biological processes operate at multiple scales to amplify or attenuate geochemical trends. Microbial reduction of U(VI) to U(IV) can immobilize uranium in sediments, but microbial cycling of iron, sulfur, and carbon also shapes redox buffering and the stability of mineral phases. Primary production and biomass turnover alter pH and oxygen dynamics in the photic zone, and biosorption and trophic transfer link dissolved and particulate uranium to ecological pathways and sedimentary burial. Hydrological connectivity, surface inflows, groundwater exchange, and atmospheric inputs sets the boundary conditions for solute supply, dilution, and long-term mass balance, so that the same limnological and biogeochemical mechanisms play out differently across catchments with different geology, climate, and hydrology.

Looking forward, effective assessment and management of uranium in abandoned mine pit lakes require site-specific characterization that couples physical lake dynamics, water and sediment geochemistry, microbial ecology, and hydrological context. Monitoring programs should target stratification patterns, redox and pH gradients, ligand chemistry, particulate fluxes, and episodic mixing events to capture the processes most likely to drive speciation and transport. Remediation and risk-management strategies must be adaptive, recognizing that interventions which stabilize one compartment (for example, promoting reductive immobilization) may be vulnerable to future disturbances that reintroduce oxidants and remobilize uranium. Integrative models that combine limnological, geochemical, and ecological modules offer the best prospects for predicting long-term behaviour, but their reliability depends on robust, site-specific data.

In closing, the conceptual picture developed in this chapter highlights the interplay of formation history, stratification, redox chemistry, pH, solid-water interactions, microbial mediation, and hydrological exchange in controlling uranium fate in open-pit lakes. While these processes provide a coherent framework for understanding potential outcomes, every open-pit lake at an abandoned uranium mine will be unique; careful, context-driven investigation is essential to translate general principles into accurate predictions and sound management decisions.

References

1. Dunbar, D., Martin, A. J., Fraser, C., & Mueller, S. *Tailings and Mine Waste Conference Proceedings*.
2. Schultze, M. (2013). In W. Geller, M. Schultze, R. Kleinmann, & C. Wolkersdorfer (Eds.), *Acidic pit lakes: The legacy of coal and metal surface mines* (pp. 11–22). Springer.
3. CAPP. (2021). In Canadian Association of Petroleum Producers (CAPP) (Ed.), (Vol. 28). Canadian Association of Petroleum Producers (CAPP).

4. Martin, A. J., & Salvador, S. In P. Stanley, C. Wolkersdorfer, K. Wolkersdorfer, & E. Mugova (Eds.), *International Mine Water Association 2023 Conference* (p. 599). International Mine Water Association.
5. Bernasconi, R., Lund, M. A., & Blanchette, M. L. (2022). Non-charismatic waterbodies and ecosystem disservices: Mine pit lakes are underrepresented in the literature. *Frontiers in Microbiology, 13*, 1063594. https://doi.org/10.3389/fmicb.2022.1063594
6. Castro, J. M., & Moore, J. N. (2000). Pit lakes: Their characteristics and the potential for their remediation. *Environmental Geology, 39*, 1254–1260. https://doi.org/10.1007/s002549900100
7. Bowell, R. J. (2003). Pit Lake systematics. *Mine Water and the Environment, 22*, 167–169.
8. Gammons, C. H., Harris, L. N., Castro, J. M., Cott, P. A., & Hanna, B. W. (2009). *Creating lakes from open pit mines: Processes and considerations—With emphasis on northern environments* (Fisheries and aquatic sciences, Vol. 106).
9. Bowell, R. J. (2002). *Mine water hydrogeology and geochemistry* (Vol. 198, pp. 159–176). Geological Society.
10. Kemanga, B., McIntyre, N., & Bulovic, N. (2024). Hydrological classification of mine pit lakes using modelling experiments. *Journal of Environmental Management, 370*, 123057. https://doi.org/10.1016/j.jenvman.2024.123057
11. Gammons, C., Harris, L., Castro, J., Cott, P., & Hanna, B. (2009). *Creating lakes from open pit mines: Processes and considerations, with emphasis on northern environments.* Canadian Technical Report of Fisheries and Aquatic Sciences (Vol. 2826).
12. Blanchette, M. L., & Lund, M. A. (2016). Pit lakes are a global legacy of mining: An integrated approach to achieving sustainable ecosystems and value for communities. *Current Opinion in Environmental Sustainability, 23*, 28–34. https://doi.org/10.1016/j.cosust.2016.11.012
13. Arcos, D., Guimerà, J., Jordana, S., Ruiz, E., & Molinero, J. In Adrian Brown, et al. (Eds.), *10th ICARD | IMWA 2015 Conference—Agreeing on Solutions for More Sustainable Mine Water Management.* IMWA—International Mine Water Association.
14. Dzimbanhete, V. L., Paulsson, O., Karlsson, T., Alakangas, L., & Martinsson, O. (2025). Identifying key uranium sources in mine water: Open pit wall rock leaching and groundwater contributions in Leveäniemi open pit. *Journal of Geochemical Exploration, 275*, 107773. https://doi.org/10.1016/j.gexplo.2025.107773
15. Doyle, G. A., & Runnells, D. D. (1998). Society for Mining, Metallurgy & Exploration.
16. Cook, P. G., et al. (2021). *Hydrological and geochemical processes and closure options for below water table open pit mines* (Report No. Final Report Project 3.3). Cooperative Research Centre for Transformations in Mining Economies Ltd.
17. Herrell, M. K., Prestia, A. L., Bowell, R., & Bosley, G. (2022). In M. Tibbett, A. B. Fourie, & G. Boggs (Eds.), *Mine closure 2022.* Australian Centre for Geomechanics.
18. Santofimia, E., Rincón-Tomás, B., López-Pamo, E., González, F. J., & Amils, R. (2023). Geochemical and limnological characterization of the Corta Atalaya Pit Lake (Riotinto Mines, Spain). *Materials Proceedings, 15*, 66.
19. España, J. S., Pastor, E. S., Toril, E. G., & Martín-Úriz, P. S. (n.d.). In R. Cidu & F. Frau (Eds.), *IMWA Symposium 2007: Water in Mining Environments.* International Mine Water Association IMWA.
20. Newman, C., Cluff, T., Gray, T., & Beale, G. (2019). Aqueous geochemistry and limnology of the Sleeper Pit Lake, Nevada, USA: Evidence for long-term subaqueous solute generation in Mine Pit Lakes. *Journal of Geoscience and Environment Protection, 7*, 64–81. https://doi.org/10.4236/gep.2019.73004
21. Dunbar, D. (2003). Society for Mining, Metallurgy & Exploration.
22. Zhang, T., et al. (2023). A review: The formation, prevention, and remediation of acid mine drainage. *Environmental Science and Pollution Research, 30*, 111871–111890. https://doi.org/10.1007/s11356-023-30220-5
23. Stewart, B. D., Amos, R. T., Nico, P. S., & Fendorf, S. (2011). Influence of uranyl speciation and iron oxides on uranium biogeochemical redox reactions. *Geomicrobiology Journal, 28*, 444–456. https://doi.org/10.1080/01490451.2010.507646

24. Kalintsev, A., et al. (2021). Uranium carbonate complexes demonstrate drastic decrease in stability at elevated temperatures. *Communications Chemistry, 4*, 120. https://doi.org/10.1038/s42004-021-00558-3
25. Lahrouch, F., et al. (2022). Uranium speciation control by uranyl sulfate and phosphate in tailings subject to a Sahelian climate, Cominak, Niger. *Chemosphere, 287*, 132139. https://doi.org/10.1016/j.chemosphere.2021.132139
26. Vopálka, D., Štamberg, K., Motl, A., & Drtinová, B. (2010). The study of the speciation of uranyl–sulphate complexes by UV–Vis absorption spectra decomposition. *Journal of Radioanalytical and Nuclear Chemistry, 286*, 681–686. https://doi.org/10.1007/s10967-010-0764-5
27. McMaster, S. A., Noller, B. N., Humphrey, C. L., Trenfield, M. A., & Harford, A. J. (2020). Speciation and partitioning of uranium in waterbodies near Ranger Uranium Mine. *Environmental Chemistry, 18*, 12–19. https://doi.org/10.1071/en20096
28. Lovley, D. R., Phillips, E. J. P., Gorby, Y. A., & Landa, E. R. (1991). Microbial reduction of uranium. *Nature, 350*, 413–416. https://doi.org/10.1038/350413a0
29. Fletcher, K. E., et al. (2010). U(VI) reduction to mononuclear U(IV) by desulfitobacterium species. *Environmental Science & Technology, 44*, 4705–4709. https://doi.org/10.1021/es903636c
30. Wall, J. D., & Krumholz, L. R. (2006). Uranium reduction. *Annual Review of Microbiology, 60*, 149–166.
31. Rogiers, T., et al. (2022). Molecular mechanisms underlying bacterial uranium resistance. *Frontiers in Microbiology, 13*, 822197. https://doi.org/10.3389/fmicb.2022.822197
32. Niida, K., Saito, T., & Tanaka, S. (2013). Comparison of uranyl adsorption on iron(III) oxyhydroxide colloids. *Chemistry Letters, 42*, 1380–1382. https://doi.org/10.1246/cl.130589
33. Ching-kuo Daniel, H., & Langmuir, D. (1985). Adsorption of uranyl onto ferric oxyhydroxides: Application of the surface complexation site-binding model. *Geochimica et Cosmochimica Acta, 49*, 1931–1941. https://doi.org/10.1016/0016-7037(85)90088-2
34. Dublet, G., et al. (2017). Partitioning of uranyl between ferrihydrite and humic substances at acidic and circum-neutral pH. *Geochimica et Cosmochimica Acta, 215*, 122–140. https://doi.org/10.1016/j.gca.2017.07.013
35. Lahrouch, F., et al. (2021). Uranium retention on iron oxyhydroxides in post-mining environmental conditions. *Chemosphere, 264*, 128473. https://doi.org/10.1016/j.chemosphere.2020.128473
36. Gronowski, A. (2013). *Effects of uranium mining on groundwater—Geochemical modeling of aqueous uranium speciation due to changing redox conditions* [Masters thesis, Royal Institute of Technology (KTH)].
37. Meinrath, G. (1998). *Freiberg on-line geosciences*. Freiberg.
38. Meinrath, G. (1997). Uranium speciation by spectroscopy. *Journal of Radioanalytical and Nuclear Chemistry, 224*, 119.
39. Smedley, P. L., & Kinniburgh, D. G. (2023). Uranium in natural waters and the environment: Distribution, speciation and impact. *Applied Geochemistry, 148*, 105534. https://doi.org/10.1016/j.apgeochem.2022.105534
40. Bird, G. A. (2012). In R. A. Meyers (Ed.), *Encyclopedia of sustainability science and technology* (pp. 11220–11262). Springer.
41. Götz, C., Geipel, G., & Bernhard, G. (2008). In B. J. Merkel & A. Hasche-Berger (Eds.), *Uranium, mining and hydrogeology* (pp. 907–914). Springer.
42. Yang, Y., et al. (2024). Systematic Raman spectroscopic study of the complexation of uranyl with fluoride. *Physical Chemistry Chemical Physics, 26*, 18584–18591. https://doi.org/10.1039/D4CP01569J
43. Faizrakhmanov, F. F. (2024). Speciation of uranium in trace concentrations in aqueous solutions of complex chemical composition. *Radiochemistry, 66*, 268–273. https://doi.org/10.1134/S1066362224020164
44. Yu, C., et al. (2024). Uranium repartitioning during microbial driven reductive transformation of U(VI)-sorbed Schwertmannite and Jarosite. *Environmental Science & Technology, 58*, 18324–18334. https://doi.org/10.1021/acs.est.4c03645

45. Malgorzata Alicja, S., et al. (2013). *47*, 12351–12358. American Chemical Society.
46. Bernier-Latmani, R., et al. (2010). Non-uraninite products of microbial U(VI) reduction. *Environmental Science & Technology, 44*, 9456–9462. https://doi.org/10.1021/es101675a
47. Meinrath, G. (1998). *Uranium Mining and Hydrogeology II. Proceedings* (Vol. 5, pp. 256–265). Sven von Loga.
48. Kumar, A., et al. (2023). A comprehensive review of uranium in the terrestrial and aquatic environment: Bioavailability, immobilization, tolerance and remediation approaches. *Plant and Soil, 490*, 31–65. https://doi.org/10.1007/s11104-023-06101-8
49. Riggle, J., & Von Wandruszka, R. (2008). The stability of uranium (VI) complexes of humates and fulvates in biphasic systems. *Annals of Environmental Science*, 1–6.
50. Lenhart, J. J., Cabaniss, S. E., MacCarthy, P., & Honeyman, B. D. (2000). Uranium(VI) complexation with citric, humic and fulvic acids. *Radiochimica Acta, 88*, 345–353.
51. Sundararajan, M., Rajaraman, G., & Ghosh, S. K. (2011). Speciation of uranyl ions in fulvic acid and humic acid: A DFT exploration. *Physical Chemistry Chemical Physics, 13*, 18038–18046. https://doi.org/10.1039/C1CP21238A
52. Labonne-Wall, N., Choppin, G. R., Lopez, C., & Monsallier, J.-M. (1999). In D. T. Reed, S. B. Clark, & L. Rao (Eds.), *Actinide speciation in high ionic strength media: Experimental and modeling approaches to predicting actinide speciation and migration in the subsurface* (pp. 199–211). Springer US.
53. Payne, T. E., & Waite, T. D. (2022). Uranium adsorption—A review of progress from qualitative understanding to advanced model development. *Radiochimica Acta, 110*, 549–559. https://doi.org/10.1515/ract-2022-0003
54. Bone, S. E., Dynes, J. J., Cliff, J., & Bargar, J. R. (2017). Uranium(IV) adsorption by natural organic matter in anoxic sediments. *Proceedings of the National Academy of Sciences, 114*, 711–716. https://doi.org/10.1073/pnas.1611918114
55. Bai, Y., et al. (2023). Exploration of the performance and mechanism of uranium adsorption by a covalent organic framework possessing the thiazole structure. *Langmuir, 39*, 16163–16173. https://doi.org/10.1021/acs.langmuir.3c02448
56. Cumberland, S. A., Douglas, G., Grice, K., & Moreau, J. W. (2016). Uranium mobility in organic matter-rich sediments: A review of geological and geochemical processes. *Earth-Science Reviews, 159*, 160–185. https://doi.org/10.1016/j.earscirev.2016.05.010
57. Lefebvre, P., et al. (2021). Diagenetic formation of uranium-silica polymers in lake sediments over 3300 years. *Proceedings of the National Academy of Sciences of the United States of America, 118*, 1–9.
58. Bone, S. E., et al. (2020). Complexation by organic matter controls uranium mobility in anoxic sediments. *Environmental Science & Technology, 54*, 1493–1502. https://doi.org/10.1021/acs.est.9b04741
59. Cologgi, D. L., Speers, A. M., Bullard, B. A., Kelly, S. D., & Reguera, G. (2014). Enhanced uranium immobilization and reduction by Geobacter sulfurreducens biofilms. *Applied and Environmental Microbiology, 80*, 6638–6646. https://doi.org/10.1128/AEM.02289-14
60. Stylo, M., Neubert, N., Roebbert, Y., Weyer, S., & Bernier-Latmani, R. (2015). Mechanism of uranium reduction and immobilization in Desulfovibrio vulgaris biofilms. *Environmental Science & Technology, 49*, 10553–10561. https://doi.org/10.1021/acs.est.5b01769
61. Li, P., et al. (2017). Highly efficient interception and precipitation of uranium(VI) from aqueous solution by iron-electrocoagulation combined with cooperative chelation by organic ligands. *Environmental Science & Technology, 51*, 14368–14378. https://doi.org/10.1021/acs.est.7b05288
62. Duff, M. C., Coughlin, J. U., & Hunter, D. B. (2002). Uranium co-precipitation with iron oxide minerals. *Geochimica et Cosmochimica Acta, 66*, 3533–3547. https://doi.org/10.1016/S0016-7037(02)00953-5
63. Veeramani, H., et al. (2013). Abiotic reductive immobilization of U(VI) by biogenic mackinawite. *Environmental Science & Technology, 47*, 2361–2369. https://doi.org/10.1021/es304025x

64. Lefebvre, P., et al. (2021). Diagenetic formation of uranium-silica polymers in lake sediments over 3300 years. *Proceedings of the National Academy of Sciences, 118*, e2021844118. https://doi.org/10.1073/pnas.2021844118
65. Roberts, H. E., et al. (2017). Uranium(V) incorporation mechanisms and stability in Fe(II)/Fe(III) (oxyhydr)oxides. *Environmental Science & Technology Letters, 4*, 421–426. https://doi.org/10.1021/acs.estlett.7b00348
66. Law, G. T. W., et al. (2011). Uranium redox cycling in sediment and biomineral systems. *Geomicrobiology Journal, 28*, 497–506. https://doi.org/10.1080/01490451.2010.512033
67. Antunes, I. M. H. R., Carvalho, P. C. S., Albuquerque, M. T. D., & Santos, A. C. T.P. Stanley, C. Wolkersdorfer, & K. Wolkersdorfer (Eds.). *IMWA 2021—"mine water management for future generations"*. International Mine Water Association.
68. Lloyd, J. R. (2003). Microbial reduction of metals and radionuclides. *FEMS Microbiology Reviews, 27*, 411–425. https://doi.org/10.1016/s0168-6445(03)00044-5
69. Shi, L., Squier, T. C., Zachara, J. M., & Fredrickson, J. K. (2007). Respiration of metal (hydr) oxides by Shewanella and Geobacter: A key role for multihaem c-type cytochromes. *Molecular Microbiology, 65*, 12–20. https://doi.org/10.1111/j.1365-2958.2007.05783.x
70. Belli, K. M., DiChristina, T. J., Van Cappellen, P., & Taillefert, M. (2015). Effects of aqueous uranyl speciation on the kinetics of microbial uranium reduction. *Geochimica et Cosmochimica Acta, 157*, 109–124. https://doi.org/10.1016/j.gca.2015.02.006
71. Ghasemi, R., Fatemi, F., & Rastkhah, E. (2025). Bioremediation performance of recombinant Shewanella azerbaijanica; considering uranium removal in the presence of nitrate. *Current Microbiology, 82*, 290. https://doi.org/10.1007/s00284-025-04228-8
72. Beller Harry, R. (2005). Anaerobic, nitrate-dependent oxidation of U(IV) oxide minerals by the chemolithoautotrophic bacterium Thiobacillus denitrificans. *Applied and Environmental Microbiology, 71*, 2170–2174. https://doi.org/10.1128/AEM.71.4.2170-2174.2005
73. Asta, M. P., Beller, H. R., & O'Day, P. A. (2020). Anaerobic dissolution rates of U(IV)-oxide by abiotic and nitrate-dependent bacterial pathways. *Environmental Science & Technology, 54*, 8010–8021. https://doi.org/10.1021/acs.est.0c01019
74. Beller, H. R., et al. (2013). Genome-enabled studies of anaerobic, nitrate-dependent iron oxidation in the chemolithoautotrophic bacterium Thiobacillus denitrificans. *Frontiers in Microbiology, 4*, 249. https://doi.org/10.3389/fmicb.2013.00249
75. Wan, J., et al. (2005). Reoxidation of bioreduced uranium under reducing conditions. *Environmental Science & Technology, 39*, 6162–6169. https://doi.org/10.1021/es048236g
76. Seder-Colomina, M., et al. (2022). Influence of microorganisms on uranium release from mining-impacted lake sediments under various oxygenation conditions. *Environmental Science: Processes & Impacts, 24*, 1830–1843. https://doi.org/10.1039/d2em00104g
77. Yu, Q., & Cui, O. (2023). Enrichment and remediation of uranium by microorganisms: A review. *Open Journal of Environmental Biology, 8*, 020–038. https://doi.org/10.17352/ojeb.000037
78. Chardi, K. J., Schenkeveld, W. D. C., Kumar, N., Giammar, D. E., & Kraemer, S. M. (2023). Effect of competing metals and humic substances on uranium mobilization from noncrystalline U(IV) induced by anthropogenic and biogenic ligands. *Environmental Science & Technology, 57*, 16006–16015. https://doi.org/10.1021/acs.est.3c01705
79. Volkov, I. V., & Polyakov, E. V. (2023). Complexation of humic acids with trace elements: Methods and approaches. *Journal of Analytical Chemistry, 78*, 1575–1602. https://doi.org/10.1134/S1061934823120213
80. Pompe, S., et al. (2000). Investigation of humic acid complexation behavior with uranyl ions using modified synthetic and natural humic acids. *Radiochimica Acta, 88*, 553–558. https://doi.org/10.1524/ract.2000.88.9-11.553
81. Stockdale, A., & Bryan, N. D. (2012). Uranyl binding to humic acid under conditions relevant to cementitious geological disposal of radioactive wastes. *Mineralogical Magazine, 76*, 3391–3399. https://doi.org/10.1180/minmag.2012.076.8.52

82. Thuéry, P. (2006). Uranyl ion complexation by citric and tricarballylic acids: Hydrothermal synthesis and structure of two- and three-dimensional uranium–organic frameworks. *Chemical Communications, 2006*, 853–855. https://doi.org/10.1039/B516191F
83. Steudtner, R., Sachs, S., Schmeide, K., Brendler, V., & Bernhard, G. (2011). Ternary uranium(VI) carbonato humate complex studied by cryo-TRLFS. *Radiochimica Acta, 99*, 687–692. https://doi.org/10.1524/ract.2011.1861
84. Pasilis, S. P., & Pemberton, J. E. (2003). Speciation and coordination chemistry of uranyl(VI)–citrate complexes in aqueous solution. *Inorganic Chemistry, 42*, 6793–6800. https://doi.org/10.1021/ic0341800
85. Mullen, L., Gong, C., & Czerwinski, K. (2007). Complexation of uranium (VI) with the siderophore desferrioxamine B. *Journal of Radioanalytical and Nuclear Chemistry, 273*, 683–688. https://doi.org/10.1007/s10967-007-0931-5
86. Gunten, K.v., et al. Biogeochemistry of U, Ni, and As in two meromictic pit lakes at the Cluff Lake uranium mine, northern Saskatchewan. *Canadian Journal of Earth Sciences, 55*, 463–474 (2018). https://doi.org/10.1139/cjes-2017-0149
87. Dzimbanhete, V. L. (2025). *Uranium sources and mobility at an iron ore mine site in Northern Sweden: A geochemical, mineralogical, and isotopic investigation* [PhD thesis, Luleå University of Technology].
88. Von Gunten, K. (2019). *Biogeochemistry of meromictic pit lakes and permeable reactive barriers at the Cluff Lake uranium mine* [PhD thesis, University of Alberta].
89. Sánchez-España, J. (2024). In L. C. Staicu & L. L. Barton (Eds.), *Geomicrobiology: Natural and anthropogenic settings* (pp. 95–121). Springer Nature Switzerland.
90. APA. (2003). The Australasian Institute of Mining and Metallurgy.
91. Skierszkan, E. K., et al. (2021). Persistence of uranium in old and cold subpermafrost groundwater indicated by linking 234U-235U-238U, groundwater ages, and hydrogeochemistry. *ACS Earth and Space Chemistry, 5*, 3474–3487. https://doi.org/10.1021/acsearthspacechem.1c00307
92. Gökçe, S., & Şengör, S. S. (2025). Reactive transport modeling of uranium in subsurface: Impact of field-scale heterogeneity and biogeochemical dynamics. *Water, 17*, 514.
93. Fuller, C., et al. (2015). *Uranium sequestration during biostimulated reduction and in response to the return of oxic conditions in shallow aquifers*. Report No. NUREG/CR-7178.
94. Suzuki, Y., Kelly Shelly, D., Kemner Kenneth, M., & Banfield Jillian, F. (2005). Direct microbial reduction and subsequent preservation of uranium in natural near-surface sediment. *Applied and Environmental Microbiology, 71*, 1790–1797. https://doi.org/10.1128/AEM.71.4.1790-1797.2005

Chapter 6
Uranium Behaviour in Abandoned In-Situ Leaching Mines

Abstract This chapter investigates the fate and transport of uranium in abandoned in situ leach (ISL) mining systems where no engineered remediation is applied. It analyses geochemical and hydrologic processes controlling uranium mobility under both acid and alkaline leaching conditions, accounting for ore mineralogy, leaching solutions, and site hydrogeology. These processes influence uranium and associated contaminant plume migration by the persistence of lixiviants in the subsurface. To assess these processes, the discussion integrates limited field observations with geochemical modelling and reactive transport simulations to project system evolution over decades to centuries, highlighting implications for groundwater quality and environmental exposure pathways in unremediated ISL sites.

Keywords Mineral dissolution · Adsorption-desorption processes · Lixiviant transport · Ion-exchange residues · Aquifer restoration · Permeability alteration · Host-rock reactivity · Solute migration · Radionuclide mobility

6.1 Introduction

In situ leaching (ISL), also known as in situ recovery (ISR) or solution mining, is a resource extraction method that dissolves ore deposits in situ. The process involves injecting chemically reactive solutions into a permeable host formation and retrieving mineral-bearing fluids through production wells [1]. The term "ISL" is more commonly used worldwide, while "ISR" is preferred in North America; therefore, ISL will be used throughout this chapter.

ISL is technically classified as a mining method, although it does not involve the classic mining concept of an ore excavation. The technique involves dissolving minerals in the ground with a chemical solution and then pumping the mineral-rich liquid to the surface for processing. It is most often used for uranium but has also been applied to potash, salt, and copper [2–4]. Unlike traditional open-pit or underground mining, ISL mobilizes uranium in place, relying on the deposit's hydrogeologic structure, the permeability and connectivity of the aquifer, and the choice of

M. Mkandawire, *Fate and Effects of Uranium in Abandoned Uranium Mines*,
https://doi.org/10.1007/978-3-032-10447-2_6

lixiviant, typically oxidizing, acidic, or alkaline fluids tailored to the host-rock mineralogy.

ISL was developed in the mid-twentieth century as a cost-effective and less surface-disturbing alternative to conventional uranium mining, with commercial applications first achieved (in the 1970s) where tabular sandstone-hosted and roll-front deposits coincided with favourable hydrologic conditions. Over decades, technical improvements advanced well construction, lixiviant chemistry, and monitoring practices [5, 6]. Despite these advances, ISL remains geographically limited because site-specific hydrogeologic and geochemical conditions tightly constrain its applicability and environmental behaviour.

This chapter examines the fate and effects of uranium in closed in situ leach (ISL) mine systems when no active remediation or engineered restoration measures are employed. It focuses on the long-term geochemical and hydrologic processes that influence uranium mobility, natural attenuation pathways, and potential exposure routes to environmental receptors following containment abandonment. Empirical data from entirely abandoned and unremediated ISL sites are scarce and often unrepresentative. As a result, our mechanistic understanding and predictive insights primarily rely on a combination of limited field observations, geochemical simulations, and reactive transport modelling. The aim is to explore system trajectories over decades to centuries.

Geochemical and reactive transport models are employed to investigate the key factors that influence post-closure behaviour. These include the redox evolution of the aquifer following the cessation of oxidant injection, the dynamics of sorption and desorption on mineral and organic surfaces, the precipitation and dissolution of secondary uranium minerals, and complexation with carbonates, sulfates, and organic ligands. Moreover, the interactions between biogeochemical reactions and evolving groundwater flow patterns are examined.

Model scenarios can produce very different outcomes. For example, one scenario may lead to a rapid decline of contaminants to near-background levels due to reducing conditions, sorptive sinks, or mineral sequestration. In contrast, another scenario might result in ongoing or episodic mobility of uranium if oxidizing conditions persist, if carbonate complexation keeps aqueous uranium in solution, or if hydraulic conditions facilitate the transport of residual lixiviant.

Readers should approach this chapter with caution, as projections rely on the model's structure, parameters, and data. As data and tools improve, projections will be refined. This chapter consolidates current understanding, presents modelled behaviours, and addresses uncertainties and monitoring strategies for unremediated sites.

6.2 ISL Procedures

6.2.1 General Principle

ISL mining extracts uranium from ore-bearing minerals by dissolving them underground with a leaching solution, then pumping the pregnant solution to the surface for uranium recovery. The general process involves three steps: (1) injection of a lixiviant solution (typically containing water mixed with oxygen and hydrogen peroxide, as well as sodium carbonate or carbon dioxide) through a series of wells into the orebody to dissolve the uranium; (2) pumping the pregnant liquor through series of recovery wells to a processing plant, where the uranium is extracted from the solution through an ion-exchange process; and (3) further purifying, concentrating, and drying to produce yellowcake. The yellowcake is then sent to a uranium conversion facility for processing to produce fuel for nuclear power reactors and other uses.

6.2.2 Conditions Favouring ISL Application

ISL viability requires a laterally continuous, hydraulically coherent ore-zone aquifer within a sedimentary sequence, bounded by laterally extensive, low-permeability confining beds that demonstrably isolate the target horizon from overlying and underlying water-bearing units; seismic quiescence and absence of through-going faults that breach confining units are preferred to minimize vertical hydraulic connectivity and lixiviant escape [3, 7–9]. The ore horizon must exhibit effective porosity and permeability sufficient to permit controlled injection and recovery at practical well spacings while maintaining advective residence times compatible with leach kinetics; transmissivity and storativity should support target recovery factors without inducing unacceptable drawdown in adjacent potable or ecological receptors [3, 10–12].

Baseline hydrochemistry must indicate geochemical conditions amenable to the chosen lixiviant (i.e., reducible uranium mineralogy, buffering capacity, and background redox and major-ion composition that limit secondary mobilization of deleterious elements), and potentiometric mapping must demonstrate predictable regional and local groundwater flow directions that allow for the engineering of hydraulic containment and recovery efficiency [2].

Thus, comprehensive pre-operational characterization is mandatory and must integrate geology, hydrology, and anticipated ISL impacts. Geological data should include regional structure and seismicity, stratigraphic cross-sections showing ore-zone geometry and confining beds, and isopach/structure mapping to define continuity and thickness. Hydrologic data must quantify potentiometric surfaces of the ore-zone and adjacent aquifers, regional and local flow directions, total and effective porosity, permeability, transmissivity, and water quality within the ore horizon and in upgradient and downgradient zones; core data, geophysical logs, and hydraulic testing must demonstrate the competence and lateral extent of confining beds. Impact assessment requires predictive modelling of lixiviant migration, mass transport, and geochemical reactions, as well as the evaluation of restoration feasibility under site-specific hydrogeologic constraints [13]. The ISL process essentially reverses the ore genesis [11].

Operational and economic limits, including depth to ore and well construction constraints, must be considered during feasibility assessment; historically, commercial ISL has concentrated on deposits at depths that permit cost-effective wellfield development and restoration, with site selection and design guided by stratigraphic correlation, hydraulic testing, tracer studies, and comprehensive baseline geochemistry to validate containment and post-operational restoration strategies.

6.2.3 ISL Uranium Mine and Process

The design of ISL wellfields varies greatly depending on local conditions, such as permeability, sand thickness, deposit type, ore grade, and distribution [14]. Figure 6.1 presents a generalized illustration of a typical ISL uranium mine system. The wellfield design typically consists of injection wells to introduce the leach solution into the orebody and extraction wells with submersible pumps to deliver a

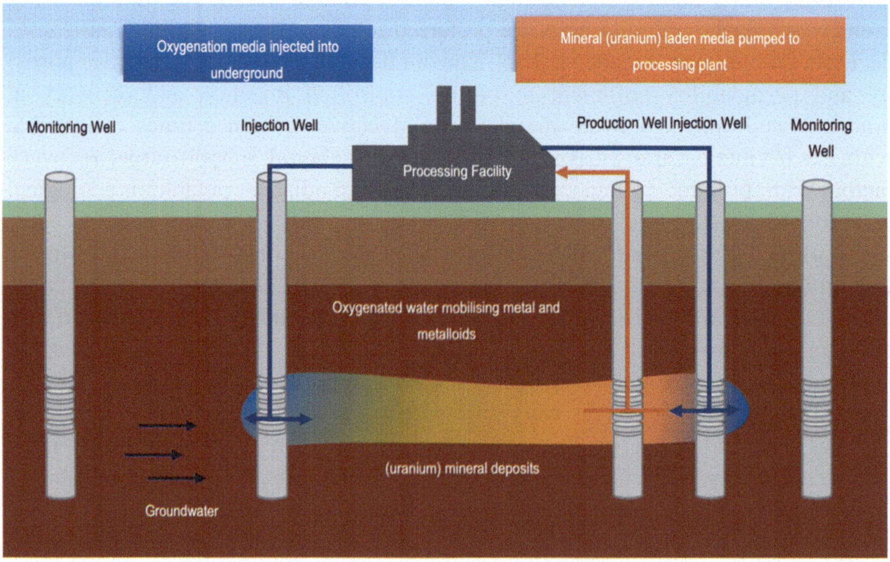

Fig. 6.1 A generalized illustration of a typical installation of an ISL uranium mine. Initially, native groundwater is extracted from the host aquifer, followed by injection of uranium-complexing reagents and an oxidant (hydrogen peroxide or oxygen) into the wellfield. The leach liquors pass through the ore to oxidize and dissolve the uranium minerals in situ

pregnant solution to the processing plant [12, 14, 15]. Wells are typically constructed as standard water bores. Due to its lower capital costs relative to conventional mining, ISL can often be a more effective method for mining low-grade uranium deposits.

Figure 6.2 illustrates a flow diagram of the generalized mineral processing operations involved in an ISL acid operation. Typically, an ISL uranium mine features a network of injection wells that inject a leach solution into the orebody zone of the aquifer. Recovery wells extract leachate containing dissolved uranium, along with various metals and metalloids commonly found with uranium mineralization, as well as elements released from the surrounding rock during the leaching process.

The leachate consists of natural groundwater that has been conditioned with acid or alkali, and it usually includes an oxidizing agent (oxidant) along with additional reagents as needed. The system of pipes connecting the injection and recovery wells includes headers and main trunk lines that lead to and from the processing plant, where uranium extraction takes place. The resulting barren solution is treated with supplementary reagents as needed to prepare it for re-injection, enabling continuous recycling of the leach solution.

Surrounding each wellfield, a series of monitoring wells and facilities is established for managing and disposing of both liquid and solid wastes. These facilities typically include storage/evaporation ponds and disposal wells. Any excess solution is reinjected into the same aquifer system, but away from areas currently undergoing mining. In some cases, particularly on an international scale, it may be directed into a different aquifer with lower-quality water. Additionally, spill containment infrastructure is typically implemented at ISL operations.

At the plant, uranium is removed from the solution. The mineral processing plant is generally located close to the wellfield; however, in situations involving intermediate processing, the wellfield and mineral processing operation can be situated several kilometres apart.

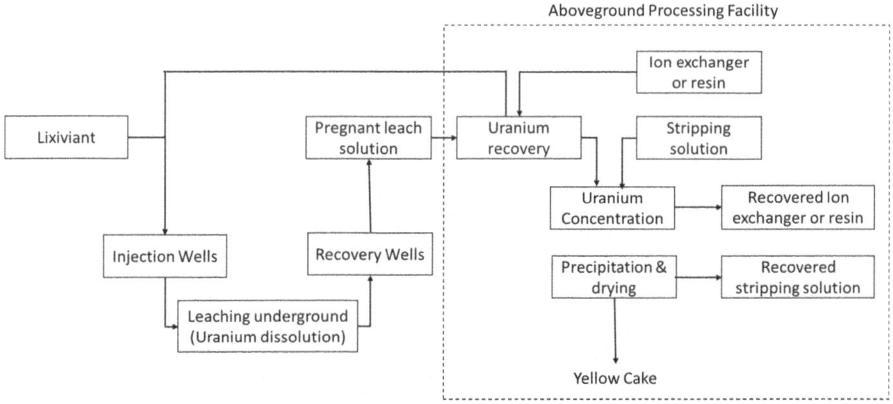

Fig. 6.2 Conceptual flow diagram of an in situ leaching uranium mining process

6.2.4 Geochemistry of Lixiviant and Oxidants

An ideal lixiviant oxidizes uranium to the soluble hexavalent state, complexes it to keep it in solution, and reacts minimally with the barren host rock. Typical lixiviants are salt solutions (i.e., sulfate, bicarbonate/carbonate, or ammonium) that form stable uranyl complexes, and an oxidant (air/oxygen, hydrogen peroxide, sodium chlorate, sodium hypochlorite, or potassium permanganate) is added to sustain oxidation and improve recovery. The principal geochemical reactions among lixiviants, ore minerals, and host rock fall into four categories: (a) oxidation and reduction, (b) dissolution, (c) re-precipitation, and (d) adsorption and ion exchange. These processes begin with lixiviant injection and may persist long after mining ends [2].

6.2.4.1 Oxidation and Reduction

Oxidation governs uranium recovery in solution mining. Tetravalent minerals such as uraninite (UO_2) and coffinite are insoluble under reducing conditions but dissolve when oxidized. Oxidants may be injected or generated by reactions with non-uranium minerals. For example, hydrogen peroxide in the presence of bicarbonate promotes oxidation:

$$UO_2 + H_2O_2 + 2HCO_3^- \rightarrow UO_2\left(CO_3\right)_2^{2-} + 2H_2O$$

Ferric iron often acts as the oxidizing agent, regenerated by chlorate oxidation of ferrous iron:

$$6Fe^{2+} + ClO_3^- + 6H^+ \rightarrow 6Fe^{3+} + Cl^- + 3H_2O$$

$$UO_2 + 2Fe^{3+} \rightarrow UO_2^{2+} + 2Fe^{2+}.$$

Once oxidized, uranium readily forms soluble complexes with sulfate or carbonate ions [16, 17].

6.2.4.2 Dissolution

Lixiviants also dissolve gangue minerals, often undesirably, as this consumes reagents and introduces contaminants. Carbonates are most susceptible, with dissolution dependent on pH:

$$CaCO_3 + 2H^+ \rightarrow Ca^{2+} + H_2O + CO_2$$

Neutral or slightly basic lixiviants minimize this effect, though localized alteration of carbonates and silicates occurs, especially where oxidation of ferrous minerals generates acidity [18].

6.2.4.3 Re-precipitation

Re-precipitation limits the mobility of dissolved species. Uranium may precipitate when the oxidizing potential decreases, such as at the leach field margins, effectively reversing leaching. Non-uranium ions, such as calcium, often reprecipitate as gypsum ($CaSO_4{\cdot}2H_2O$) or secondary calcite. Uranium can co-precipitate with vanadate or arsenate minerals under certain conditions [19, 20].

6.2.4.4 Adsorption and Ion Exchange

Metals are commonly removed by adsorption onto hydrous iron oxides. Ion exchange between lixiviant and clays, especially montmorillonite, is significant. Calcium and magnesium may be replaced by sodium, while ammonium fixation occurs with ammonium bicarbonate lixiviants, resulting in ammonium contents up to twice those of pre-leach levels in clays [21].

6.2.5 Operating Regimes for ISL Uranium Mining

There are two established operating regimes for ISL uranium mining, determined by geology, groundwater, and environmental requirements. These regimes are (a) alkali leaching and (b) acid leaching. The choice between acid and alkaline leaching is determined primarily by the geology of the orebody and the chemistry of the surrounding aquifer. The main controlling factor is the carbonate content of the orebody. Table 6.1 summarizes the factors that influence the choice between acid and alkali leaching regimes. If the carbonate content of the orebody exceeds 2%, alkaline leaching is necessary; otherwise, acid leaching is preferred for higher recovery and lower costs. Acid leaching enables faster, more efficient uranium recovery. Environmental and regulatory considerations are also taken into account.

6.3 Uranium in Alkali ISL

When the orebody has significant quantities of acid-consuming minerals, mainly calcium, such as more than 2% limestone or gypsum, alkaline leaching with carbonate is used to give a uranyl-carbonate [22–24]. The alkaline leaching technique is used

Table 6.1 The choice between acid and alkaline leaching for ISL uranium mining

Factor	Acid leaching	Alkaline leaching
Geological composition	Typically used for ores with low carbonate mineral content, often less than 2%	Mandated for ores with significant amounts of acid-consuming minerals like limestone and gypsum
Recovery efficiency	Achieves higher uranium recovery rates, typically 70–90%	Has lower recovery rates, usually in the range of 60–70%
Reaction kinetics	Offers faster leaching kinetics, meaning the uranium dissolves more quickly	Features slower leaching kinetics, requiring more time to dissolve the uranium
Reagent cost	Uses sulfuric acid, which is corrosive but often has lower overall operating costs compared to alkaline methods	Uses less corrosive but more expensive reagents, such as bicarbonate or ammonium carbonate, and often has higher operating costs
Groundwater impact	Mobilizes a broader range of dissolved solids, including metals like cadmium and lead, which can make post-mining groundwater remediation more difficult and expensive	It is more selective for uranium, mobilizing fewer metals and reducing overall water contamination This generally facilitates groundwater restoration more easily
Pyrite content	Not inhibited by the presence of pyrite and other sulfides	Unsuitable for deposits with significant pyrite content, as the carbonate can accelerate pyrite oxidation and form complex iron-carbonate compounds

mainly in the USA and to a lesser extent in Kazakhstan and other Eastern European countries [9, 24, 25]. The technique has a uranium recovery efficiency of 60–70%.

6.3.1 Alkali ISL Uranium Mining Process and Mechanisms

Alkaline in situ leaching mobilizes uranium by oxidizing U(IV) minerals to soluble uranyl-carbonate complexes under high carbonate/bicarbonate alkalinity. The process is controlled by redox fronts, mineral buffering, sorption/desorption kinetics, and advective transport (with diffusive exchange) in heterogeneous sandstone aquifers.

Alkaline ISL of sandstone-hosted uranium deposits uses a bicarbonate/carbonate-based lixiviant, typically sodium bicarbonate or sodium carbonate solutions, with an oxidant (commonly dissolved oxygen, sometimes supplemented by more potent oxidants) to convert reduced uranium phases into soluble uranyl-carbonate complexes that are recovered by pumping and ion exchange. The operational chemistry is deliberately designed to favour the formation of highly stable anionic complexes such as $UO_2\left(CO_3\right)_3^{4-}$ and mixed Ca-uranyl-carbonates like $Ca_2UO_2(CO_3)_3$, which minimize adsorption to mineral surfaces and maximize mobility in porewater under alkaline pH and elevated carbonate concentrations [6].

The geochemical mechanism begins at the mineral scale, where oxidation of U(IV) (uraninite and coffinite) to U(VI) (uranyl) is the rate-limiting step for mobilization. Under alkaline carbonate conditions, the net reaction can be represented conceptually as:

$$UO_2(s) + \frac{1}{2}O_2 + 3CO_3^{2-} \rightarrow UO_2(CO_3)_3^{4-}$$

The net reaction involves concomitant proton and electron transfers, buffered by carbonate/bicarbonate equilibria. Oxidants used in practice include dissolved oxygen, hydrogen peroxide, sodium hypochlorite or other chlorine-releasing compounds, and permanganate in some enhanced applications; the choice and dose are dictated by the reducing capacity of the host rock (organic matter, pyrite, and Fe(II)-bearing silicates) and by kinetic constraints on oxygen transfer in low-permeability zones [26].

At the hydrogeochemical scale, redox fronts migrate with the injected lixiviant, producing spatially variable zones of mobilization and attenuation. Where the oxidant persists and carbonate activity is high, uranyl-carbonate complexes dominate and are transported by advection to recovery wells. Where reducing minerals or microbial activity consume oxidant, U(VI) is reduced and may precipitate as uraninite or incorporate into secondary phases (e.g., coffinite and U-bearing phosphates) or be immobilized by sorption to Fe-oxyhydroxides and clay surfaces. Competitive complexation and ion pairing with Ca^{2+} and Mg^{2+} alter speciation and charge, producing neutral or less negatively charged species that change sorption behaviour and mobility. Thermodynamic modelling shows that speciation is highly sensitive to carbonate activity, pH, and Ca/Mg concentrations, and that small changes in these parameters can shift dominant species between highly mobile anionic complexes and less mobile neutral complexes [27, 28].

In abandoned ISL fields, the long-term fate of uranium is governed by residual alkalinity, remaining oxidant, diffusion from low-permeability strata, and the re-establishment of reducing conditions. Rebound phenomena, where dissolved uranium concentrations rise after pumping ceases, are explained by back-diffusion from matrix porosity, slow re-oxidation of previously reduced zones when oxygen or other oxidants infiltrate, and desorption from mineral surfaces as geochemical conditions evolve. Understanding and predicting post-closure behaviour, therefore, requires coupled reactive transport models that include kinetic oxidation/reduction, carbonate complexation equilibria, sorption/desorption kinetics, and mineral precipitation/dissolution pathways, calibrated to site-specific mineralogy and hydrogeology.

6.3.2 *Selection of Lixiviants for Alkali ISL*

Choosing an alkaline lixiviant is a balance between thermodynamic speciation that favours stable uranyl-carbonate complexes, kinetic constraints on oxidant delivery and consumption, and site-specific environmental conditions that increasingly favour bicarbonate/CO_2 systems over ammonium-based reagents.

Selection of an alkaline lixiviant begins with the fundamental geochemical objective: to maintain a solution chemistry that stabilizes uranium as soluble uranyl-carbonate species while minimizing secondary mobilization of deleterious constituents and facilitating post-closure restoration. In practice, this means maintaining elevated carbonate activity and pH in the presence of an effective oxidant so that U(VI) predominates as complexes such as $UO_2(CO_3)_3^{4-}$ or Ca-bearing neutral complexes under the prevailing Ca^{2+} and Mg^{2+} activities. The choice of carbonate source, dissolved CO_2, sodium bicarbonate, sodium carbonate, or ammonium carbonate, controls not only carbonate activity but also ionic strength, cation exchange effects, and clay swelling tendencies, all of which feed back on permeability and flow patterns in the ore horizon [6].

Operational constraints drive the second axis of lixiviant selection. Oxidant compatibility is critical: dissolved oxygen alone may be sufficient in low reducing-capacity systems, but where organic matter, pyrite, or Fe(II) minerals consume oxidant rapidly, more potent oxidants or higher oxygen delivery rates are required. The lixiviant must therefore be evaluated for its ability to carry and sustain oxidant in the pore network while limiting undesirable side reactions such as carbonate-driven dissolution of gangue carbonates or excessive mobilization of trace metals. Laboratory column and batch tests that couple oxidation kinetics with carbonate complexation and sorption experiments remain indispensable for predicting field behaviour and for sizing oxidant doses and injection schedules [5, 6].

Environmental considerations increasingly influence the choice of lixiviant. Historically, ammonium carbonate was used because of its strong complexation and favourable kinetics, but concerns about ammonium persistence, nitrification, and aquifer restoration have led regulators and operators to prefer sodium bicarbonate or dissolved CO_2 in many jurisdictions; dissolved CO_2 is often the least expensive option, but it is not universally effective across deposit types and hydrogeochemical settings. Post-closure rebound risk and the potential for long-term mobilization of co-contaminants require that the selected lixiviant be evaluated for its reversibility: how readily carbonate activity and cation composition can be returned to baseline, and how sorbed or matrix-diffused uranium will respond as redox and carbonate conditions evolve [29].

Lixiviant selection for alkali ISL is a multidisciplinary optimization problem that integrates speciation thermodynamics, redox and kinetic constraints, hydrogeologic transport, and environmental closure criteria. Site-specific mineralogy and native groundwater chemistry determine whether sodium bicarbonate, sodium carbonate, dissolved CO_2, or ammonium carbonate is the most appropriate choice, and rigorous laboratory-to-field testing is required to translate chemical theory into reliable field performance (Table 6.2).

Table 6.2 Common lixiviants used in alkaline ISL uranium mining, their compatibility, advantages, and limitations

Lixiviant	Typical composition	Oxidant compatibility	Advantages	Limitations
Sodium bicarbonate	$NaHCO_3$ solution; moderate ionic strength	Compatible with O_2; supports H_2O_2, permanganate	Limits ammonium issues; controllable pH	Sodium exchange; clay swelling; cost vs CO_2
Sodium carbonate	Na_2CO_3 or mixed $Na_2CO_3/NaHCO_3$; higher carbonate activity	Compatible with common oxidants	Strong carbonate buffering Promotes uranyl-carbonate formation	Higher pH effects; carbonate precipitation with Ca; permeability impacts
Dissolved CO_2	CO_2 gas dissolved toformHCO$_3^-$/CO$_3^{2-}$	Works with O_2; limited with some strong oxidants	Low reagent cost; minimal cation exchange; easy to strip	Not effective in all deposits; gas handling complexity
Ammonium carbonate	$(NH_4)_2CO_3$ or NH_4HCO_3 solutions	Compatible with O_2 and H_2O_2	Strong complexation; effective in many ores	Restoration concerns; nitrification risk; ammonium persistence

6.3.3 Uranium Behaviour in Post-Alkali ISL

In alkaline ISL uranium operations, ore is mobilized by injecting oxidants (typically dissolved oxygen or hydrogen peroxide) together with carbonate/bicarbonate to form highly soluble uranyl-carbonate complexes; after cessation, a residual plume of lixiviant and reaction products persists in the ore zone and adjacent aquifer unless fully restored to baseline. Restoration is frequently incomplete, with trace metals, anions (especially carbonate and sulfate), and elevated total dissolved solids lingering and capable of migrating along connected hydrostratigraphic pathways if gradients and barriers are not properly managed. These concerns collectively emphasize the need to understand post-closure plume behaviour under alkaline conditions and the hydrogeochemical controls on natural attenuation versus persistence [1, 12, 30, 31].

Under the pH, alkalinity, and ionic strength typical of alkaline ISL systems, uranium is predominantly present as U(VI) in solution, complexed by carbonate to form species such as $[UO_2(CO_3)_2]^{2-}$ and $[UO_2(CO_3)_3]^{4-}$, and, critically in Ca-bearing waters, the neutral ternary complexes $CaUO_2(CO_3)_3$ and $Ca_2UO_2(CO_3)_3$ that substantially increase mobility by suppressing electrostatic sorption and enhancing advective transport. The existence and environmental dominance of these Ca-uranyl-carbonate complexes is supported both experimentally and by modelling, and their formation is characteristic of calcite-rich, carbonate-buffered mine waters [32, 33].

Sorption and mineral interactions under alkaline, carbonated conditions generally disfavour immobilization. Ferrihydrite/goethite and other Fe-(hydr)oxides are potent U(VI) sorbents in low-carbonate waters, but sorption edges are shifted and capacities diminished as carbonate complexes form; the presence of Ca further reduces adsorption and promotes persistence of neutral aqueous complexes. Nonetheless, secondary retention pathways exist: Fe(II)-induced transformation of ferrihydrite to goethite can incorporate U into the crystal lattice, and mixed retention via sorption plus partial reduction can occur in Fe-rich, mildly reducing zones. These mechanisms compete with carbonate complexation and are sensitive to Ca and bicarbonate concentrations, sediment reactivity, and the extent of Fe(II) availability [34–36].

Redox evolution following closure is pivotal. As oxidants are consumed by reaction with reduced minerals (e.g., pyrite and organic matter), Eh declines and U(VI) can be reduced to sparingly soluble U(IV), commonly precipitating as nanocrystalline uraninite or forming non-crystalline U(IV) phases bound to Fe/organic aggregates. Field and laboratory studies demonstrate that downgradient redox barriers akin to ore-forming roll fronts can attenuate ISL-derived U plumes, yielding stabilization within years to decades at many sites. However, the kinetics and completeness of reduction are site-specific. Attenuation behaviour has been quantified in historical case analyses and in cross-hole tracer tests in restored and unrestored ISR fields, showing significant U losses over tens of metres when water encounters reduced sands [37, 38].

An important caveat is that the reductive pathway does not guarantee immobility: U(IV) can persist as mobile colloids associated with natural organic matter and Fe-rich nanoparticles, enabling transport under anoxic conditions despite nominal reduction. U(IV) is associated with 1–3 kDa organic colloids and 5–50 nm Fe-bearing particles, highlighting a colloid-facilitated transport vector that must be considered in risk assessments for abandoned ISL systems [38–40].

Hydrogeologic architecture strongly governs plume migration. Sandstone-hosted roll-front deposits are typically heterogeneous and anisotropic, with layered permeability, local clay barriers, and dual-porosity behaviour in which matrix-fracture or macro−/micropore exchanges control residence times and tailing. Plumes preferentially follow high-K (i.e., high hydraulic conductivity) pathways because these zones offer the path of least resistance to fluid flow in porous media. Vertical anisotropy and compartmentalization can offset oxidation fronts and create complex plume geometries, while mischaracterized faults or connected units can serve as conduits to adjacent aquifers. These features complicate restoration and post-closure predictions, necessitating site-specific characterization of hydraulic gradients, heterogeneity, and interporosity transfer during and after pumping cessation [12].

Carbonate system dynamics, especially calcite saturation, modulate U fate. Where evaporation or CO_2 degassing drives calcite precipitation, co-precipitation and occlusion can trap U(VI) within growing $CaCO_3$; conversely, sustained high bicarbonate levels keep U complexed and soluble, inhibiting the formation of low-solubility hydrated uranyl oxides (e.g., schoepite) under ambient conditions. Thus, immobilization via calcite co-precipitation is more likely in zones of carbonate supersaturation and limited bicarbonate replenishment. In contrast, plume persistence is favoured in well-buffered, Ca-HCO_3 waters, precisely those produced by alkaline ISL [41, 42].

Temperature dependence of carbonate complex stability is a secondary, usually minor, factor in aquifers (ambient to ~30 to 40 °C), but experiments at elevated temperatures (>100 °C) underscore that uranyl-carbonate complexes lose stability, implying that thermal perturbations (e.g., deep circulation or geothermal gradients) would tend to reduce carbonate-mediated mobility; such conditions are atypical for the near-surface aquifers affected by ISL [37]. From a restoration perspective, monitored natural attenuation (MNA) frameworks evaluate whether destructive (i.e., reduction) and non-destructive (i.e., dilution, dispersion, and sorption) processes will meet remedial goals on reasonable timescales. They emphasize plume stability and protection of points of exposure. Field programs at ISR sites have combined groundwater sweep/reverse osmosis with geochemical interventions (abiotic reductants such as dithionite and biostimulation) and have investigated downgradient attenuation capacity to justify relaxation of in-zone targets when natural barriers are effective; nevertheless, federal and state reviews caution that residual constituents often remain above baseline after restoration, reinforcing the need for long-term surveillance.

In closed alkaline ISL settings, migration is promoted by high alkalinity and calcium, which stabilize weakly charged to neutral uranyl-carbonate complexes; by elevated ionic strength, which suppresses electrostatic sorption; and by connected high-permeability pathways and anisotropy, which channel flow and shorten residence times. Conversely, immobilization is favoured when oxidant depletion allows Eh to fall, enabling reduction to U(IV) coupled to Fe-mediated incorporation or precipitation; when carbonate is consumed or diluted, strengthening adsorption to Fe-(hydr)oxides; and when calcite co-precipitation occludes pores and co-traps uranyl. Each of these controls is sensitive to site-specific variables: bicarbonate and Ca concentrations, availability of reactive Fe(II), microbial activity, grain-scale surface area and mineral coatings, and hydrodynamic exchange between mobile and immobile domains. The contemporaneous presence of mobile U(IV) colloids in NOM/Fe dispersions can partially offset reductive immobilization, extending a low-level tail beyond the redox barrier.

6.3.4 Potential Receiving Environments

Downgradient receptors include adjacent potable aquifers and wellfields, springs, streams, wetlands, and hyporheic zones where plume discharge mixes with surface waters; carbonate-rich recharge (e.g., irrigated landscapes) can further mobilize U and drive young, U-affected water deeper into supply horizons. The risk pathways from ISR sites to hydraulically connected units emphasize the need to define points of compliance and exposure under aquifer-exemption regimes. In wetlands and organic-rich surface environments, cycling between U(VI) and colloidal U(IV) is plausible, with seasonal redox variability affecting accumulation and release. From a management and assessment perspective, in abandoned alkaline ISL mines, plume prognosis hinges on carbonate-calcium speciation, redox trajectories, and aquifer architecture. Effective long-term management integrates (i) thermodynamic/kinetic speciation models that include Ca-uranyl-carbonates, (ii) surface complexation/incorporation models for Fe phases, and (iii) transport models that represent heterogeneity and dual-domain exchange, all grounded by field tests that interrogate downgradient attenuation capacity and colloid dynamics.

6.3.5 *Plume Migration in Post-Alkali ISL*

Abandoned alkaline ISL uranium mines develop long-lived, chemically driven plumes controlled by residual carbonate-alkaline lixiviant, redox gradients, mineral buffering, and coupled advective-dispersive transport with kinetically limited sorption and precipitation/dissolution reactions. Predictive descriptions begin with the hydrogeochemical state left by alkaline ISL: elevated pH, high carbonate/bicarbonate concentrations, elevated Ca and Na, and residual complexing agents that stabilize uranyl in solution as carbonate complexes, such as $UO_2(CO_3)_n^{2-2n}$. These conditions favour uranium mobility because carbonate complexation increases aqueous uranium solubility and reduces adsorption to mineral surfaces. Over time, natural groundwater recharge, oxygen ingress, and microbial activity shift redox conditions, producing spatially variable zones where U(VI) can be reduced to U(IV) and immobilized as uraninite or incorporated into secondary phases and coatings on Fe-Mn oxides; conversely, re-oxidation can remobilize uranium. Accurate prediction, therefore, requires explicit representation of carbonate complexation equilibria together with redox-controlled precipitation/dissolution reactions and surface processes.

Transport is commonly described by the reactive advection-dispersion equation, with source and sink terms accounting for aqueous speciation, sorption, and mineral reactions. The mass balance for dissolved uranium species C in a porous medium can be written as

$$\frac{\partial(\phi C)}{\partial t} + \nabla \cdot (qC) - \nabla \cdot (\phi D \nabla C) = -\rho_b \frac{\partial S}{\partial t} + R_{\mathrm{aq}} + R_{\mathrm{min}}$$

where ϕ is porosity, q is Darcy flux, D is the hydrodynamic dispersion tensor, ρ_b is bulk density, S is sorbed concentration, R_{aq} represents aqueous reaction rates (including complexation and redox transformations), and R_{min} represents mineral precipitation/dissolution rates. For first-order approximations, sorption is often represented by a linear partition coefficient ,K_d with $S = K_d C$, but this simplification fails to capture the strong pH- and carbonate-dependence of uranium surface interactions; surface complexation models that explicitly represent site speciation and electrostatic effects are required to capture competitive adsorption with carbonate and other anions.

Aqueous speciation is controlled by equilibrium constants for carbonate complexation and by pH and ionic strength. Representative reactions include the formation of carbonate complexes such as

$$UO_2^{2+} + CO_3^{2-} \rightleftharpoons UO_2CO_3$$

$$UO_2^{2+} + 2CO_3^{2-} \rightleftharpoons UO_2\left(CO_3\right)_2^{2-}$$

These reactions have stability constants that strongly favour multi-carbonate complexes at high carbonate alkalinity and elevated pH. These complexes reduce the fraction of free UO_2^{2+} available for inner-sphere adsorption and increase mobility.

Redox transformations are central to immobilization and remobilization. Reduction of U(VI) to U(IV) can be represented by kinetically controlled reactions that depend on the availability of electron donors, mineral catalysts, and microbial activity. A commonly used kinetic form for biotic or abiotic reduction is

$$R_{\text{red}} = k_{\text{red}} C_{\text{U(VI)}} \frac{B}{K_B + B}$$

where k_{red} is a rate constant, $C_{\text{U(VI)}}$ is the aqueous U(VI) concentration, B is an electron donor or microbial biomass proxy, and K_B is a half-saturation constant; simpler first-order forms $R_{\text{red}} = k_{\text{red}}C_{\text{U(VI)}}$ are sometimes used for conceptual models. Oxidative dissolution of U(IV) solids follows analogous kinetics and is strongly influenced by oxygen and nitrate ingress. Coupled Fe-oxide redox cycling is important because Fe(II) produced during microbial Fe(III) reduction can abiotically reduce U(VI), while Fe(III) oxides provide sorption sites and can sequester uranium through surface complexation or co-precipitation.

Mineral precipitation and dissolution provide additional sinks and sources. Precipitation of uranyl-carbonate minerals or incorporation into secondary Fe- and Mn-oxyhydroxides can immobilize uranium, while dissolution of these phases under changing pH or redox conditions can release uranium back to solution. Mineral reaction rates are often described by transition-state theory or affinity-dependent kinetic laws of the form

$$R_{\min} = k_{\min} A_{\text{reactive}} \left(1 - \frac{Q}{K_{\text{eq}}}\right)$$

where $k_{\min}$ is a rate constant, A_{reactive} is the reactive surface area, Q is the ion activity product, and K_{eq} is the equilibrium constant for the mineral phase.

Heterogeneity and multi-scale transport processes control plume geometry and longevity. Preferential flow in high-permeability channels accelerates advective transport, carrying carbonate-stabilized uranyl complexes far from the source, while low-permeability matrix blocks act as long-term sources through diffusion and slow desorption. Diffusion-limited release from sorbed pools and from mineral coatings produces long tails in concentration-time histories. The hydrodynamic dispersion tensor D combines molecular diffusion and mechanical dispersion and is commonly parameterized by longitudinal and transverse dispersivities that scale with travel distance; however, scale-dependent dispersivity and non-Fickian transport behaviour are frequently observed in fractured or highly heterogeneous ISL settings.

Reactive transport models that couple flow, multi-component aqueous speciation, surface complexation, kinetic redox reactions, and mineral kinetics are required to capture the nonlinear feedbacks that control plume persistence and attenuation. Numerical implementation must resolve the coupling between fast equilibria (e.g., carbonate complexation) and slower kinetic processes (e.g., microbial reduction and mineral dissolution). It must represent mass transfer limitations between mobile and immobile porosities. Essential parameters for model calibration and monitoring include background carbonate alkalinity, pH, Eh/pE profiles, dissolved oxygen and nitrate, concentrations of Ca, Na, and other major cations, reactive surface area and abundance of Fe-Mn oxides, sorption parameters (surface complexation constants or K_d ranges), effective dispersivity, porosity, and hydraulic conductivity contrasts. Sensitivity and uncertainty analyses are essential because kinetic rates, reactive surface areas, and microbial activity are often poorly constrained yet exert first-order control on long-term behaviour.

In abandoned alkali ISL sites, uranium plumes reflect carbonate-stabilized aqueous speciation, redox-driven immobilization and remobilization, mineral buffering through precipitation/dissolution, and heterogeneous advection-dispersion with diffusion-limited release from low-permeability zones. The governing equations couple advection-dispersion with sorption and reaction terms, equilibrium speciation for fast complexation, and kinetic laws for redox and mineral reactions, producing a strongly coupled, nonlinear system that requires reactive-transport modelling to predict plume evolution and evaluate remediation or natural attenuation.

6.4 Uranium in Acid Leaching ISL

When the orebody contains either low-acid-consuming or alkaline minerals, acid leaching with sulfate is generally preferred [3, 24, 25]. In this case, the leach solution has a pH of 2.5–3.0, yielding uranium recovery of 70–90%. The operating costs of acid leaching are approximately half those of alkaline leaching. Accordingly, acid leaching is the most widely applied ISL technique worldwide. In Australian ISL mines, the oxidant is hydrogen peroxide, and the complexing agent is sulfuric acid, yielding uranyl sulfate [3]. In Kazakhstan, oxidants are generally not used in ISL operations, and instead, higher acid concentrations are used in the circulating solutions [3, 7].

6.4.1 Acid ISL Uranium Mining Process

Acid in situ leaching (ISL) mobilizes uranium by creating an engineered subsurface geochemical environment in which oxidants and low pH lixiviants dissolve reduced uranium minerals and maintain uranium in soluble hexavalent complexes; the process is controlled by coupled redox, complexation, sorption, and hydrodynamic transport mechanisms that determine both extraction efficiency and long-term post-abandonment behaviour.

Acid ISL is implemented by establishing an injection-production wellfield through permeable, roll-front, or tabular sandstone-hosted uranium deposits and circulating an acidic, oxidant-bearing leach solution through the ore-bearing aquifer to solubilize uranium and transport it to recovery wells. Operational lixiviants are typically sulfuric acid with an oxidant, such as dissolved oxygen, hydrogen peroxide, or ferric iron generated in situ; the acid lowers pH to promote mineral dissolution, while oxidants convert tetravalent uranium (U(IV)) in uraninite or coffinite to the more soluble uranyl ion (UO_2^{2+}). The engineered flow field is maintained by controlling injection pressures and extraction rates to create advective flow paths through the ore zone and to limit bypassing or channelling; hydraulic gradients, anisotropic permeability, and matrix diffusion into low-permeability facies govern residence times and mass transfer between mobile and immobile porosity. The wellfield hydrodynamics, therefore, set the spatial pattern of geochemical reaction fronts and the temporal evolution of aqueous composition.

Geochemically, the primary dissolution step can be represented in simplified form as the oxidation and protonation of UO_2 (uraninite):

$$UO_2(s) + \frac{1}{2}O_2 + 2H^+ \rightarrow UO_2^{2+} + H_2O$$

Once formed, the uranyl ion is stabilized by complexation with sulfate under acidic conditions, forming species such as UO_2SO_4(aq) and higher-order sulfate complexes, which increase uranium mobility and reduce adsorption to mineral surfaces. Concurrent iron and manganese redox cycling strongly influences oxidant availability and secondary mineral formation: ferrous iron released from dissolution or reductive dissolution of Fe(III) oxides can be reoxidized, producing ferric oxyhydroxysulfates (e.g., schwertmannite and jarosite) that scavenge arsenic, trace metals, and can co-precipitate or adsorb uranyl species, thereby creating spatially heterogeneous sinks and sources for uranium during and after leaching. Acid attack also dissolves carbonate and aluminosilicate phases, releasing Ca^{2+}, Mg^{2+}, and silica; in carbonate-bearing systems, carbonate complexation competes with sulfate complexation and can buffer pH, altering uranium speciation and extraction kinetics.

Sorption and ion-exchange processes on clay minerals and Fe-Mn oxyhydroxides modulate aqueous uranium concentrations. Under low pH and high sulfate, cation exchange sites are occupied by H^+ and competing cations (Na^+ and Ca^{2+}), reducing sorptive capacity for U(VI) and favouring transport. Conversely, during neutralization or natural attenuation phases, pH rebound and reductive conditions can promote U(VI) sorption, reduction to U(IV), precipitation of secondary uranium minerals such as uranophane or coffinite, or incorporation into iron oxide matrices, complicating restoration and creating long-term diffuse sources.

Hydrogeologically, the ISL process imposes transient hydraulic gradients that drive advective-dispersive transport, with longitudinal dispersion, transverse mixing, and diffusion into low-permeability strata controlling plume shape and mass recovery. Post-abandonment, residual acidity, oxidant remnants, and secondary mineral assemblages established during leaching continue to evolve as oxygen is consumed, redox fronts migrate, and groundwater chemistry re-equilibrates; these processes determine whether uranium remains mobile, is attenuated by sorption or precipitation, or is remobilized by later changes in redox or pH. Field observations and monitoring indicate pronounced spatiotemporal changes in leachate composition during and after acid ISL, underscoring the need to couple reactive transport modelling with detailed hydrogeologic characterization to predict long-term uranium behaviour in abandoned ISL sites.

Conditions that favour the application of acid leaching are therefore those that maximize uranium solubility and controllable flow while minimizing neutralizing buffers and undesirable side reactions. Practically, acid ISL is favoured in permeable, well-sorted sandstone aquifers with high transmissivity and continuous orebodies that permit uniform sweep and predictable wellfield hydraulics; low carbonate content or limited carbonate buffering capacity is advantageous because carbonate minerals consume acid and promote carbonate complexation, which can reduce sulfate-driven mobilization. A host matrix with limited reactive iron oxide coatings and moderate clay content reduces acid neutralization and excessive sorption losses, while an aquifer with a confined or semi-confined hydrogeologic setting and low vertical connectivity limits off-zone migration of acidic solutions. Groundwater with low natural reducing capacity and low dissolved organic carbon concentrations supports sustained oxidant activity and uranyl stability in solution, whereas high natural reductants or abundant sulfides favour rapid attenuation and complicate recovery.

Operationally, sites where surface water and potable groundwater are hydraulically separated or where containment and restoration infrastructure can be implemented are preferred because they reduce environmental risk and facilitate post-mining remediation. These criteria, when combined with economic considerations, define the technical envelope in which acid ISL is the most effective extraction method.

6.4.2 Acid ISL Lixiviants and Selection Factors

Acidic lixiviants for ISL are dominated by sulfuric acid combined with oxidants to produce soluble uranyl-sulfate complexes; selection of a lixiviant is governed by host-rock mineralogy, carbonate buffering, redox capacity, permeability, and operational constraints that together control dissolution kinetics, complexation equilibria, and advective-diffusive transport.

The principal lixiviant employed in acid in situ leach (ISL) uranium operations is sulfuric acid, typically delivered as dilute H_2SO_4 solutions that are augmented with an oxidant to convert U(IV) to U(VI) and to sustain oxidative dissolution of uranium-bearing phases; the canonical stoichiometry for primary oxidative dissolution of uraninite under acidic conditions can be written as:

$$UO_2\left(s\right)+½O_2+2H^+ \rightarrow UO_2^{2+}+H_2O$$

Following oxidation, uranyl forms sulfate complexes such as UO_2SO_4(aq) that dominate speciation and mobility in low-pH, high-sulfate porewaters. The efficacy of sulfuric acid arises from its ability to both supply protons that attack silicate and carbonate matrices and to generate high-sulfate activity that stabilizes uranyl in solution, thereby minimizing adsorption to mineral surfaces and enhancing sweep efficiency in wellfields.

Oxidant selection is a co-determinant of lixiviant performance because oxidant availability controls the rate of U(IV) oxidation and the persistence of mobile U(VI). Dissolved oxygen is the simplest oxidant but is limited by low solubility and slow kinetics in some systems; hydrogen peroxide provides rapid, high-potential oxidation but requires careful dosing to avoid decomposition and localized oxidant spikes; ferric iron can be used either as an injected oxidant or generated in situ by acid-mediated dissolution of Fe(III) minerals and subsequent redox cycling, producing ferric sulfate species that both oxidize U(IV) and precipitate as ferric oxyhydroxysulfates under evolving pH and Eh conditions. These oxidant pathways are accompanied by secondary reactions that can sequester uranium (co-precipitation and adsorption) or consume oxidant and acid. Accordingly, oxidant choice is made with attention to kinetics, reagent stability, and by-product formation.

Mineralogical and hydrochemical constraints strongly influence lixiviant selection. High carbonate content in the host rock imposes a significant acid demand because calcite and dolomite react rapidly with H^+ (e.g., $CaCO_3 + 2H^+ \rightarrow Ca^{2+} + CO_2 + H_2O$), buffering pH and promoting carbonate complexation of U(VI) that can either enhance or reduce extractability depending on ionic strength and Ca^{2+} activity. Where carbonate buffering is significant, non-acidic lixiviants such as bicarbonate/CO_2 systems are often considered, but in deposits where carbonate is low and acid consumption minimal, sulfuric acid remains the most cost-effective and kinetically favourable lixiviant. Clay content and iron oxide coatings also modulate acid neutralization and sorption capacity, affecting reagent consumption and uranium recovery efficiency.

Hydrogeological processes determine how a chosen lixiviant performs in situ because permeability, anisotropy, and connectivity control advective sweep, while dispersion, transverse mixing, and diffusion into the low-permeability matrix control contact time and mass transfer. Low transmissivity or high heterogeneity increases reagent residence time. It can lead to localized acid exhaustion or oxidant depletion, whereas high transmissivity favours rapid sweep but may reduce contact time and recovery of uranium from less permeable facies. Operational parameters, well spacing, injection pressure, and flow rates, are therefore optimized to balance advective transport with reaction kinetics, and post-leach rebound and attenuation are predicted by coupling reactive transport models that include acid neutralization, redox evolution, complexation equilibria, and matrix diffusion processes.

6.4.3 Uranium Behaviour in Post-acid ISL

Post-acid ISL groundwater evolution is initially dominated by low pH, high sulfate, and oxidizing conditions that stabilize U(VI) as soluble uranyl-sulfate complexes; subsequent fate is governed by pH rebound, redox evolution, sorption/desorption dynamics, secondary mineral precipitation/dissolution, and hydrogeologic transport processes that together determine whether uranium is attenuated in situ or remobilized over decadal timescales.

Immediately following cessation of acid in situ leaching, the ore-bearing aquifer typically exhibits elevated sulfate, low pH, high redox potential, and high dissolved U(VI) concentrations as a consequence of the prior circulation of sulfuric acid and oxidants. Under these conditions, uranium speciation is dominated by uranyl-sulfate complexes; the principal aqueous equilibrium can be represented as $UO_2^{2+} + SO_4^{2-} \rightleftharpoons UO_2SO_4(aq)$, with higher-order complexes forming at elevated sulfate activity. The persistence of these complexes suppresses adsorption to mineral surfaces. It enhances advective transport, so the initial post-leach plume geometry largely reflects the engineered wellfield hydraulics and the distribution of high-permeability pathways. Numerical reactive-transport studies of natural attenuation following acid ISL emphasize the coupled roles of mixing, pH buffering, adsorption, and redox reactions in controlling U(VI) decline and spatial redistribution.

As the aquifer evolves, acid neutralization reactions and consumption of oxidants drive pH upward and lower Eh in many locales, producing conditions that favour uranium attenuation. Acid neutralization is commonly controlled by carbonate minerals and silicate weathering; calcite buffering follows the stoichiometry, which both consumes protons and releases Ca^{2+} that can influence uranyl complexation equilibria. The reaction is:

$$CaCO_3 + 2H^+ \rightarrow Ca^{2+} + CO_2 + H_2O$$

The pH rebound increases the propensity for uranyl sorption to Fe- and Mn-oxyhydroxides and to clay exchange sites, and it promotes reductive pathways when electron donors are available. Microbially mediated or abiotic reduction of U(VI) to U(IV) is a key attenuation mechanism, which can be summarized as:

$$UO_2^{2+} + 2e^- \rightarrow UO_2(s)$$

The kinetics and permanence of this reduction depend on the availability of labile organic carbon, reactive Fe(III) phases and sulfide, and on the stability of the precipitated U(IV) phase (e.g., uraninite versus more disordered U(IV) phases).

Secondary mineral formation during and after leaching creates both sinks and potential long-term sources. Ferric sulfate minerals such as schwertmannite and jarosite precipitate under acidic, sulfate-rich conditions and can incorporate or adsorb uranyl species; upon neutralization, these phases may transform to goethite or haematite, releasing or re-sequestering associated trace metals and uranium. Carbonate and silicate secondary phases (e.g., uranophane and autunite-group minerals) may precipitate under higher pH and elevated Ca^{2+} or PO_4^{3-} conditions, immobilizing uranium but potentially being metastable under future geochemical perturbations. The interplay among sorption, co-precipitation, and incorporation into secondary minerals produces heterogeneous, kinetically controlled attenuation, complicating the prediction of long-term mobility.

Hydrogeological controls remain decisive: advective transport, longitudinal dispersion, transverse mixing, and matrix diffusion into low-permeability facies set contact times for geochemical reactions and control plume persistence. Heterogeneity and vertical connectivity determine whether neutralizing fronts and reducing zones can migrate through the ore zone or whether isolated pockets of acidic, oxidizing water persist as long-term sources. Residual saturation in the vadose zone, wellbore integrity, and legacy reagent mass in sorbed or mineral-bound forms can sustain remobilization events during seasonal recharge or anthropogenic perturbation. Effective prediction of post-acid ISL uranium fate, therefore, requires coupled reactive transport modelling that resolves mineral kinetics, redox microenvironments, and multi-scale hydrogeologic heterogeneity to assess both natural attenuation trajectories and the risk of future remobilization.

6.4.4 Plume Migration in Post-acid ISL

Post-acid ISL plumes are distinguished from alkali ISL plumes by persistent low pH and high sulfate that stabilize uranyl-sulfate complexes, by acid-driven mobilization and redistribution of iron, trace metals, and carbonate buffers, and by acid-induced changes to porosity and permeability that alter advective and diffusive transport; these processes produce distinct attenuation pathways, secondary mineral sinks, and rebound risks not present in alkaline systems.

Following acid ISL, plume migration is governed by processes that are chemically and hydrogeologically distinct from those in alkali ISL. Under acidic, sulfate-rich conditions, the dominant aqueous uranium species are uranyl-sulfate complexes (e.g., $UO_2SO_4(aq)$ and polymeric sulfate complexes), which confer lower sorption affinity to mineral surfaces than the uranyl-carbonate complexes typical of alkaline systems; consequently, retardation factors are generally lower in acid plumes, producing more advectively mobile uranium under comparable hydraulic gradients. The acid environment also promotes extensive dissolution of Fe-oxyhydroxides and aluminosilicates, releasing Fe^{2+}, Al^{3+}, $SiO_2(aq)$, and associated trace elements. Oxidation of released ferrous iron regenerates ferric species that precipitate as ferric-sulfate phases (e.g., schwertmannite and jarosite) according to reactions such as the generalized formation of jarosite and schwertmannite-type precipitation that sequesters sulfate and iron while creating metastable solids that can later transform and release sorbed constituents during neutralization. The reaction that forms jarosite is:

$$K^+ + 3Fe^{3+} + 2SO_4^{2-} + 6H_2O \rightarrow KFe_3\left(SO_4\right)_2\left(OH\right)_6 + 6H^+$$

These acid-specific secondary minerals act as kinetically controlled sinks for uranium and co-contaminants during the acidic phase; however, they are prone to transformation upon pH rebound, potentially releasing previously sequestered uranium.

An acid attack on carbonate phases consumes proton capacity and produces elevated Ca^{2+} and dissolved inorganic carbon, altering complexation equilibria via the reaction:

$$CaCO_3 + 2H^+ \rightarrow Ca^{2+} + CO_2 + H_2O$$

Unlike alkaline ISL, where carbonate complexation dominates and can enhance mobility, in acid systems, carbonate buffering can locally neutralize acidity and shift speciation towards mixed carbonate-sulfate complexes; however, the net effect is spatial heterogeneity in uranium speciation and mobility that depends on the distribution of carbonate lenses and acid consumption rates.

Hydrogeologically, acid leach solutions can increase effective porosity and permeability by dissolving cementing phases and mobilizing fine particles, thereby altering preferential flow paths and increasing longitudinal connectivity relative to pre-mining conditions. Enhanced permeability can accelerate the advective transport of uranium, while increased matrix porosity can both increase matrix diffusion uptake and create new low-permeability storage domains for sorbed or precipitated uranium. Matrix diffusion into clays and siltstones under low pH is complicated by proton exchange on clay surfaces, which reduces cation exchange capacity for U(VI) and modifies diffusion-retardation coupling compared with alkaline systems.

Redox evolution in post-acid plumes is controlled by the consumption of oxidants and by the availability of organic carbon and reduced minerals; microbial reduction of U(VI) to U(IV) can occur where pH rebounds, and electron donors are present, but the kinetics and permanence of reduction are often limited by the instability of U(IV) phases formed within acid-altered matrices. Seasonal recharge, well integrity issues, or anthropogenic perturbations that introduce alkalinity or reductants can trigger transformation of ferric-sulfate phases and remobilization of uranium, producing rebound events that are more pronounced and chemically complex than those observed in alkali ISL contexts. Predictive modelling of acid ISL plume migration, therefore, requires coupled reactive transport formulations that explicitly represent acid-driven mineral dissolution/precipitation kinetics, sulfate-iron cycling, evolving porosity/permeability, and multi-scale diffusion processes unique to acidic environments

6.5 Perspectives

6.5.1 Important Aspect of ISL

The ISL process is characterized by three key aspects. The most important aspect is its potential to produce uranium at a lower cost from low-grade ore, with lower capital investment and operational expenses than conventional mining methods. The second key point is that the ISL method is applicable for certain types of uranium deposits, mainly roll-front deposits in permeable sandstone that is saturated with water and functions as an aquifer. As a result, ISL is very effective under specific geological and hydrogeological conditions but is not practical or feasible for most other types of uranium deposits, which require more costly and environmentally disruptive traditional mining methods to access much of the world's uranium resources.

The third significant aspect is the inherent difficulty in predicting the performance of an ISL project. There have been substantially more unsuccessful ISL attempts than successful ones. This method involves competing chemical reactions underground within a natural uranium deposit, leading to significant variations in chemical and hydrogeological parameters that affect flow rates, reaction rates, and uranium production. These factors can differ significantly, not only from one ore deposit to another but also within the same deposit, reflecting spatial heterogeneity and evolving geochemical conditions during operation.

Other aspects include that ISL mining inherently releases less radiation and fewer contaminants than conventional mining and milling because ore is not excavated, transported, or stored. ISL produces no waste rock or mill tailings, generates no windblown particulates, and can often achieve zero discharge of process water. Consequently, potential radiation doses at ISL operations are generally lower than those from conventional uranium mining and milling, since many exposure pathways present in conventional operations are absent.

6.5.2 Post-operational Concerns

The first concern in post-operational ISL mines is the recovery of lixiviants, the leaching solutions injected underground. A substantial amount of these leaching liquids often remains in the ground long after mining activities have ended. At present, there are no effective large-scale methods for removing residual leachate from the deposit or entirely arresting the ongoing leaching of uranium and other contaminants.

Moreover, no lixiviant is completely inert towards the minerals typically found in uranium deposits. This results in various chemical interactions among the lixiviants, uranium, and associated secondary minerals that persist long after mining operations cease. As a result, uranium from the host rock is not the only major contaminant; other metals, metalloids, and co-released elements are also present. These may include radioactive isotopes and progeny of uranium, thorium, radium, and radon, as well as non-radioactive elements, such as arsenic, vanadium, zinc, selenium, and molybdenum. These contaminants can remain underground as long as the leaching solution persists. Because leaching occurs within water-bearing formations, residual solutions also pose a continuing risk to surrounding groundwater systems.

Following the cessation of in situ leach (ISL) mining operations, several interrelated release pathways and environmental risks exist. The primary concern is contamination of aquifers. Residual leaching solutions and mobilized constituents can persist in the ore zone and surrounding strata, allowing dissolved metals and process chemicals, including uranium, to migrate into groundwater through advection, diffusion, or preferential flow along wellbores and fractures. Long-term plume persistence is possible where geochemical conditions slow attenuation, making groundwater quality restoration and monitoring essential components of closure.

A second major concern is the release of radioactive substances from residual leach fluids and disturbed materials. Radionuclides can stay dissolved or sorbed in the subsurface and may be released into the vadose zone or shallow aquifers as fluids move or evaporate; radon gas can emanate from contaminated sediments and migrate to the surface or into buildings if pathways are available. These radiological pathways require evaluation of both dissolved and gaseous forms, as well as potential exposure routes to receptors after abandonment. Beyond uranium, ISL operations can mobilize other dissolved metals and metalloids, such as vanadium, selenium, and arsenic, which may become more mobile under the oxidizing, acidic, or complexing conditions created by leaching solutions. Changes in redox state, pH, or competing-ion chemistry during and after operations can remobilize previously immobile species, complicating natural attenuation and remediation efforts and potentially creating long-term water-quality liabilities.

6.5.3 Real Environmental Footprint

ISL mining is often portrayed as a low-impact, environmentally benign method of uranium extraction due to its subterranean nature. This perception is reinforced by the absence of large-scale surface disruption and the relatively limited generation of airborne emissions and solid waste. However, this image can be misleading. The true environmental risk lies not in what is visible aboveground, but in the persistent, potentially hazardous contamination that can occur underground, particularly in groundwater systems.

While ISL operations are designed with safeguards such as monitoring wells and controlled lixiviant injection rates to prevent contamination from migrating beyond the orebody, these measures are not foolproof. Lixiviants can remain active underground long after injection ceases. If not fully recovered, these chemicals may migrate with groundwater flow, forming contaminant plumes that extend beyond the intended recovery zone. This plume migration poses a significant threat to surrounding aquifers and ecosystems, especially if excursions are not promptly detected or effectively remediated.

ISL facilities attempt to mitigate these risks through rigorous monitoring and restoration protocols. Monitoring wells are strategically placed beyond the wellfield to detect excursions early, and injection volumes are carefully managed to maintain a slight hydraulic inward gradient that draws fluids towards the recovery zone. Nevertheless, the effectiveness of these controls depends heavily on timely detection, accurate hydrogeological modelling, and consistent operational discipline. Even with these precautions, the long-term behaviour of residual lixiviants in complex subsurface environments remains uncertain.

Upon closure, ISL sites are required to restore groundwater to pre-mining baseline conditions, a process that involves extensive sampling and analysis of a wide range of water quality parameters. These include major ions, trace elements, radionuclides, and general indicators, such as pH and conductivity. Restoration efforts are guided by baseline data collected before operations begin. However, achieving full restoration is technically challenging and often protracted, with no guarantee that all contaminants can be removed or neutralized.

Thus, while ISL may appear clean and contained due to its underground footprint, the potential for long-term groundwater contamination reveals a more complex and concerning environmental reality. The persistence of lixiviants and the risk of contaminant plume migration underscore the need for cautious oversight and transparent evaluation of ISL's true environmental impact.

6.5.4 Potential Trends

ISL is expected to expand towards deeper orebodies as demand for cost-effective uranium extraction grows. ISL remains attractive for low-grade deposits due to its lower capital and operating costs compared to conventional mining, and several new projects are under development globally. However, technical and economic challenges persist for deep ISL operations beyond approximately 500–600 m. While commercial success has been achieved at depths of ~150 to 200 m, pilot tests at 600 m have demonstrated technical feasibility but limited economic viability. Key constraints include well deviation control, high drilling and completion costs, and water quality issues associated with saline aquifers at greater depths. Significant technological innovation, such as advanced directional drilling, improved well integrity, and enhanced leaching chemistry, will be required before deep ISL becomes a reliable, low-cost option [6, 43, 44].

Globally, ISL has proven competitive with higher-grade conventional mining when applied to suitable sandstone-hosted roll-front deposits. These deposits dominate ISL operations because they offer the necessary permeability and confinement for controlled leaching. Table 6.3 summarizes major ISL, both active and closed.

Table 6.3 Global distribution of operating and closed uranium ISL mines by 2025

Country	Operation location	Operation period	Total production
USA	Wyoming; Texas, Nebraska; new projects proposed in Texas, Wyoming, New Mexico	Operational (data point cited for 1996)	1684 t U (1996 ISL production)
Eastern Germany	Königstein near Dresden	Operated until the end of 1990 (converted underground → ISL)	18,000 t U total; ~5400 t U from ISL (~30%)
Czech Republic	Stráž pod Ralskem, North Bohemia	Large-scale ISL production was reported in 1994	13,835 t U (1994 production)
Bulgaria	Multiple ISL fields (15 fields; total ISL area ~6 km^2)	ISL from 1979; intensive through early 1990s	5175 t U (total ISL production to 1994)
Ukraine	Devladove; Bratske; Safonovskoye	ISL used 1966–1983	Not specified in the source
Russia	Dalmatovkoye (Western Siberia)	Proposed	Not specified/proposed only
Kazakhstan	Kandjugan; Uvanas; Mynkuduk; Karamurun; Muyunkum and Inkay planned	Operational by 1994; new projects planned	19,961 t U (total ISL to 1994); 1580 t U (1994 ISL production)
Uzbekistan	Uchkuduk; Zarafabad; Nurabad (area ~13 km^2)	Since 1995, all production from ISL	~3050 t U per year (since 1995)
China	Tengchong; Yining	ISL in use (period not specified)	Not specified in the source
Australia	Beverley; honeymoon (South Australia) proposed/advanced projects	New projects proposed; Beverley developed later	Not specified in the source

However, their scattered distribution limits large-scale production growth. Future expansion will depend on the availability of amenable orebodies. Countries with substantial reserves suitable for ISL include Australia, South Africa, Gabon, Zambia, Malawi, and Brazil, alongside major producers such as Kazakhstan and Uzbekistan. Australia alone holds over 3.6 million tonnes of uranium resources, while Brazil and South Africa each exceed 900,000 tonnes, offering significant potential for ISL deployment.

Market forecasts indicate that ISL will remain the dominant uranium extraction method, accounting for over half of global production, driven by its environmental advantages and cost efficiency. The global ISL uranium market is projected to grow steadily through 2030, supported by nuclear energy expansion and technological improvements in leaching systems and monitoring frameworks.

6.6 Conclusion Perspectives

As interest in small modular reactors grows and clean-energy strategies diversify, ISL is poised to re-emerge as a pragmatic route to secure uranium supplies. ISL operations typically have lower capital requirements and smaller operating footprints, making low-grade deposits economically viable and shortening the timeline from discovery to production, which is attractive where rapid, modular nuclear deployment is being considered. This chapter offers a high-level generalization of the issues that follow ISL operations: it synthesizes insights from field studies, laboratory experiments, and computational modelling, but it is necessarily framed by limited field data drawn from a relatively small number of ISL sites. Consequently, the actual behaviour and fate of uranium and co-mobilized contaminants in abandoned ISL mines are highly site-specific, governed by local geology, climate, and hydrology and by the particular reagents and operational histories used at each site.

Because lixiviants introduced during ISL are difficult to recover completely, residual reagents, dissolved uranium, and associated metals can remain mobile in the subsurface after closure. Post-closure groundwater flow, heterogeneous permeability, and evolving redox conditions can transport these constituents away from the wellfield as plumes, with transport and attenuation controlled by complex, spatially and temporally varying hydrogeochemical processes. The challenge of predicting plume behaviour and of recovering lixiviants therefore represents a central risk, as post-mining mobility can create long-term water-quality liabilities that undermine the short-term economic advantages of ISL.

Given these realities, ISL project planning must treat post-mining uranium behaviour as a core design criterion rather than an afterthought. Robust site characterization, conservative pilot testing, adaptive wellfield design, and long-term monitoring are essential to constrain uncertainty. Closure strategies should prioritize containment, active or passive treatment where feasible, and legally enforceable commitments to long-term stewardship. Regulators, operators, and communities must recognize that economic efficiency does not obviate environmental responsibility; credible expansion of ISL to support modular reactor deployment will depend on integrating rigorous science, transparent risk assessment, and enforceable closure and remediation plans. Only by embedding post-mining behaviour into the earliest stages of project design can ISL contribute to clean-energy goals without leaving unacceptable legacy contamination.

References

1. Hall, S. (2009). *Groundwater restoration at uranium in-situ recovery mines, South Texas Coastal Plain*. U.S. Department of the Interior & U.S. Geological Survey.
2. IAEA. (2001). *Manual of acid in situ leach uranium mining technology. Report No. IAEA-TECDOC-1239, 294*. International Atomic Energy Agency, Vienna.
3. IAEA. (2016). *In Situ Leach Uranium Mining: An Overview of Operations. Report No. NF-T-1.4, 60*. International Atomic Energy Agency, Vienna.
4. IAEA. (1989). *In situ leaching of uranium: Technical, environmental and economic aspects*. International Atomic Energy Agency.
5. Khan, Y., Shah, S. S., & Siddiq, M. (2012). Selection of Lixiviant system for the alkaline in-situ leaching of uranium from an Arkosic type of sandstone and measuring the dissolution behaviour of some metals and non-metals. *Journal of the Chemical Society of Pakistan, 34*, 826–840.
6. Li, G., & Yao, J. (2024). A review of In Situ Leaching (ISL) for uranium mining. *Mining, 4*, 120–148.
7. NEA. (2014). *Uranium 2014: Resources, production and demand. Report No. NEA No. 7209, 508*. OECD Nuclear Energy Agency (NEA).
8. Seidel, D. C. (1981). Extracting uranium from its ores. *IAEA Bulletin, 23*, 24–28.
9. Committee on Uranium Mining in Virginia, & Committee on Earth Resources National Research Council. (2012). *Uranium mining in Virginia: Scientific, technical, environmental, human health and safety, and regulatory aspects of uranium mining and processing in Virginia. 342*. National Academy of Sciences.
10. IAEA. (2001). *Vol. IAEA-TECDOC-1239 (ed International Atomic Energy Agency Nuclear Fuel Cycle and Materials Section)*. International Atomic Energy Agency.
11. IAEA. (1989). *Vol. IAEA-TECDOC-1239 (ed International Atomic Energy Agency Nuclear Fuel Cycle and Materials Section)*. International Atomic Energy Agency.
12. . (2016). IAEA. In *Situ leach uranium mining: An overview of operations. 60*. International Atomic Energy Agency.
13. Ruzicka, V. (1993). Vein uranium deposits. *Ore Geology Reviews, 8*, 247–276. https://doi.org/10.1016/0169-1368(93)90019-U
14. Ward, J. R. (1983). Well design and construction for in situ leach uranium extraction. *Groundwater Monitoring & Remediation, 3*, 79–85. https://doi.org/10.1111/j.1745-6592.1983.tb00865.x
15. Qiu, W., et al. (2025). A deep-learning-based multiobjective optimization for the design of in-situ uranium leaching system under multiple uncertainties. *Journal of Hydrology, 651*, 132576. https://doi.org/10.1016/j.jhydrol.2024.132576
16. Venter, R. & Boylett, M. 2009. The evaluation of various oxidants used in acid leaching of uranium. *Hydrometallurgy Conference 2009*, The Southern African Institute of Mining and Metallurgy, 2009.
17. Ho, E., & Ring, R. *ALTA 2007 uranium conference*. ALTA Metallurgical Services.
18. Chen, S., et al. (2022). Weak acid leaching of uranium ore from a high carbonate uranium deposit. *Journal of Radioanalytical and Nuclear Chemistry, 331*, 2583–2596. https://doi.org/10.1007/s10967-022-08323-z

19. Liu, X., et al. (2025). Spatiotemporal evolutionary patterns of mineral dissolution and precipitation (formation) amount in acid in situ leaching of uranium. *Journal of Radioanalytical and Nuclear Chemistry, 334*, 2949–2962. https://doi.org/10.1007/s10967-025-10005-5
20. Tokunaga, T. K., Kim, Y., & Wan, J. (2009). Potential remediation approach for uranium-contaminated groundwaters through potassium uranyl vanadate precipitation. *Environmental Science & Technology, 43*, 5467–5471. https://doi.org/10.1021/es900619s
21. IAEA-TECDOC-492. *Proceedings of a technical committee meeting organized by The International Atomic Energy Agency.* International Atomic Energy Agency.
22. Coral, T., et al. (2018). Microbial communities associated with uranium in-situ recovery mining process are related to acid mine drainage assemblages. *Science of the Total Environment, 628–629*, 26–35. https://doi.org/10.1016/j.scitotenv.2018.01.321
23. Mkandawire, M. (2013). Biogeochemical behaviour and bioremediation of uranium in waters of abandoned mines. *Environmental Science and Pollution Research, 20*, 7740–7767. https://doi.org/10.1007/s11356-013-1486-3
24. NEA. (2014). In Nuclear Energy Agency (Ed.), *Organisation for Economic Co-operation and Development* (Vol. NEA No. 7062). OECD.
25. Senes, C. L. (2008). (ed Alberta environment) 57. .
26. IAEA. (2001). In International Atomic Energy Agency (Ed.), *International Atomic Energy Agency* (Vol. IAEA-TECDOC-1239). International Atomic Energy Agency.
27. Liao, W., Que, W., Wang, L. & Du, Z. (2020).
28. Shen, N., Li, J., Guo, Y., & Li, X. (2020). Thermodynamic modeling of in situ leaching of sandstone-type uranium minerals. *Journal of Chemical & Engineering Data, 65*, 2017–2031. https://doi.org/10.1021/acs.jced.9b01152
29. Tweeton, D. R., & Peterson, K. A. (1981). *Selection of lixiviants for in situ uranium leaching. Information circular*. Bureau of Mines/Twin Cities Research Center.
30. EPA. Aquifer Restoration after Uranium Recovery: Evaluation of aquifer restoration at sample uranium in-situ recovery sites. 2 (U.S. Environmental Protection Agency, 2017).
31. Davis, J. A., & Curtis, G. P.U. S. Geological Survey (Ed.). *U.S. Geological Survey* (Vol. NUREG/CR-6870). U.S. Geological Survey, Menlo Park, CA.
32. Stewart, B. D., Mayes, M. A., & Fendorf, S. (2010). Impact of uranyl−calcium−Carbonato complexes on uranium(VI) adsorption to synthetic and natural sediments. *Environmental Science & Technology, 44*, 928–934. https://doi.org/10.1021/es902194x
33. Bernhard, G., et al. (2001). Uranyl(VI) carbonate complex formation: Validation of the Ca2UO2(CO3)3(aq.) species. *Radiochimica Acta, 89*, 511–518.
34. Um, W., Serne, R. J., Brown, C. F., & Rod, K. A. (2008). Uranium(VI) sorption on iron oxides in Hanford Site sediment: Application of a surface complexation model. *Applied Geochemistry, 23*, 2649–2657. https://doi.org/10.1016/j.apgeochem.2008.05.013
35. Payne, T. E., & Waite, T. D. (2022). Uranium adsorption—A review of progress from qualitative understanding to advanced model development. *Radiochimica Acta, 110*, 549–559. https://doi.org/10.1515/ract-2022-0003
36. Massey, M. S., et al. (2014). Competing retention pathways of uranium upon reaction with Fe(II). *Geochimica et Cosmochimica Acta, 142*, 166–185. https://doi.org/10.1016/j.gca.2014.07.016
37. Kalintsev, A., et al. (2021). Uranium carbonate complexes demonstrate drastic decrease in stability at elevated temperatures. *Communications Chemistry, 4*, 120. https://doi.org/10.1038/s42004-021-00558-3
38. Wang, Y., et al. (2013). Mobile uranium(IV)-bearing colloids in a mining-impacted wetland. *Nature Communications, 4*, 2942. https://doi.org/10.1038/ncomms3942
39. Dublet, G., et al. (2019). Colloidal size and redox state of uranium species in the Porewater of a Pristine Mountain wetland. *Environmental Science & Technology, 53*, 9361–9369. https://doi.org/10.1021/acs.est.9b01417

40. Deb, D., & Chakma, S. (2023). Colloid and colloid-facilitated contaminant transport in subsurface ecosystem, a concise review. *International Journal of Environmental Science and Technology, 20*, 6955–6988. https://doi.org/10.1007/s13762-022-04201-z
41. Reeder, R. J., et al. (2001). Coprecipitation of uranium(VI) with calcite: XAFS, micro-XAS, and luminescence characterization. *Geochimica et Cosmochimica Acta, 65*, 3491–3503. https://doi.org/10.1016/S0016-7037(01)00647-0
42. Kitano, Y., & Oomori, T. (1971). The coprecipitation of uranium with calcium carbonate. *Journal of the Oceanographical Society of Japan, 27*, 34–42. https://doi.org/10.1007/BF02109313
43. Hidayat, M. (2025). In-situ leach uranium mining: Innovative extraction transforming global production. *Discovery Alert*.
44. Estay, H., Díaz-Quezada, S., Arancibia, E., & Vargas, T. (2023). Economic assessment of an in situ leaching operation with Ore preconditioning using sublevel stoping techniques. *Mining, Metallurgy & Exploration, 40*, 493–504. https://doi.org/10.1007/s42461-023-00736-y

Chapter 7
Uranium and Contaminant Dispersion from Abandoned Mining Sites

Abstract This chapter presents a comprehensive framework for predicting the dispersion of uranium, co-occurring metals, and radionuclides from abandoned mining and processing sites into surface and groundwater systems. It reviews the key geochemical and hydrogeological controls governing contaminant mobility, including advection, dispersion, sorption, redox transformations, and mineral interactions. Conceptual and numerical models, both deterministic and stochastic, are developed to support site-specific assessments of fate and transport. Challenges associated with reactive-transport modelling in heterogeneous media, limitations in geochemical databases, and the importance of calibration, validation, and uncertainty analysis are examined in detail. The insights provided here inform monitoring strategies, risk assessment, and remediation planning for uranium-contaminated environments. Portions of this chapter overlap with material published in *Nano and Bio-Based Technologies for Wastewater Treatment: Prediction and Control Tools for the Dispersion of Pollutants in the Environment*, in the chapter "Geochemical Prediction of Metal Dispersion in Surface and Groundwater Systems", and are reproduced with kind permission from John Wiley and Sons (Mkandawire, Nano and Bio-Based Technologies for Wastewater Treatment: Prediction and Control Tools for the Dispersion of Pollutants in the Environment, Wiley-Scrivener, p. 504, 2019).

Keywords Contamination plumes · Phreatic processes · Vadose processes · Geochemical prediction · Reactive transport · Mine geochemistry · Numerical modelling · Conceptual models · Abandoned mines

7.1 Introduction

The dispersion of uranium, associated metals, metalloids, and radionuclides in aquatic systems, particularly from abandoned uranium mining and milling sites, poses a complex challenge for environmental geochemistry and hydrogeological modelling. These legacy sites act as persistent sources of contamination, releasing uranium (U), thorium (Th), radium (Ra), and other decay-series radionuclides,

M. Mkandawire, *Fate and Effects of Uranium in Abandoned Uranium Mines*,
https://doi.org/10.1007/978-3-032-10447-2_7

along with trace metals, such as arsenic (As), selenium (Se), and lead (Pb), into surrounding surface water and groundwater systems. The mobilization and transport of these contaminants are governed by a combination of physical processes (i.e., advection, dispersion, and diffusion) and geochemical interactions, including redox transformations, aqueous complexation, sorption-desorption, and mineral precipitation-dissolution dynamics.

In abandoned uranium mining sites, the fate and transport of uranium and its daughter products are highly sensitive to site-specific hydrogeochemical conditions, including pH, redox potential (Eh), ionic strength, and the presence of complexing ligands, such as carbonate, phosphate, and sulfate. These parameters influence the speciation, solubility, and mobility of radionuclides, often resulting in spatially and temporally variable contaminant plumes. Furthermore, the heterogeneity of aquifer materials, preferential flow paths, and fluctuating hydrological regimes, exacerbated by climatic variability, complicates the prediction of contaminant dispersion.

Given the long half-lives and radiotoxicity of uranium-series isotopes, accurate prediction of their dispersion is critical for long-term risk assessment, regulatory compliance, and the design of monitoring and remediation strategies. Numerical models that couple hydrological transport with geochemical reactions are essential tools for simulating the evolution of contaminant plumes in both saturated and unsaturated zones. However, these models must contend with uncertainties in input parameters, limitations in thermodynamic and kinetic databases, and the computational complexity of simulating reactive transport in heterogeneous media.

This chapter discusses the fundamental principles and methodologies for designing effective geochemical prediction models for the dispersion of uranium, metals, and radionuclides in both surface and groundwater systems. It outlines the conceptual and mathematical frameworks, modelling approaches (deterministic vs. stochastic; steady-state vs. transient), and numerical techniques (e.g., finite-difference and finite-element methods) necessary to simulate uranium fate and transport. Emphasis is placed on integrating site-specific data, addressing model limitations, and applying predictive tools to inform environmental management and remediation of uranium-contaminated sites.

7.2 Contamination Plume Sources

7.2.1 Fundamental Processes

All metals are always in a dynamic cycle. Natural systems maintain the metal geochemical cycle in equilibrium, which controls the accumulation of metals in the environment or within a single ecological range. However, their normal geochemical cycle is altered or accelerated by mining and ore processing [1]. Mining and processing metal ores are significant sources of metal contaminants in water systems. Although some mine water from abandoned uranium mines may contain relatively low amounts of metals and radionuclides, long-term loading can eventually result in radionuclide and metal accumulation in the receiving aquatic environments. The most common metal and metalloid pollutants in abandoned uranium mines are arsenic, cadmium, chromium, copper, nickel, lead, and mercury. Airborne metals include stack or duct emissions of air, gas, or vapour streams, as well as fugitive emissions such as dust from storage areas or waste piles. Metals from airborne sources are generally released as particulates contained in the gas stream. Some metals, such as Cd and Pb, as well as metalloid As, can also volatilize during high-temperature processing.

7.2.2 *Source Types*

Uranium and most radionuclides and related metals are very soluble in acid solutions, which can disperse them into the receiving environment, including surface and groundwater. In abandoned uranium mines, these contaminants can enter the aquatic system from (a) point sources (i.e., localized pollution), where pollutants come from single, identifiable sources (e.g., overflow from a flooded mine shaft, leakage from a yellowcake storage site or waste rock pile) or (b) nonpoint sources, where pollutants come from dispersed (and often difficult to identify) sources.

Nonpoint sources are not as easily quantifiable or localizable; hence, it is challenging to develop effective geochemical methods for nonpoint-source pollutants. Thus, geochemical prediction of metal dispersion involves considerable uncertainty, especially from area-wide impacts from nonpoint sources such as mine wastes and spills during the transportation of mine material. In addition to uncertainties inherent in measuring water quality, there are uncertainties in models used to predict the dispersion of uranium and related metals, especially in their transport through different media. However, pollutant loadings from nonpoint sources in abandoned uranium mines can be much higher than those from point sources.

7.3 Uranium Dispersion Mechanism and Processes

The metal dispersion process in surfaces and subsurfaces is zone-related and complementary between phreatic and vadose zones. The movement of metal solutes and particles can be against gravity; therefore, a medium (water, air, and mineral particles) and a force (concentration, electrical, temperature, or pressure gradients) are necessary to drive the transfer via diffusion or advection [2].

7.3.1 Phreatic Process

Processes in the phreatic zone are underpinned by groundwater characteristics, including hydraulic gradients (i.e., water flow), convection, dilatancy, bubbles, diffusion, and electromigration [3]. A hydraulic gradient leads to water flow, which can transport and disperse dissolved metals in the groundwater or from the ground or solid sources to the surface water bodies [4]. Subsurface metal-contaminated water can be transferred to the surface via faults due to seismic or dilatancy pumping. However, this mechanism is limited to low-rainfall and neo-tectonic areas with regular seismic activity [2, 4]. Another scenario is where an oxidizing sulfide orebody acts as a point or diffusive heat source, resulting in buoyancy-driven currents and convective currents, potentially inducing solute migration in a particular direction. Metal dispersion can also occur through electrochemical processes involving the movement of charged ions within an electrical field. Electrochemical forces arise due to spontaneous potentials around an orebody and result in excess cation concentrations at the sulfide ore's upper edges with subsequent diffusion of cations along a concentration gradient.

The presence of redox anisotropy between reducing conditions (e.g., caused by a buried sulfide body) and oxidizing conditions in the water table, resulting in a redox gradient, induces the migration of reduced metal species and their subsequent oxidation. Further, migration of microscopic gas bubbles (dominantly composed of carrier gases—CO_2), formed within the groundwater due to overpressure or exsolution and released from the water table, can transport uranium and some specific metals (e.g., Cu, Zn, Pb, Hg, and actinides). Ultra-fine particles (e.g., clays, oxides, and microorganisms) can attach to the surfaces of gas bubbles, especially if the bubbles have an organic coating acquired from trace organics in groundwater [3]. All the phreatic processes are supported by groundwater and are limited by the water table's height, except that of seismic pumping [2, 3].

7.3.2 Vadose Processes

The vadose zone is the shallow zone of unsaturated porous media, roughly between the land surface and groundwater. This zone is responsible for metal dispersion, mainly through capillary migration, gaseous transport, and biological transfer mechanisms. Capillary or tension forces at and above the water table can induce solute migration, depending on the texture of the material, aquifer grain size, and evaporation rates, especially in finer-grained materials. Further, evaporation in the root zone can create suction pressures that would transfer a thin film of water and dissolved metals [5, 6].

Gases are responsible for dispersing possibly volatile metals like Hg, I, As, and Se. Metallic uranium is not considered volatile, but uranium hexafluoride (UF_6) is a volatile, white solid used in uranium enrichment. The diffusion of gases is a viable transport mechanism for volatile metals, but the temporal and geochemical stability of individual gases limits it. When subsurface barometric pressure changes significantly, atmospheric pumping occurs, driving rapid air migration through pores and conduits. Atmospheric pumping is restricted to fractured media. Most of the gases may be formed during weathering. The heat from oxidizing sulfides can create a convective gas flow in which oxygen diffuses into the oxidizing orebody and is released by advection.

The other critical vadose zone processes are biological, namely vegetation transfer, plant physiological uptake, and bioturbation. Vegetation can take up metals from the subsurface and release them to the surface via litter. Some plants have evolved novel methods to biotransform, store, and release toxic metals to the surface. Additionally, hydraulic lift, redistribution of deeper water acquired by sinker roots to near-surface soil horizons used by laterals, can rapidly transfer water and possibly metals within the overburden. Metals can also be dispersed through bioturbation by soil fauna, potentially transferring anomalous material from deep within the biosphere beyond the biomantle [7, 8]. The microbial metabolism affects the kinetics of many hydrochemical processes, especially sulfide oxidation and other redox transfers, and impacts the generation of gases [2, 6].

7.4 Fundamentals of Geochemical Prediction

7.4.1 *Basic Principles*

Predicting metal dispersion in aqueous systems, such as surface water and groundwater, at a given point in time and space requires a model to simulate the arrival concentration of a metal at a downgradient receptor from the source. All geochemical predictions of metal dispersion use models based upon mass conservation (i.e., mass balance), which applies the first law of thermodynamics (Fig. 7.1). Mass is neither created nor destroyed in the system but transferred between solid, aqueous, and gaseous phases. The input equals the output unless mass is stored or converted to different phases within the same system. Thus, most of these prediction models

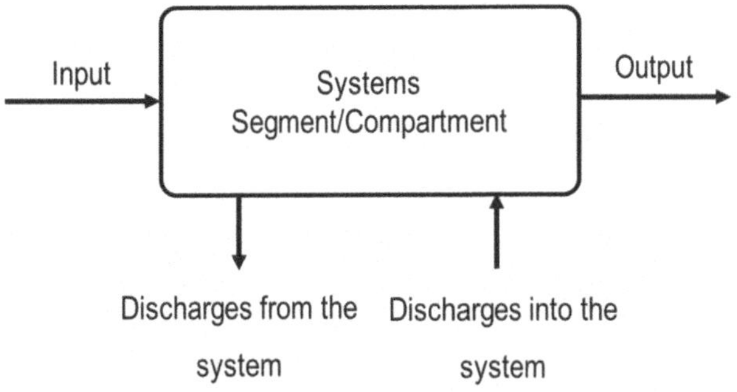

Fig. 7.1 Conceptual mass balance system used as a basis for the development of geochemical predictions of uranium and metal dispersions in surface and groundwater segments

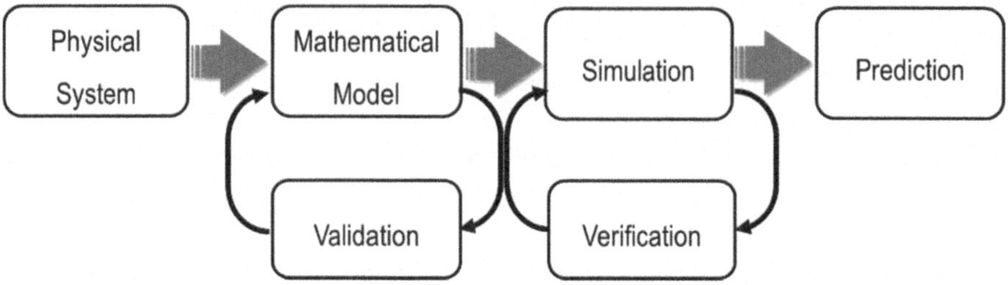

Fig. 7.2 General elements in the development of prediction modelling

simulate the behaviour of soluble metal concentrations in a specific system segment (e.g., ground or surface water) over a discrete period [9–11]. The components of the mass balance for a segment include: (a) changes by transport into and out of the segment; (b) changes by physical or chemical processes occurring within the mine system; and (c) changes by sources/discharges to or from the segment.

The model development progression is illustrated in Fig. 7.2. These models can generally be divided into two distinct categories based on how they incorporate transport processes. Models that do not consider transport processes are called geochemical reaction models or batch models. Models that consider transport processes and geochemical reactions are called coupled transport and reaction models [12, 13]. The geochemical reactions between dissolved metal ions and the minerals composing the host aquifer affect the behaviour of the metal contaminant of concern. Geochemical modelling involves combining knowledge of chemical components' reactions with the components that describe the hydrogeological conditions. The dispersion of metals in water is affected by several factors, including:

(a) Chemical reactions which determine the partitioning of contaminants among different phases and species.
(b) Advective transport, which moves dissolved species and causes mixing (dispersion).
(c) Biological processes which mediate reactions, transform compounds, and often control the rate of kinetically controlled reactions.
(d) Groundwater flow through an aquifer may be affected by precipitation or dissolution of the solid aquifer phase.
(e) Heat transport affects the reaction rate.

7.4.2 Types of Geochemical Prediction Models

Dispersion of metals in the surface and groundwater is determined by processes, such as groundwater flow and solute transport. Thus, geochemical prediction models for metal dispersion can use either a single equation or a set of governing equations that represent these processes. The models can be analytical or numerical, deterministic or stochastic, steady-state or transient. Depending on the site conditions, properties, and the prediction goal, the models can be one-, two-, or three-dimensional [14]. For instance, one-dimensional can be used in sites where a potential receptor is immediately downgradient of a contaminant source; two-dimensional can be used for predictions in sites with one or more water sources or receiving environments in the direction of water flow in two dimensions [14, 15]. For example, contaminant migration can affect transverse dispersion, and the contaminant plume can be either lateral or vertical. Three-dimensional is where the site has well-known hydrogeological conditions and multiple groundwater zones [16].

7.4.2.1 Analytical Model

An analytical model is simply a mathematical equation that describes relationships among variables in a historical data set [17]. In essence, an analytical model draws a "line" through a set of data points to predict outcomes. As an initial assessment of field conditions or screening of remedial alternatives, field monitoring is conducted, and data on the most important parameters, especially those that characterize flow and solute transport processes, are collected. In analysing this field data, exact solutions to one- or two-dimensional water flow or transport equations can be deduced and used to extrapolate or interpolate [12]. Thus, geochemical analytical models require simplifying the flow system, including a horizontal aquifer base, uniform hydraulic and chemical reaction properties, and simple flow or chemical reaction boundaries [17]. Further, analytical models are typically steady-state and one-dimensional. However, selected groundwater flow models are two-dimensional (e.g., analytical element models), and some contaminant transport models assume one-dimensional groundwater flow conditions and one-, two- or three-dimensional transport conditions [17, 18].

7.4.2.2 Numerical Models

Mostly, a given site's geochemical characteristics are governed by interactions of several parameters (e.g., groundwater flow directions, hydrogeologic or geochemical conditions, and hydraulic or chemical sources or sinks with variable time and space). Consequently, the mathematical formulation of the model describing the sites yields a system of nonlinear algebraic equations that can be solved numerically. A numerical model solves the equations by approximating them using matrix algebra and discretizing the modelled domain. In discretization, the model domain is represented by a network of grid cells or elements, and time steps represent the simulation time [12, 19, 20]. The accuracy of numerical models depends on the model input data, the spatial and temporal discretization, and the numerical method used to solve the model equations. Generally, a numerical model can simulate and predict the system with greater accuracy when sufficient data are available.

7.4.3 *Modelling Procedures and Approaches*

7.4.3.1 Deterministic and Stochastic Modelling Approaches

In geochemical predictions of metal dispersion, deterministic or stochastic approaches can be used. In geochemical predictions of metal dispersion, a deterministic approach is appropriate when all input data are single, "best estimate" values and result in single-value outputs [20, 21]. Simply because deterministic approaches use a mathematical model in which outcomes are precisely determined by known relationships among states and events, there is no room for random variation. Thus, the deterministic approach is generally appropriate for predicting metal dispersion at a site-specific scale when extensive data have been collected, and spatial characterization is well established. With proper calibration, sensitivity analysis, and history matching, deterministic models can adequately represent metal dispersion in the real hydrogeologic system.

In most cases, not enough field data is collected before or during contamination; sometimes, the available data is limited to a local area, yet a larger regional scale is involved. In such a scenario, metal dispersion can be predicted using a stochastic approach (e.g., Monte Carlo simulations). A stochastic model can estimate probability distributions of potential outcomes by allowing for random variation in one or more inputs over time. The random variation is usually based on historical data fluctuations for a selected period using standard time-series techniques. Thus, a stochastic approach to geochemical predictions of metal dispersion characterizes parameter uncertainty by incorporating it into the parameters and databases used in the simulations [22, 23]. When data are scarce and uncertainty is high, calibration and additional historical matching can be lengthy. The stochastic approach maintains uncertainty throughout the modelling process, potentially enabling more realistic interpretations of results by providing a range of scenarios applicable to the entire system [21–24].

7.4.3.2 Steady-State and Transient Modelling Approaches

Metal dispersion in surface or groundwater can be simulated using either steady-state or transient flow models. In steady-state systems, inputs and outputs are in equilibrium, so there is no net change in the system over time, thereby providing average, long-term results [25]. In transient simulations, the inputs and outputs are not in equilibrium, so there is a net change in the system with time [24, 26]. Transient models are appropriate in the contemporary era of climate change, where the water regime and other factors, such as temperature, vary over time.

7.4.4 Rationale for Geochemical for Predicting Dispersion

The dwindling availability of portable water resources has increased measures to protect water sources, especially from pollution. However, the uranium mining and processing legacy has negatively affected the environment, leading to elevated metal levels. The mere presence of high levels of metals at abandoned uranium mining and processing sites is not a concern unless there is a reasonable risk of human exposure or unless environmental degradation where the metals are located is likely to occur. Therefore, the dissolution of metals from contaminant sources and their dispersion into receiving water sources, where they become exposure pathways, is an issue. Consequently, monitoring and predicting the dispersion of metal contaminants into the environment is essential for both water management and pollution control.

Classic procedures for monitoring and predicting the dispersion of metal contamination involve regular water-quality sampling. For surface water, it involves setting representative sampling stations, usually arranged in a gradient, on the water bodies of interest; for groundwater, it also involves monitoring wells. Hydrochemical monitoring data can indicate the potential for metals to be enriched in a system from their sources. However, this information is temporal and spatial, as it does not indicate metal dispersion in real time. Given the complexity and time involved, it is often impossible to conduct sufficiently realistic sampling and laboratory experiments to predict the long-term behaviour of the uranium, especially regarding dispersion into water systems [27, 28]. Geochemical models can interpret and predict processes that may occur over timescales not directly achievable in sample analysis and experiments. Consequently, there has been increased development of geochemical models that predict the dispersion of pollutants and metals into the environment. This development has generally gone hand in hand with advances in numerical techniques for solving complex mathematical problems, improvements in calculation speed and capability, and the general accessibility of computers [16, 29].

The accuracy in predicting metal dispersion into surface and groundwater systems has significant implications in water resource management and pollution control, and it can significantly impact the reasonability and scientific significance of pollution control strategies [30]. Developing effective geochemical prediction tools for metal dispersion requires a sound understanding of the prevailing hydrogeochemical processes and the uranium behaviour in the aquatic system. Uranium dispersion depends solely on its inherent chemical properties vis-à-vis chemical form and speciation in the water, as facilitated by the prevailing hydrogeochemical processes. For instance, the complex nature of sediment-soil-water interactions across different hydrodynamic zones can produce a manifold of effects, including the mobilization, concentration, and dispersion of metals at short- and long-time scales [31].

Chemical erosion due to variations in natural climate and hydrodynamic conditions significantly influences the uranium concentration dissolved in water. Depending on their dissolved form and redox status, some metals form free or complex cations when dissolved in water, while others, including some metalloids, are present as anions in their dissolved form [32]. Differences among metal groups have significant consequences for the partitioning of uranium among several dissolved and particulate phases, generally described as sorption onto particulates, precipitation in minerals, and complexation in solution.

The partitioning is significant for the dispersion of metals and divides the total amount of a pollutant into a "dissolved" fraction and several "adsorbed" fractions. Every hydrogeological process influences the fraction of a metal which is adsorbed onto geological material in the ground. Similarly, the fraction of metal adsorbed on the particulate matter is subjected to any process which affects particulates, such as settling and resuspension in surface waters. In groundwater systems, metals are initially adsorbed via fast reactions, followed by slower adsorption reactions, and are then redistributed into different chemical forms with varying mobility [33]. This distribution is controlled by reactions of metals such as (i) mineral precipitation and dissolution, (ii) ion exchange, adsorption, and desorption, (iii) aqueous complexation, and (iv) biological immobilization and mobilization.

Given the number of parameters to be considered in model development, more and more geochemical prediction models have been developed using various algorithms [34]. Consequently, there are often notable differences between modelling results (i.e., predicting metal dispersion) due to different theories and algorithms on which these models are based; different models lead to inconsistent prediction results. Therefore, harmonizing geochemical prediction tools for metal dispersion models is essential. This can be achieved by having a set of fundamental "must-be-included" parameters that each modelling code should include.

7.5 Prediction Modelling

7.5.1 Steps in Model Development

Models are simplified representations of complex natural systems, which provide insight and understanding. Consequently, the first step in developing a metal dispersion prediction model is to obtain biological information to gain insight into the contamination source history and characteristics. Once this information is collected, it is used to develop a conceptual understanding of what is happening at the site. With this understanding of the site characteristics and the development of a conceptual model, the information is used to simulate scenarios that predict the geochemical behaviour and dispersion of the metal in the system (see Fig. 7.3 for a summarized flow diagram of the steps). Depending on the question(s) that need to be answered in a particular system, models can be applied at varying levels of complexity and sophistication. However, all geochemical models suffer from uncertainty stemming from the natural system's understanding, the model formulation, and the variability of input parameter values [35].

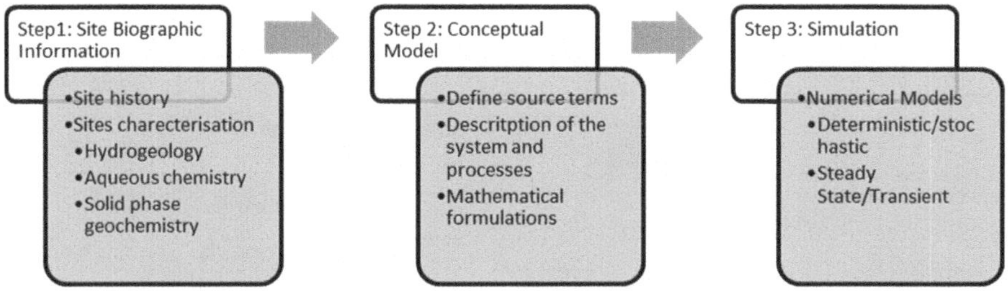

Fig. 7.3 Generalized steps in the development of effective geochemical prediction models for metal dispersion in surface and groundwater systems

7.5.2 Conceptual and Mathematical Formulations

7.5.2.1 Conceptual Model Development

In predicting metal dispersion at any site, the first step should be to develop a conceptual model. A conceptual model is a simplified description and schematic that outlines the system's components to be modelled. The model must be based on an understanding of site hydrogeologic conditions [36, 37]. The conceptual model should include the geologic and hydrologic framework, hydraulic properties, recharge and discharge areas (sources and sinks), boundary and initial conditions, transport processes, and spatial and temporal dimensions [4, 12, 25]. For metal dispersion prediction modelling, additional factors that may affect dispersion should be incorporated, such as contaminant sources, affected media, concentration distributions, the physical and chemical properties of the metal contaminants, and the plume. Figure 7.4 is a basic conceptual model of a contaminated site, outlining the processes that determine how mobile uranium and metals are dispersed into a receiving environment, which is usually a surface or groundwater body.

7.5.2.2 Mathematical Formulations

Once the conceptual model is drawn with the basic process outlined in Fig. 7.4, a mathematical description of the processes is formulated. The formulation of the reaction model for metal dispersion can be based upon the distribution of geochemically reactive species in an aqueous solution [12, 39]. The reaction types crucial to the accuracy of predicting metal dispersions include complexation, ion exchange, redox reactions, precipitation/dissolution, surface complexation, and other forms of adsorption. In all geochemical models, the reactions that describe the aqueous composition are defined in a set of basic components, which are the minimum set of fundamental species required to explain all the free and derived species (complexes) present in the aqueous solution.

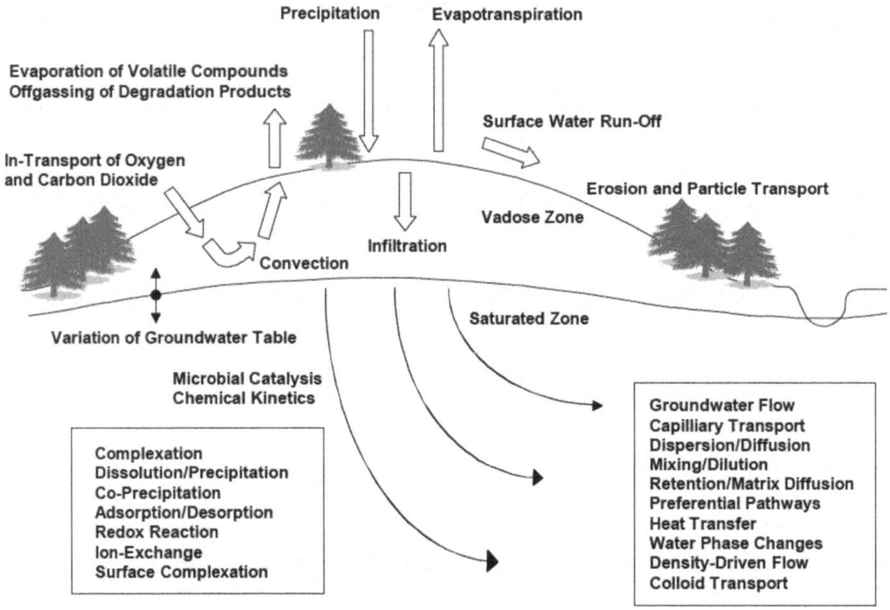

Fig. 7.4 Conceptual model of essential processes in the dispersion of contaminants. (Adapted from Crawford (1999) [38])

Analytical Model Formulation

An analytical model of metal dispersion from the source to the receiving environment would be the same as the one reported in numerous pieces of literature as a geochemical conceptual model. This involves using historical data and feeding in mathematical descriptions of the conceptual model's processes. Several factors influence the generation, transport, and dispersion of metals in the subsurface. However, the most significant aspects in developing prediction models are the fundamentals of how plumes develop, such as:

- Physical and chemical characteristics of the contaminants.
- Geometry and temporal variations in the contaminant source zone.
- Transport mechanisms (advection and dispersion).
- Reactions (destructive and non-destructive).

To illustrate this, let us consider mine and waste rock pile (WRP) sites as an example. In a WRP, the development of a contaminant plume is a function of oxygen ingress and net percolation, responsible for pyrite oxidation and generating acid mine drainage (AMD). AMD dissolves metal from rock material and transports it into the receiving environment. In brief, acidity can arise from two forms: stored or potential acidity. Stored acidity is the previously generated acids that can readily be transported to the receiving environment by water movement (net percolation). Potential acidity is unoxidized material that can be easily turned into stored acidity

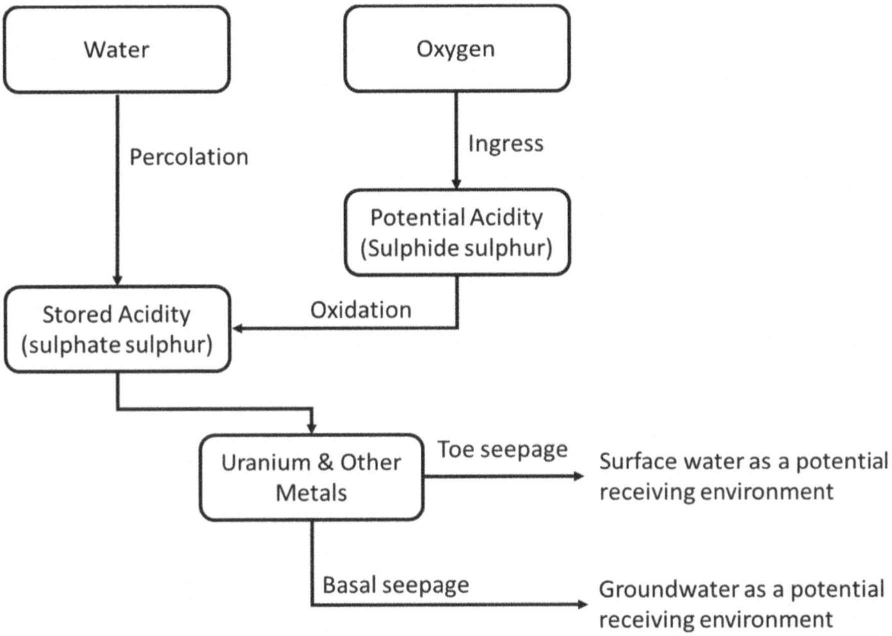

Fig. 7.5 Schematic WRP conceptual model, illustrating the essential process for developing a geochemical prediction model for metal dispersion from an abandoned mining site. It can be a waste pile, tailings, or a mine

when exposed to oxygen [40]. Schematic flow diagram of the conceptual model is presented in Fig. 7.5.

Calculating Oxygen Ingress The oxygen flux into the underlying waste rock generates oxidation products and acidity. Estimating the size of metal sources can be based on site data on the volume and mass of stored waste material, as well as the area covered. The amount of oxygen responsible for a chemical reaction in the waste pile, degrading acid- and metal-containing materials, is estimated by summing ingress from diffusion, advection, and dissolved oxygen.

Diffusion is the movement of molecules, atoms, or ions from a region of higher concentration to one of lower concentration due to random Brownian motion. Fick's Law can describe the diffusion of oxygen through the material:

$$J_{\text{diff}} = -D \times \frac{\Delta C}{\Delta x}$$

where J is the oxygen flux per unit area per unit time [mol/m^2/s], D is the diffusivity coefficient [m^2/s], ΔC is the change in concentration [mol/m^3], and Δx is the change in the distance [m]. The diffusion of oxygen is driven by the concentration gradient across the waste material and by the oxygen diffusivity coefficient, which is specific to soil material and texture. The value of the soil material's oxygen diffusivity coefficient at the site varies with material and requires the hydraulic conductivity and porosity of the material to be determined. Diffusion is generally the dominant mechanism [16, 36].

An influx of oxygen through advection occurs when spatial variations in temperature, gas consumption, or changes in atmospheric pressure create pressure gradients. Oscillations mostly drive advection in barometric pressure due to the heating and cooling of the atmosphere [12, 41]. An increase in atmospheric pressure drives new, oxygen-enriched surface air into the soil, while a decrease results in oxygen-depleted soil air venting upward into the atmosphere. A general form of Darcy's law describing airflow is modified in that, instead of the hydraulic head, the pneumatic head h drives the gas flow [41, 42]. Consequently, a one-dimensional system of Darcy's Law is written as:

$$q = -Ki = -\frac{k\rho g}{\mu} \times \frac{\Delta h}{\Delta l}$$

where k is the effective air permeability, ρ is the density of air, g is the gravitational acceleration, μ is the viscosity of air, Δh is the gradient in the pneumatic head, and Δl is the distance of flow. For an ideal gas, the pneumatic head h (m) is expressed as follows [37]:

$$h = z + \frac{RT}{gm} \times \ln \frac{p}{P_0}$$

where z is the elevation relative to a reference level, R is the gas constant (8.314 Pa m^3/mol K), T is the absolute temperature (K), m is the molar mass of the gas phase (kg/mol), p is the gas pressure (Pa), and P_0 is the reference gas pressure (Pa). According to the expression of the pneumatic head, gas flow can occur under gradients in (i) pressure p, (ii) temperature T, and (iii) composition (affecting molar mass m) [24]. Generally, the advective oxygen flux is negligible.

Dissolved oxygen that infiltrates the waste through net percolation is calculated from the estimated water ingress and concentration of dissolved oxygen in the water as follows:

$$\mathrm{DO} = \frac{0.678(P - p)}{35 + t}$$

where DO is the dissolved oxygen [mg/L], P is the barometric pressure [torr] (note: 1 torr = 133.32 Pa), t is the water temperature [°C], and p is the water vapour pressure [torr].

The total oxygen flux estimate is the sum of ingress through diffusive, advective, and dissolved oxygen. The total oxygen flux to the waste is determined by multiplying the obtained oxygen flux by the surface area.

Estimating Potential Acidity As a conservative rule, it is assumed that all sulfide is present as pyrite, oxygen is available for pyrite oxidation, and no oxygen is consumed by the oxidation of organic material in the cover system or carbonaceous material in the waste rock, thus creating the maximum amount of acidity per mole of oxygen. Therefore, oxidation of 1 mole of pyrite generates 2 moles of acid per 3.75 moles of oxygen present, as shown in the following equation:

$$FeS_2 + \frac{7}{2}H_2O + \frac{15}{4}O_2 \rightarrow Fe(OH)_3 + 2H_2SO_4$$

The oxygen flux is converted to an acidity load based on the H_2SO_4 equivalent (mol/m^2/year). Then, the calculated total oxygen flux can predict potential moles of acid to be generated, assuming a conservative rule that acidity is not neutralized by any carbonates or silicate minerals present.

Using these acidity concentrations, the total extrapolated catchment acidity load over the area of the cover system was determined in $CaCO_3$:

$$AL_{\mathrm{Catch}} = \mathrm{Acid} \times \mathrm{MW} \times \mathrm{Area}$$

where AL_{Catch} is acidity loading in the catchment, Acid is the acidity generated due to oxygen flux [mol/m^2/year], MW is the molar weight of $CaCO_3$ [100 g of $CaCO_3$ equivalent/mol], and Area is the surface area of the WRP [m^2]. Thus, the acidity concentration multiplied by the molar weight of $CaCO_3$/mol and the site surface area provides the catchment acidity load per year (as $CaCO_3$).

Calculating Water Ingress Net percolation is critical in developing a contamination plume from a WRP as a source. Net percolation into the underlying waste rock is primarily a function of water and energy balances and will occur when storage in the cover material is overwhelmed.

In simple terms, net percolation is the water that enters the waste pile, which reaches the waste and transports the acid. It is estimated from the water balance equation as follows:

$$\mathrm{NP} = \mathrm{PPT} - (R + \mathrm{AET} + \Delta S + \mathrm{LP})$$

where PPT is the total annual precipitation received at the site, R is the total annual amount of water leaving the site through runoff, AET is the total annual loss of water through evapotranspiration, ΔS is the change of moisture storage capacity of the material covering the waste, and LP is the lateral percolation and interflow. Using the mean annual net percolation and the pile footprint area, estimates of the downward vertical flow rate, the minimum and maximum annual flow rates, and total water ingress through the pile can be made. Numerically, net percolation estimates can also be obtained using software such as the VADOSE/W modelling program.

Estimating Groundwater Flow Rate The primary mechanism by which contaminants, including metal contaminants, disperse from source zones and into the receiving environment is advection through groundwater flow. Many factors affect groundwater flow, including climate (especially recharge from precipitation), watershed characteristics, geology, hydraulic conditions (e.g., water table slope and groundwater potential), and hydrogeologic boundary conditions, such as discharge or "no-flow" locations [43–45]. The velocity of the groundwater can be estimated using Darcy's law from the hydraulic gradient data obtained from two monitoring wells in line with the groundwater flow direction, as follows:

$$q = \frac{Q}{A} = K\frac{\Delta h}{\Delta x}$$

where q is the groundwater flux [m/s], Q is the discharge [m^3/s], A is the area, K is the hydraulic conductivity [m/s], Δh is the difference in water level in the wells, and Δx is the distance between the wells. Given this, a theoretical groundwater velocity and field test (e.g., dye tracer test) would be required to confirm the actual velocity. Further, it is limited to lateral groundwater flow and assumes a homogeneous, isotropic hydrostratigraphic unit. To obtain the groundwater flow rate, the velocity is multiplied by the cross-sectional area through which the groundwater flows. This cross-sectional area is equal to the height/thickness of the subaqueous waste, and the covered area's width is perpendicular to the flow direction.

Acidity and Metal Loads Stored acidity is mobilized to function as net percolation, like acidity released in the paste pH test, a simple static test employed to identify the presence of acid salts and reactive minerals. Sulfate-sulfur represents both acid-forming and non-acid-forming minerals, but, in most cases, it is assumed that all sulfate-sulfur is present as acid-forming minerals and that 2 moles of H^+ will be produced per mole of sulfur. This is a conservative assumption and assumes a maximum stored acidity load. It is expected that the total sulfur (wt%) is equal to the sulfide-sulfur (wt%) and sulfate-sulfur (wt%). Stored acidity is also reflected in the acid neutralization potential (ANP), which is simply the amount of acid consumed by minerals in a rock or soil. The ANP is measured using static acid-base accounting procedures and is usually expressed in calcium carbonate equivalence units.

Most water acidity arises from free protons, and the mineral acidity from dissolved iron, aluminium, and manganese [46, 47]. Thus, acidity is calculated from its pH and the sum of the milliequivalents of the dissolved acidic metals, as follows [48]:

$$\text{Acidity} = 50\left(\left(\frac{2\times[\text{Fe}]}{56}\right)+\left(\frac{3\times[\text{Al}]}{27}\right)+\left(\frac{2[\text{Mn}]}{55}\right)+\left(1000\times10^{-\text{pH}}\right)\right)$$

where the concentrations of the metals iron (Fe), aluminium (Al), and manganese (Mn) are given in units of mg/L, and 50 is the equivalent weight of $CaCO_3$, which converts the acidity in units of meq/L into units of mg/L of $CaCO_3$ equivalent. Water with a pH > 4.5 has acid-neutralizing capacity and is said to have alkalinity. The main alkalinity in water is dissolved carbonate, which can exist in bicarbonate and other carbonate forms. Alkalinity and acidity are not mutually exclusive terms. When water contains both mineral acidity and alkalinity, a comparison between the two measurements results in a determination as to whether the water is net alkaline (alkalinity > acidity) or net acidic (acidity > alkalinity). Net alkaline water contains enough alkalinity to neutralize the mineral acidity represented by dissolved iron and manganese. Net acidic water means that the mineral acidity, plus the acidity generated by the oxidation and precipitation of metals, exceeds the initial alkalinity. The net acidity is calculated as follows:

$$\text{Net Acidity} = \text{Acidity}_{\text{calc}} - \text{Alkalinity}$$

where alkalinity is measured in mg/L as $CaCO_3$.

From the ABA analysis, when the mean paste pH is low, it suggests that the waste is already acid-forming with reasonably low pH values. It is acceptable to assume that this is due to geochemical reactions within the waste, including carbonate/silicate mineral dissolution, sulfate-reducing bacteria generating alkalinity, and other reactions. Regarding the groundwater source term, when iron concentration is high, it is likely due to a low-oxygen environment that prevents Fe(II) from precipitating. Aluminium is expected to be colloidal at pH levels greater than 4.2–5.0, remaining in solution as fine particulate, and any interaction with lower pH drainage waters could redissolve the $Al(OH)_3$. In most cases, the elevated concentrations of these metals are assumed to be a function of pyrite oxidation, where these trace metals are contained within the pyrite mineral and are released upon oxidation. However, other sulfide species may be present. Thus, the loading of metals is due mainly to acid leaching from the minerals in the waste material. As with acidity, it is assumed that the concentrations are uniform throughout the waste material. To identify the enrichment of any elements in the waste material, the Geochemical Abundance Index (GAI), which compares the actual concentration of each component of the waste material sample with the average crustal abundance, is calculated according to Förstner et al. [49] and is defined as:

$$\text{GAI} = \log_2\left[\frac{C}{1.5 \times B}\right]$$

where C is the measured concentration of the specific element in sample B and is the background concentration of the particular component (e.g., average crustal abundance).

Conceptual Flow Model Formulation

To understand the sources and the dispersion of contamination in a given area, a conceptual flow model should be developed to understand contaminant transport from the source to the receiving environment. Source terms for downgradient groundwater and surface water receptors are derived using a mass balance, confirming that the mass per unit time entering the system equals the mass per unit time exiting the system. Thus, the source term loading can be calculated as follows:

$$C_{d} \times Q_{d} = \left(C_{u} \times Q_{u}\right) + \left(C_{s} \times Q_{s}\right)$$

where C_d is the concentration downgradient of the contamination source [mg/L], Q_d is the downgradient flow rate [L/s], C_u is the upgradient concentration [mg/L], Q_u is the upgradient flow rate [L/s], C_s is the concentration exiting the contaminant source [mg/L], and Q_s is the exiting flow rate [L/s].

The various flow components and associated flow rates are then combined in the geochemical conceptual model to determine the loading of the receiving environment. The flow rate through the unsaturated portion of the waste material is equal to the annual net percolation rate, which can contribute to both toe seepage (which leads to dispersion of metal contaminants to the surface water) and basal seepage (which leads to dispersion into the underlying groundwater and is related to the downward net percolation). Since it is assumed that the net percolation is the sole contributor to both toe and basal seepage, the basal seepage is equal to the mean net percolation minus the observed mean toe seepage rate. To obtain the minimum basal seepage rate possible, the maximum toe seepage rate is subtracted from the minimum net percolation rate, while the maximum basal seepage rate is estimated by subtracting the minimum toe seepage from the maximum net percolation rate.

It is assumed that 100% of the basal seepage from the site mixes with upgradient groundwater. The groundwater flow modelling provides specific insight into the contaminant plume as it migrates downgradient of the source. Following the derivation of the source terms for upgradient, WRP, and downgradient water, a mass balance concept is applied to evaluate and confirm that the mass per unit time entering the system is equal to the mass per unit time exiting the system:

$$\left(C_{UP} \times Q_{UP}\right) + \left(C_{Site} \times Q_{Site}\right) = C_{d} \times Q_{d}$$

where C_{UP} is the acidity concentration upgradient, Q_{UP} is the flow rate upgradient [L/s] of the pile of the site, which is equal to the groundwater flow rate in the till material, C_d is the acidity concentration exiting the waste heap, and Q_d is the flow rate [L/s]. It is also taken as the basal seepage rate, and C_d is the acidity concentration downgradient of the site. Following the derivation of the source terms for seepage, groundwater recharge, and upstream and downstream water, a mass balance concept is applied to evaluate and confirm that the mass per unit time entering the system is equal to the mass per unit time exiting the system:

$$\left(C_{\mathrm{UP}} \times Q_{\mathrm{UP}}\right)+\left(C_{\mathrm{Toe}} \times Q_{\mathrm{Toe}}\right)+\left(G_{\mathrm{GWR}} \times Q_{\mathrm{GWR}}\right)=C_{\mathrm{d}} \times Q_{\mathrm{d}}$$

where C_{TOE} and Q_{TOE} are the acidity concentration and the flow rate from the toe seep, respectively, while C_{GWR} and Q_{GWR} are the acidity concentration and flow rate from groundwater recharge, respectively. In all these calculations, the primary assumption is that porewater in the waste material has a constant concentration. However, bacterial activity (e.g., reducing bacteria) contributes to alkalinity and processes, such as dispersion, aquifer recharge, retardation, neutralization, and redox reactions, thereby affecting water quality, including the metal content plume. Further, the flow modelling simulations predict the groundwater plume's spatial and temporal variability as its metal disperses in the water and migrates downstream.

7.5.3 Numerical Simulations

7.5.3.1 Mathematical Basics for Simulations

The conceptual model and its mathematical formulations help predict the potential loading of metal from the course to the receiving surface and groundwater environment. It is a very conservative approach, whereas the actual scenario of metal contaminant dispersion in the environment involves multicomponent processes that couple transport and reaction. Coupled transport and reaction models, based upon the principle of mass conservation, are used to simulate how a geochemical system evolves along a fluid flow path in one, two, or even three dimensions. The coupling of hydrologic transport and geochemical reaction processes into a mathematical model typically results in a combination of partial differential equations and nonlinear algebraic equations [39, 43]. The partial differential equations are non-steady-state mass balances that relate changes over time in the total concentrations of basic components to the hydrologic transport components in space [12]. The mass balance includes all dissolved, sorbed, and precipitated species. The nonlinear algebraic equations are mass action equations that define the equilibrium chemistry of the system. When kinetic processes are included in the model, most of the mass action equations are replaced by partial differential equations that describe the rate-limited transformation of species by chemical reactions [12, 39]. There are many ways to solve the coupled hydrologic transport and geochemical reaction equations. Except for some straightforward cases, analytical solutions are not available for generalized, coupled transport and reaction problems. Consequently, most solution procedures are based on numerical methods. Such numerical methods are suited to mixed sets of nonlinear algebraic and ordinary differential equations.

7.5.3.2 Factors Prerequisite to Practical Simulations

Critical to a reliable predictive simulation is a detailed comprehension of the site's hydrogeological and geochemical properties being modelled. The models require the development of a calibrated flow model or an accurate determination of (a) the flow velocity to simulate the movement of contaminants by advection, diffusion, and dispersion; (b) the influence and impact of removal or release of metal contaminants by sorption or desorption from soil or waste material; and, (c) alteration of contaminants by biological or physical processes, or by chemical reactions [39, 46]. Further, the characterization of horizontal and vertical distributions, direction, and magnitude of average linear groundwater velocity is very important. All these should be obtained from field data, although heterogeneity is challenging to capture (e.g., preferential flow paths, fracture zones, regions of variable hydraulic conductivity and porosity, and stagnant zones).

Field data, augmented with data estimated either directly from experimental measurements or by calibration, is the input into the simulation model. Calibration involves adjusting key model parameters to match measured field data. Calibration does not always guarantee that the model will accurately represent a groundwater system and should be used with care. Thus, field data should, at a minimum, provide an initial distribution of solute, location, history and mass loading rate of chemical sources or sinks, and effective porosity and soil bulk density. The outputs from model simulations are contaminant concentrations in equilibrium with the groundwater flow system and the geochemical conditions of the modelled area.

7.5.3.3 Common Techniques Used in Prediction Simulations

Several numerical simulation techniques can be used to predict metal dispersion in surface and groundwater. However, the most commonly used are the (a) advection-dispersion-reaction (ADR) equation and (b) finite-difference (FD) and finite-element (FE) techniques.

(a) *The ADR equation*, based upon the assumption of transport within a homogeneous porous medium with a constant flow velocity, is the conservation of mass of the transported contaminant concentration. Consider the illustration in Fig. 7.6 with the concentration of metal. The ARD equations are:

$$\frac{\partial C}{\partial t} = -v\frac{\partial C}{\partial x} + D_{\mathrm{L}}\frac{\partial^2 C}{\partial x^2} - \frac{\partial q}{\partial t} \tag{7.1}$$

where C is the concentration in water (mol/kg), t is time (s), v is the porewater flow velocity (m/s), x is the distance (m), D_{L} is the hydrodynamic dispersion coefficient [m^2/s, $D_{\mathrm{L}} = D_{\mathrm{e}} + \alpha_{\mathrm{L}}v$, with D_{L} as the effective diffusion coefficient and α_{L} as the dispersivity (m)], and q is the concentration in the solid phase

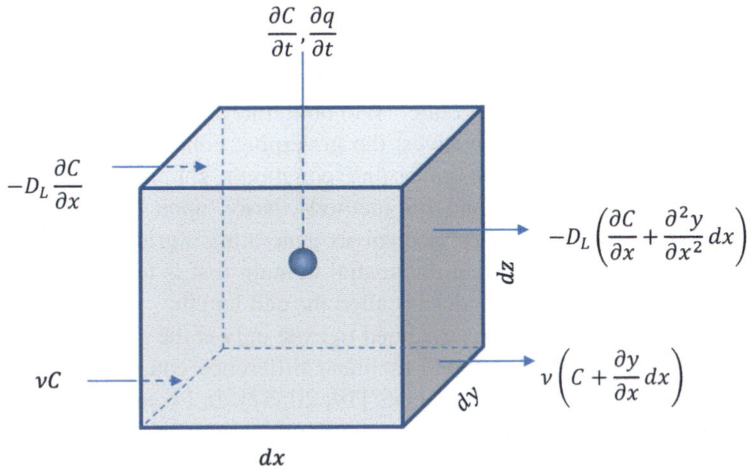

Fig. 7.6 An illustration shows the source of terms in the advection-reaction-dispersion equation

(expressed as mol/kg in the pores). The term $-v\frac{\partial C}{\partial x}$ represents advective transport, $D_L\frac{\partial^2 C}{\partial x^2}$ represents dispersive transport, and $\frac{\partial C}{\partial t} = \frac{\partial q}{\partial t}$ is the change in concentration in the solid phase due to reactions (q in the same units as C). The usual assumption is that v and D_L are equal for all solute species so that C can be the total dissolved concentration for an element, including all redox species.

The transport part of the equation is solved using an explicit finite-difference scheme that is forward in time, central in space for dispersion, and upwind for advection. For the chemical interaction term $\frac{\partial q}{\partial t}$ of each element, the transport part is calculated separately at each time step and is the sum of all equilibrium and non-equilibrium reaction rates. The numerical approach follows the basic components of the ARD equation in a split-operator scheme [50, 51]. With each time step, advective transport is calculated first, followed by all equilibrium and kinetically controlled chemical reactions. After that, dispersive transport is calculated, followed by the calculation of all equilibrium and kinetically controlled chemical reactions [12]. The scheme differs from most other hydrogeochemical transport models in that kinetic and equilibrium chemical reactions are calculated after both the advection and dispersion steps. This reduces numerical dispersion and the need to iterate between the chemistry and transport models. One conceptual problem with the ADR equation is that it assumes (and predicts) scale-independence of dispersivity. However, in real groundwater systems, heterogeneities lead to dispersion characteristics that vary with measurement scale. In addition, the flow field in a real system may vary considerably depending on local conditions of porosity and hydraulic conductivity in the medium.

(b) *The FD and FE Techniques:* Due to the nonlinear physical processes associated with concentration changes (gradients) in both time and space, and the presence of both time and space derivatives, the governing conservation equations are complex. Consequently, the governing equation's solution is sought using numerical techniques, FD and FE methods, based upon discretization techniques [19, 20]. In both cases, it involves generating a grid or mesh of points (nodes) distributed throughout the spatial domain that is to be modelled. The distance between adjacent nodes is called the cell length.

In FD methods, the fluid is assumed to exist only at the grid nodes, and spatial derivatives are approximated as linear difference equations based on the concentrations at neighbouring nodes [19, 20, 52]. In FE methods, the fluid is assumed to occupy the regions between grid nodes, and fluid concentrations are represented by interpolating polynomials based on concentrations at neighbouring nodes. Spatial derivatives are then approximated as the derivatives of these interpolating functions. In both FD and FE methods, time derivatives are approximated as finite-difference equations based upon a discretized time-frame. The numerical solution to the unsteady-state problem is obtained by solving the spatially discretized equation system with an appropriate algorithm and then advancing the solution forward using discrete time steps [19, 52]. The FD method requires that the discretization mesh consists only of rectangular polygons, limiting its flexibility [20, 22]. In comparison, FE methods can handle non-rectangular discretization meshes, allowing the simulation of unusual geometries in two- and three-dimensional space.

7.5.3.4 Simulation Model Development

The choice of a numerical simulation model is based on its applicability to the conceptual model, the availability of input data, and the purpose or objective of the modelling effort. Therefore, a prerequisite to simulation development is the availability of field data, sometimes augmented with literature values. Here are the steps followed in the development of the simulation models:

- *Setting Model Boundary Conditions:* For the simulation, the chosen model should satisfy fundamental assumptions of the area's boundary and initial conditions to be modelled [35]. Consequently, the first step is to set boundary conditions for the site, which defines the area where the model is both applicable and valid. Boundaries of a site can be surface water bodies, rivers, geologic structures, injection barriers, and groundwater divides. The selection of boundary conditions profoundly affects model simulation results, yielding biased or erroneous results. Ideally, natural hydrogeologic boundaries are the best choice. Still, these features are widely spaced and applicable only to analytic element models and to large regional numerical models with distant flow boundaries. Most small, site-specific numerical models use grid systems that require an artificial boundary at the edge of the grid [20, 22, 35]. In this instance, there are a few strategies to set the most appropriate boundary conditions, which include the following:

- Setting the grid boundaries sufficiently remote from the area of interest so that the artificial boundary does not significantly impact the predictive capabilities of the model [20, 32, 50].
- Using a stepwise or telescopic refinement modelling approach, either a coarser regional numerical model or a regional analytic element model is developed based on natural hydrogeologic boundaries. The model results define appropriate boundary conditions for a smaller-scale, more detailed numerical model.
- If the set artificial boundaries impact the model's predictive capabilities, a larger-scale, simpler analytic element model can be developed to test the influence of various boundary conditions. A simpler analytical element model can also be developed to test the influence of boundary conditions on the interest areas before developing a more complex numerical model.

• *Designing an Aerial Grid:* Most numerical methods require the development of an aerial grid overlay. The input parameters and grid form the database on which the groundwater system is defined [12, 19, 20]. The formation and input of this database are specific to the chosen computer code. Closely spaced grid patterns produce more accurate results; however, the finer the grid pattern, the longer the computational run time. A general guideline for developing the grid is to locate a node corresponding to a field feature, such as near pumping wells or near the centre of a wellfield, and to locate boundaries accurately. Grid spacing should be appropriate to the problem [16, 33, 35]. For instance, grid spacing should be closer in areas with significant spatial changes in transmissivity or hydraulic head. The grid should then be aligned with the anisotropies' major directions (i.e., orienting the grid with significant trends). Multiple planes, layers, or nodules should accommodate strong vertical gradients within a single saturated zone. As a rule of thumb, when expanding finite-difference grids beyond the interior nodes to the boundaries, grid spacing should not be more than 1.5 times the previous nodal spacing [19, 33, 53]. It may help develop an analytic element model first, determine appropriate boundary conditions, and then create a numerical model based on the information gained from the analytic element model [18].
• *Model Calibration:* This is adjusting the input parameters to match field conditions within acceptable criteria (i.e., goodness-of-fit). Thus, the calibration requires properly characterized field conditions. Similarly, field parameters necessary to produce a functional calibration should reflect actual field conditions, such as groundwater flow and direction, hydraulic heads and gradients, site water balance, infiltration rates, soil moisture content, metal contaminant migration rates and direction, and concentrations. Some inputs (e.g., hydraulic conductivity, transmissivity, and dispersivity) are highly variable, sometimes suspect, and the data are limited. Consequently, the values are typically adjusted and extrapolated iteratively until an acceptable "match" is achieved [18, 19, 33, 53]. As the

calibration proceeds, data gaps often become evident. The conceptual model may be redefined based on the calibration results, and additional data may be collected. When the best-calibrated match is achieved, a final input data set is established and demonstrated to be reasonable and realistic. The accuracy and precision of the match should be determined by how well the model simulates the dispersion of the metal into the water system. Although there are no universally accepted "goodness-of-fit" criteria, it is essential to minimize the difference between model-simulated and field conditions.

- *Model Verified:* To ensure that the modelled results accurately simulate the observed field data, the model undergoes field verification. Field verification is the process of testing the calibrated model to reproduce a set of field observations independent of what is used in the model calibration (e.g., historical matching) [53, 54]. The degree of verification required depends on the modelling purpose, the type of model, the sensitivity analysis, and the site complexity. If the model cannot be adequately field-verified, more emphasis is placed on the sensitivity and uncertainty analyses. Sensitivity analysis involves varying inputs within a reasonable range of parameter uncertainty and observing the relative change in the model response. The sensitivity analysis aims to demonstrate the simulations' sensitivity by varying input values. For instance, if a change in a parameter or boundary condition causes significant changes in the output, then the model is sensitive to that parameter or boundary [55]. Suppose the modelled hydraulic conductivity is varied, and the heads in the model do not vary significantly. The particular model is not sensitive to K. Sensitivity analyses also help guide future data collection activities. An uncertainty analysis is conducted by assigning distributions to parameters that exhibit the most significant variability in the field and are most sensitive to the model output [55]. There are several methods for introducing uncertainty into the models and the modelling process. One approach is to employ Monte Carlo methods in which the various possibilities are represented in a large number of simulated realizations [21, 24]. Another approach is to construct stochastic models in which the various coefficients are represented as probability distributions rather than deterministic values [23].
- *Prediction Simulation:* Upon completing calibration, sensitivity analysis, and field verification, the model is ready to predict both best- and worst-case scenarios. The simulations combine the conceptual model, the hydrogeological and geochemical input parameters, and the numerical model algorithms. Thus, the parameter estimates and calculations in the conceptual model, along with other site-specific data, should be used as inputs for simulations. Predictive simulations are often conservative. Given the uncertainty in model input parameters and their corresponding uncertainties, model input values have been selected to yield a "worst-case" simulation [56, 57]. Site-specific data should be used to support a more reasonable worst-case scenario to limit the range of uncertainty in predictive models. If long-term action is necessary, the model must be refined and updated as additional data are collected and new trends emerge.
- *Model Validation:* The geochemical prediction model uses historical data to predict future scenarios. Thus, even when a model is sufficiently calibrated and

field-validated, it is still challenging to predict the magnitude, location, and duration of future developments [54, 58]. Consequently, predictive simulations are monitored to verify whether their predictions align with reality. Thus, additional field data are collected after model development is complete to evaluate prediction accuracy. The new data is used to recalibrate the model to update and improve the simulation.

7.6 Limitations and Challenges

Despite advancements in computational methods and model development, geochemical prediction models for metal dispersion remain limited in accurately predicting. Here, some of the crucial limitations and challenges are listed and briefly discussed:

1. To execute geochemical simulation models, predictive calculation codes are widely used to estimate potential metal dispersion in water. These codes are often used within elaborate models to explain the source, mixing, and fate of such geochemical systems [59]. However, there remain many unknowns that influence the ability of the users to provide realistic predictions. Foremost among these are the inherent limitations of the thermodynamic databases and the lack of published kinetic databases that define the reactions. There is not enough data on the complex behaviour of the metals and secondary/amorphous phases encountered under field conditions to support the model predictions.
2. In numerical solution procedures, the mathematical formulation of most geochemical models involving equilibrium constants often yields a system of nonlinear algebraic equations that are solved numerically [19, 24, 51]. Most programs use a modified Newton-Raphson method to solve the system of equations [60]. Being an iterative method, convergence problems may arise in the numerical Newton-Raphson method if the initial values of unknown variables are not sufficiently close to the equilibrium values. Almost all field systems contain multiple phases (i.e., containing gases or minerals and water), and the number of phases exceeds that allowed by the Gibbs' Phase Rule [61–63]. This leads to a singular matrix in the model's mathematical formulation. Computer programs will generally fail to find a numerical solution for systems defined with inconsistent or physically unrealistic parameters.
3. One of the most significant issues in the numerical solution of metal dispersion is the coupling of the reaction and transport terms in the finite-difference or finite-element formulation of the system [12, 19]. There are many methods for solving the coupled transport and reaction problem, including the mixed differential-algebraic equation approach, direct substitution, sequential non-iterative, and sequential iterative approaches, to mention a few. Most of these approaches are not widely used for two- and three-dimensional systems owing to their excessive RAM requirements during computation. Further, they are based

on approximating real functions with discrete difference equations or interpolating functions. Consequently, errors may arise in the numerical solution due to the grid resolution (i.e., truncation errors) and the accuracy of the computer used to perform the calculations (i.e., rounding errors) [19, 24, 51].

4. The finite-difference and finite-element methods encounter numerical problems when simulating systems in which advection dominates over dispersion and diffusion. In this situation, the numerical solution can exhibit non-physical oscillations in the vicinity of a concentration front (i.e., where the concentration changes rapidly over a short distance). This problem occurs when a dimensionless number, called the grid Peclét number, exceeds a value of about two [1] [64]. The grid Peclét number is grid cell length and advective flux divided by the dispersion/diffusion coefficient. Therefore, this limitation can be partially eliminated by increasing the grid resolution and decreasing cell length. However, increasing the grid resolution in the spatial domain is very computationally expensive; accordingly, it can be done only selectively at locations where sharp reaction fronts occur. As reaction fronts tend to migrate over time, this often requires special front-tracking techniques to adapt the grid as needed to avoid the problem [20, 22, 50, 64]. A related problem that may occur near sharp reaction fronts in advectively dominated systems is numerical dispersion, also called numerical diffusion, which can smooth concentration profiles in a non-physical manner [64].
5. Furthermore, different implementations of the finite-difference and finite-element methods are susceptible to numerical oscillations and dispersion to varying degrees [20, 22, 50, 64]. In systems where dissolution and precipitation processes are considered, a program may need to take millions, or sometimes billions, of time steps to simulate small changes in the spatial distribution of metals in the flow system. This is one of the most significant problems encountered when coupled transport and reaction programs simulate diagenetic processes over long-time scales.
6. Many hydrological model extensions assume purely homogeneous reaction systems. However, the realistic scenario is heterogeneous, and some programs have been developed to simulate such systems. These models account for changes in the distribution of metals in the system driven by reactive transport processes. The mathematical formulation of models for heterogeneous reaction systems is much more complicated than that for homogeneous systems, as dissolution and precipitation zones form and slowly advance [20, 35, 39]. One challenge in simulating heterogeneous reaction systems is tracking the positions of these mineral reaction fronts over time. Simulating heterogeneous reaction kinetics accurately is challenging. Kinetic interactions with solid-phase materials are usually entirely dependent upon the mineral surface area exposed to porewater and the residence time of water in the random pores and fractures that characterize most geological media [12]. The exposed mineral surface area and the porosity of the medium change during diagenesis because of the precipitation and dissolution of various metals in a process referred to as armouring. Local changes in the medium's porosity may create preferential flow paths [20]. Armouring and porosity change processes are virtually impossible to predict.

7. Another challenge is to develop a geochemical prediction model with unique solutions. A numerical solution to a general multicomponent equilibrium problem is unique under ideal conditions (i.e., dilute concentrations) when formulated as a mass balance [12, 61]. However, geochemical prediction models often combine mass balance constraints (constraints on fluid bulk composition) with mass action constraints [12]. Mass action constraints include fixed pH, pE, and individual species activity and assumptions of gas and metal equilibrium. Solutions to problems involving mixed mass balance and mass action constraints, such as those in metal dispersion into surface and groundwater, are not always unique even in thermodynamically ideal systems. There are occasionally multiple solutions that, although satisfying the problem equally well in a mathematical sense, are not necessarily physically realistic and are often referred to as metastable equilibria. A computer program will not always converge to the most physically realistic solution; therefore, some care is needed when interpreting simulation results [65].
8. Finally, the major challenge in geochemical prediction simulation is incorporating transient behaviour into the model. In steady-state mass-balance systems, factors influencing metal dispersion, such as climate change, are rarely incorporated into model development. As a result, even during model validation, most models cannot incorporate transient behaviour and make accurate predictions beyond a 3-year time horizon.

7.7 Concluding Remarks

Geochemical modelling can predict uranium dispersion from the contamination source into surface and groundwater systems. Although by no means a substitute for classical monitoring, geochemical modelling and simulation are valuable predictive tools that bridge the gap between laboratory experiments, field observations, and the long-term behaviour of geochemical systems. They prove to be practical tools for simulating and predicting pollutant transport in water, helping reduce the cost of labour and materials for a large number of chemical experiments and field monitoring. Moreover, it is the only means to explain or predict the fate and transport of uranium at the abandoned mining sites that are inaccessible to sampling or experiments. Therefore, geochemical models have become an essential tool for predicting the final fate of uranium and its behaviour in aqueous environments.

Nevertheless, geochemical models have limitations and challenges, ranging from the methods themselves to the complexity of capturing, in a mathematical sense, all the parameters needed for high processing power and computing memory. Therefore, for metal dispersion, it is essential to assess the modes of occurrence and distribution patterns across different contexts, including soil, sediment, and water, to determine total concentrations and reactive fractions. Such geochemical signatures reflect the complex influence of sources and the combined effects of natural and anthropogenic processes on concentrations and dispersion within the respective abandoned uranium mining and processing sites. Further, contaminant transport modelling should include a simulation of advective flow, which is typically the major contaminant transport component. Mechanical dispersion and diffusion also play a role, and these parameters are often lumped into a single dispersion value. Sorption and transformation processes (e.g., biotransformation and hydrolysis) can alter contaminants' physical or chemical states. Thus, all applicable transport processes should be considered when modelling uranium, contaminant movement, and dispersion from abandoned uranium mines.

References

1. Wuana, R. A., & Okieimen, F. E. (2011). Heavy metals in contaminated soils: A review of sources, chemistry, risks and best available strategies for remediation. *ISRN Ecology, 2011*, 20. https://doi.org/10.5402/2011/402647
2. Aspandiar, M. F., Anand, R. R., & Gray, D. J. (2008). *Geochemical dispersion mechanisms through transported cover: Implications for mineral exploration in Australia* (Vol. 84). CRC LEME.
3. Pingitore, N. E. (1976). Vadose and phreatic diagenesis; processes, products and their recognition in corals. *Journal of Sedimentary Research, 46*, 985–1006. https://doi.org/10.1306/212F70B8-2B24-11D7-8648000102C1865D
4. Ronen, D., Scher, H., & Blunt, M. (1997). On the structure and flow processes in the capillary fringe of phreatic aquifers. *Transport in Porous Media, 28*, 159–180. https://doi.org/10.1023/a:1006583410617
5. Heinse, R., & Link, T. E. (2013). Vadose zone processes: A compendium for teaching interdisciplinary modeling. *Journal of Contemporary Water Research & Education, 152*, 22–31. https://doi.org/10.1111/j.1936-704X.2013.03163.x
6. Anand, R. R., Aspandiar, M. F., & Noble, R. R. P. (2016). A review of metal transfer mechanisms through transported cover with emphasis on the vadose zone within the Australian regolith. *Ore Geology Reviews, 73*, 394–416. https://doi.org/10.1016/j.oregeorev.2015.06.018
7. Tschinkel, W. R. (2015). Biomantling and bioturbation by colonies of the Florida harvester ant, *Pogonomyrmex badius*. *PLoS One, 10*, e0120407. https://doi.org/10.1371/journal.pone.0120407
8. Mehring, A. S., & Levin, L. A. (2015). Potential roles of soil fauna in improving the efficiency of rain gardens used as natural stormwater treatment systems. *Journal of Applied Ecology, 52*, 1445–1454. https://doi.org/10.1111/1365-2664.12525
9. Zhang, Y., Jiang, J., & Chen, M. (2008). MINTEQ modeling for evaluating the leaching behavior of heavy metals in MSWI fly ash. *Journal of Environmental Sciences, 20*, 1398–1402. https://doi.org/10.1016/S1001-0742(08)62239-1
10. Ishchenko, V. (2018). Prediction of heavy metals concentration in the leachate: A case study of Ukrainian waste. *Journal of Material Cycles and Waste Management, 20*, 1892–1900. https://doi.org/10.1007/s10163-018-0740-7
11. Monteith, H. D., et al. (1993). Modeling the fate of metals in municipal water pollution control plants. *Water Environment Research, 65*, 129–137.
12. Smith, S. L., & Jaffé, J. P. (1998). Modeling the transport and reaction of trace metals in water-saturated soils and sediments. *Water Resources Research, 34*, 3135–3147. https://doi.org/10.1029/98WR02227
13. Danesi, P. R. (1984). A simplified model for the coupled transport of metal ions through hollow-fiber supported liquid membranes. *Journal of Membrane Science, 20*, 231–248. https://doi.org/10.1016/S0376-7388(00)82001-3

14. Martinez, B. C., DeJong, J. T., & Ginn, T. R. (2014). Bio-geochemical reactive transport modeling of microbial induced calcite precipitation to predict the treatment of sand in one-dimensional flow. *Computers and Geotechnics, 58*, 1–13. https://doi.org/10.1016/j.compgeo.2014.01.013
15. Carranza, E. J. M. (2009). In E. J. M. Carranza (Ed.), *Handbook of exploration and environmental geochemistry* (Vol. 11, pp. 3–21). Elsevier Science B.V..
16. Power, C., Ramasamy, M., & Mkandawire, M. (2018). Performance assessment of a single-layer moisture store-and-release cover system at a mine waste rock pile in a seasonally humid region (Nova Scotia, Canada). *Environmental Monitoring and Assessment, 190*, 186. https://doi.org/10.1007/s10661-018-6555-0
17. Gokhale, S. S., & Trivedi, K. S. (1998). *Encyclopedia of distributed systems*. Kluwer Academic Publishers.
18. Strack, O. D. L. (2003). Theory and applications of the analytic element method. *Reviews of Geophysics, 41*, 1–16. https://doi.org/10.1029/2002RG000111
19. Duretz, T., May, D. A., Gerya, T. V., & Tackley, P. J. (2011). Discretization errors and free surface stabilization in the finite difference and marker-in-cell method for applied geodynamics: A numerical study. *Geochemistry, Geophysics, Geosystems, 12*. https://doi.org/10.1029/2011GC003567
20. Radu, F. A., Muntean, A., Pop, I. S., Suciu, N., & Kolditz, O. (2013). A mixed finite element discretization scheme for a concrete carbonation model with concentration-dependent porosity. *Journal of Computational and Applied Mathematics, 246*, 74–85. https://doi.org/10.1016/j.cam.2012.10.017
21. Liberti, L., & Kucherenko, S. (2005). Comparison of deterministic and stochastic approaches to global optimization. *International Transactions in Operational Research, 12*, 263–285. https://doi.org/10.1111/j.1475-3995.2005.00503.x
22. Zeitoun, D. G., & Baker, R. (1992). A stochastic approach for settlement predictions of shallow foundations. *Géotechnique, 42*, 617–629. https://doi.org/10.1680/geot.1992.42.4.617
23. Alberto, M., & Armando, B. (2004). A stochastic approach for assessing the uncertainty of rainfall-runoff simulations. *Water Resources Research, 40*. https://doi.org/10.1029/2003WR002540
24. Alberto, G., Nisha, R., & Newman, P. N. S. (2003). Three-dimensional steady state flow to a well in a randomly heterogeneous bounded aquifer. *Water Resources Research, 39*. https://doi.org/10.1029/2002WR001443
25. Sayer, C., & Giudici, R. (2002). A comparison of different modeling approaches for the simulation of the transient and steady-state behavior of continuous emulsion polymerizations in pulsed tubular reactors. *Brazilian Journal of Chemical Engineering, 19*, 89–104.
26. Clement, T. P., Wise, W. R., Molz, F. J., & Wen, M. (1996). A comparison of modeling approaches for steady-state unconfined flow. *Journal of Hydrology, 181*, 189–209. https://doi.org/10.1016/0022-1694(95)02904-4
27. Wang, Q., Li, S., Jia, P., Qi, C., & Ding, F. (2013). A review of surface water quality models. *The Scientific World Journal, 2013*, 231768. https://doi.org/10.1155/2013/231768
28. Ayangbenro, A., & Babalola, O. (2017). A new strategy for heavy metal polluted environments: A review of microbial biosorbents. *International Journal of Environmental Research and Public Health, 14*, 94–110.
29. Oke, S., & Vermeulen, D. (2017). Geochemical modeling and remediation of heavy metals and trace elements from artisanal mines discharge. *Soil and Sediment Contamination: An International Journal, 26*, 84–95. https://doi.org/10.1080/15320383.2017.1241216
30. Unsworth, E. R., et al. (2006). Model predictions of metal speciation in freshwaters compared to measurements by in situ techniques. *Environmental Science & Technology, 40*, 1942–1949. https://doi.org/10.1021/es051246c
31. Arakel, A. V., & Hongjun, T. (1992). Heavy metal geochemistry and dispersion pattern in coastal sediments, soil, and water of Kedron brook floodplain area, Brisbane, Australia. *Environmental Geology, 20*, 219–231. https://doi.org/10.1007/bf01706165

32. Mohammed, A. S., Kapri, A., & Goel, R. (2011). In M. S. Khan, A. Zaidi, R. Goel, & J. Musarrat (Eds.), *Biomanagement of metal-contaminated soils* (pp. 1–28). Springer.
33. Galitskaya, I. V., et al. (2017). Assessment of soil and groundwater contamination by heavy metals and metalloids in Russian and Indian megacities. *Procedia Earth and Planetary Science, 17*, 674–677. https://doi.org/10.1016/j.proeps.2016.12.180
34. Cohen, D. R., Kelley, D. L., Anand, R., & Coker, W. B.B. Milkereit (Ed.). *Proceedings of Exploration 07: Fifth Decennial International Conference on Mineral Exploration* (pp. 3–18).
35. King, K. S. (2005). *Hydrogeological assessment tools for modelling transport of metals in groundwater* (Vol. 28). Ministry of Environment.
36. Power, C., et al. (2017). Five-year performance monitoring of a high-density polyethylene (HDPE) cover system at a reclaimed mine waste rock pile in the Sydney coalfield (Nova Scotia, Canada). *Environmental Science and Pollution Research International, 24*, 26744–26762. https://doi.org/10.1007/s11356-017-0288-4
37. Lefebvre, R., Hockley, D., Smolensky, J., & Gélinas, P. (2001). Multiphase transfer processes in waste rock piles producing acid mine drainage: 1: Conceptual model and system characterization. *Journal of Contaminant Hydrology, 52*, 137–164. https://doi.org/10.1016/S0169-7722(01)00156-5
38. Crawford, J. (1999). *Geochemical Modelling—A review of current capabilities and future directions* (p. 39). Swedish Environmental Protection A gency.
39. Guoxiang, Z., Zuoping, Z., & Jiamin, W. (2005). Modeling reactive geochemical transport of concentrated aqueous solutions. *Water Resources Research, 41*. https://doi.org/10.1029/2004WR003097
40. Power, C., Ramasamy, M., MacAskill, D., & Mkandawire, M. (2016). *RemEAST conference 2015*. RemEAST.
41. Phillips, F. M., & Castro, M. C. (2003). In H. D. Holland & K. K. Turekian (Eds.), *Treatise on geochemistry* (pp. 451–497). Pergamon.
42. Marcati, P., & Milani, A. (1990). The one-dimensional Darcy's law as the limit of a compressible Euler flow. *Journal of Differential Equations, 84*, 129–147. https://doi.org/10.1016/0022-0396(90)90130-H
43. Ko, K.-S., Chang, H. W., Kim, T., & Lee, K.-K. (2002). Factors affecting the groundwater system around an underground LPG storage cavern. *Quarterly Journal of Engineering Geology and Hydrogeology, 35*, 279–290. https://doi.org/10.1144/1470-9236/2000-23
44. Barackman, M., & Brusseau, M. L. (2002). In J. F. Artiola, I. L. Pepper, & M. L. Brusseau (Eds.), *Environmental monitoring and characterization* (pp. 121–139). Academic Press.
45. Maltby, E. (Ed.). (2009). *Functional assessment of wetlands* (pp. 241–e216). Woodhead Publishing.
46. Kairies, C. L., Capo, R. C., & Watzlaf, G. R. (2005). Chemical and physical properties of iron hydroxide precipitates associated with passively treated coal mine drainage in the bituminous region of Pennsylvania and Maryland. *Applied Geochemistry, 20*, 1445–1460. https://doi.org/10.1016/j.apgeochem.2005.04.009
47. Watzlaf, G. R., Schroeder, K. T., Kleinmann, R. L. P., Kairies, C. L., & Nairn, R. W. (2014). *The passive treatment of coal mine drainage* (Vol. 72). US Department of Energy.
48. Park, W.-P., et al. (2016). Dissolution of metals from biosolid-treated soils by organic acid mixtures. *Applied and Environmental Soil Science, 2016*, 14. https://doi.org/10.1155/2016/9858437
49. Förstner, U., Ahlf, W., & Calmano, W. (1993). Sediment quality objectives and criteria development in Germany. *Water Science and Technology, 28*, 307.
50. Li, Y. S., & Chen, C. P. (1989). An efficient split-operator scheme for 2-D advection-diffusion simulations using finite elements and characteristics. *Applied Mathematical Modelling, 13*, 248–253. https://doi.org/10.1016/0307-904X(89)90083-8
51. Li, C. W., & Zhang, F. (1996). Three-dimensional simulation of thermals using a split-operator scheme. *International Journal of Numerical Methods for Heat & Fluid Flow, 6*, 25–35. https://doi.org/10.1108/09615539610113073

52. Yang, Y., Webb, G. I., & Wu, X. (2005). In O. Maimon & L. Rokach (Eds.), *Data mining and knowledge discovery handbook* (pp. 113–130). Springer US.
53. Ohio EPA. (2007). *Ground water flow and fate and transport modeling* (Vol. 30). Ohio Environmental Protection Agency.
54. Maloszewski, P., & Zuber, A. (1993). Principles and practice of calibration and validation of mathematical models for the interpretation of environmental tracer data in aquifers. *Advances in Water Resources, 16*, 173–190. https://doi.org/10.1016/0309-1708(93)90036-F
55. Vardoulakis, S., Fisher, B. E. A., Gonzalez-Flesca, N., & Pericleous, K. (2002). Model sensitivity and uncertainty analysis using roadside air quality measurements. *Atmospheric Environment, 36*, 2121–2134. https://doi.org/10.1016/S1352-2310(02)00201-7
56. Lahmira, B., & Lefebvre, R. (2015). Numerical modelling of transfer processes in a waste rock pile undergoing the temporal evolution of its heterogeneous material properties. *International Journal of Mining, Reclamation and Environment, 29*, 499–520. https://doi.org/10.1080/17480930.2014.889362
57. Beaudry, C., Lefebvre, R., Rivard, C., & Cloutier, V. (2018). Conceptual model of regional groundwater flow based on hydrogeochemistry (Montérégie Est, Québec, Canada). *Canadian Water Resources Journal/Revue canadienne des ressources hydriques, 43*, 152–172. https://doi.org/10.1080/07011784.2018.1461579
58. Nordstrom, D. K. (2012). Models, validation, and applied geochemistry: Issues in science, communication, and philosophy. *Applied Geochemistry, 27*, 1899–1919. https://doi.org/10.1016/j.apgeochem.2012.07.007
59. Nordstrom, D. K. (2017). *Goldschmidt 2017*. Goldschmidt.
60. Chun, C., & Neta, B. (2008). Some modification of Newton's method by the method of undetermined coefficients. *Computers & Mathematics with Applications, 56*, 2528–2538. https://doi.org/10.1016/j.camwa.2008.05.005
61. Rumble, D. (1974). Gibbs phase rule and its application in geochemistry. *Journal of the Washington Academy of Sciences, 64*, 199–208.
62. Fegley, B. (Ed.). (2013). *Practical chemical thermodynamics for geoscientists* (pp. 225–286). Academic Press.
63. Bleam, W. (2017). In W. Bleam (Ed.), *Soil and environmental chemistry* (2nd ed., pp. 189–251). Academic Press.
64. El-Kadi, A. I., & Ling, G. (1993). The courant and Peclet number criteria for the numerical solution of the Richards equation. *Water Resources Research, 29*, 3485–3494. https://doi.org/10.1029/93WR00929
65. Mkandawire, M. (2019). In E. Fosso-Kankeu (Ed.), *Nano and bio-based technologies for wastewater treatment: Prediction and control tools for the dispersion of pollutants in the environment* (p. 504). Wiley-Scrivener.

Chapter 8
Uranium Toxicology and Ecotoxicology in Abandoned Mining Sites

Abstract This chapter introduces the toxicological and ecotoxicological hazards posed by abandoned uranium mining and processing sites, with a primary focus on how hazards transfer from mine wastes into human and ecological receptors. It synthesizes chemical and radiological toxicity, exposure routes, toxicokinetics, genotoxicity, food-chain transfer, and vegetation impacts to frame risk assessment and remediation priorities.

Keywords Abandoned uranium mines · Hazard transfer pathways · Human health risks · Environmental exposure · Trophic transfer · Radionuclide contamination · Toxic metals

8.1 Introduction

Abandoned uranium mining and processing sites pose persistent and complex hazards to human health and the environment because they concentrate both chemical contaminants and radioactive materials in waste rock, tailings, and process waters that remain exposed to weathering and transport processes. Although mining methods vary, the fundamental hazards are similar: mobilizable uranium species, decay-chain radionuclides, and co-occurring metals and metalloids that can be liberated by physical erosion, chemical weathering, and hydrological movement [1]. Over time, these source materials become vectors for hazard transfer, releasing contaminants into air, surface water, groundwater, soils, and biota and creating pathways by which people and ecological receptors are exposed [2, 3].

Hazard transfer from abandoned sites is governed by the physical and chemical characteristics of wastes, site hydrology and geomorphology, climatic and seasonal dynamics, and the presence or absence of vegetation and engineered containment. Solubility and mineralogy determine whether uranium and associated elements remain bound in sediments or dissolve and migrate; wind and dust generation mobilize particulates and adsorbed radionuclides; and runoff and groundwater flow carry dissolved contaminants downstream or into aquifers. Land use, landscape

M. Mkandawire, *Fate and Effects of Uranium in Abandoned Uranium Mines*, https://doi.org/10.1007/978-3-032-10447-2_8

disturbance, and the lack of long-term management amplify these processes, so that even remote legacy sites can become sources of localized contamination with far-reaching ecological and human health consequences.

Human exposure occurs primarily through ingestion of contaminated water and food, inhalation of dust and radon progeny, and dermal contact with contaminated media. Toxicological effects reflect a combination of chemical and radiological mechanisms: chemically mediated nephrotoxicity and other organ effects from soluble uranium compounds, and stochastic radiogenic risks associated with internalized radionuclides that deposit in bone and soft tissues [4–6]. Toxicokinetic behaviour links external exposure to internal dose and target organ burden. The behaviour mentioned includes absorption influenced by chemical form and route, distribution to the kidney and bone, limited biotransformation, and elimination mainly via urine. Evidence for genotoxicity, oxidative stress, and DNA damage from uranium and some decay products further informs assessments of carcinogenic potential and long-term health risk [7].

Ecotoxicological impacts follow parallel pathways. In aquatic systems, dissolved uranium and radionuclides move with water and accumulate in sediments, where benthic organisms, invertebrates, and fish may be exposed; in terrestrial systems, contaminated soils and dust affect microbes, plants, and herbivores and can alter community composition and vegetation distribution [4–6]. While uranium typically shows limited biomagnification through food webs, it can bioconcentrate in particular taxa and sediments, and trophic transfer of co-contaminants can create hotspots of ecological risk. Vegetation both responds to and influences contamination: plant uptake and toxicity can change species composition and impede natural revegetation, while established plant cover can reduce erosion and dust emissions, thereby limiting further hazard transfer [8, 9].

A comprehensive assessment of abandoned uranium sites, therefore, requires integrating source characterization, transport mechanisms, exposure routes, toxicokinetics, and food-web dynamics to identify sensitive receptors and prioritize remediation. Understanding the interplay of chemical toxicity and radiological effects, the factors that facilitate contaminant mobilization, and the ecological consequences for aquatic and terrestrial communities is essential for designing effective monitoring, risk assessment, and long-term management strategies that protect human health and restore ecosystem function.

8.2 Toxicology Vs. Ecotoxicology

Toxicology and ecotoxicology both investigate how harmful agents affect living systems, but toxicology emphasizes individual-level responses, whereas ecotoxicology emphasizes population, community, and ecosystem-level consequences. Both fields rely on natural processes such as uptake, transport, transformation, and ecological interactions to explain how contaminants cause harm.

In this chapter, the terms toxicology and ecotoxicology are deliberately constrained to the context of uranium mining legacies, specifically uranium and its decay-series radionuclides (e.g., ^{226}Ra, 230th, and short-lived progeny) together with the suite of metal contaminants commonly associated with ore and waste (such as arsenic, lead, nickel, and copper). Within that scope, toxicology is treated primarily as the discipline concerned with human health: it addresses how these specific contaminants enter the body (inhalation of dust and radon, ingestion of contaminated water or food, and dermal contact), how they are distributed and retained in tissues, the internal dose delivered to critical organs, and the dose-response relationships that underlie clinical and subclinical outcomes (for example, radiogenic cancer risk from radon and internalized radionuclides, and nephrotoxicity or neurotoxicity from metals).

Ecotoxicology in this chapter is ecosystem-scale science that traces the fate and effects of contaminants across all ecosystem compartments (i.e., air, surface water, groundwater, soils, sediments, and biota), examining processes of environmental transport and transformation, bioavailability and trophic transfer, bioaccumulation and biomagnification, and the resulting impacts on populations, community structure, and ecosystem functions. This delimitation clarifies that, although the underlying mechanisms of harm (e.g., oxidative stress, genotoxicity, and disruption of physiological processes) may be shared, toxicology here emphasizes human exposure and health endpoints, while ecotoxicology emphasizes contaminant behaviour and ecological consequences across interconnected environmental compartments.

8.3 Hazard Transfer from Abandoned Uranium Mines

8.3.1 *Hazards of Concern*

The principal hazard at abandoned uranium mines is the mobilization of uranium, radionuclides, and associated heavy metals from tailings and waste rock into human and environmental exposure pathways, driven by physical destabilization and chemical alteration of ore residues. These processes convert formerly immobile decay-series elements into respirable aerosols, dissolved contaminants, and gaseous radon that enter human and ecological exposure pathways.

Abandoned uranium mine wastes contain minerals in which radioactive elements and heavy metals are initially held in a near-secular equilibrium within the ore matrix. Mining and milling disrupt that equilibrium by comminution, chemical processing, and exposure to atmospheric conditions, producing finely divided solids and chemically reactive surfaces that are far more susceptible to dissolution and transport than intact ore [10]. The mechanical breakup of rock during extraction and milling increases specific surface area and creates fresh mineral surfaces that oxidize when exposed to air and water. Oxidative weathering of sulfide and uranium-bearing minerals (e.g., pyrite and uraninite) generates acidic microenvironments that enhance the solubility of uranium, radium, thorium progeny, and a suite of heavy metals. Once solubilized, these species partition into porewaters and runoff and can be transported off-site by overland flow or infiltration into groundwater.

Physical destabilization of tailings and waste-rock piles is a primary vector for hazard transfer. Wind erosion of desiccated tailings produces respirable radioactive aerosols and dust that can be dispersed to adjacent lands, depositing radionuclide-bearing particulates on soils and vegetation and creating inhalation and ingestion exposure routes. Surface erosion and mass wasting expose deeper, previously isolated waste layers to oxygen and water, accelerating chemical weathering and increasing the flux of dissolved contaminants. Infiltration and seepage through loosely consolidated tailings mobilize dissolved radionuclides and metals into shallow groundwater; near-surface groundwater contamination can re-emerge into surface water bodies via springs, baseflow, or ephemeral channels, thereby linking subsurface contamination to aquatic and atmospheric exposure pathways.

Radon emanation and extreme-event mobilization are additional, mechanistically distinct hazards. ^{222}Rn, a gaseous decay product of ^{226}Ra, is released from tailings as the physical matrix becomes more permeable and as radium-bearing solids are exposed; radon diffuses through pore spaces and is emitted to the atmosphere, where it contributes to the inhalation dose for nearby populations. Extreme hydrological or geotechnical events, such as floods, dam failures, and seismic shaking, can produce episodic, large-magnitude releases of both particulate and dissolved contaminants, rapidly expanding the spatial footprint of contamination and overwhelming natural attenuation processes. Over longer timescales, diagenetic consolidation, root penetration, and freeze-thaw cycles alter heap porosity and permeability, changing redox conditions and thereby the chemical speciation and mobility of radionuclides and metals; these coupled physical-chemical evolutions convert legacy wastes into persistent sources of environmental and human exposure unless actively managed or remediated [11].

8.3.2 *Hazards Transfer Routes*

Hazard transfer from abandoned uranium mines occurs along interconnected air, surface-water, groundwater, and food-chain pathways, driven by the physical disaggregation of ore residues, oxidative and aqueous geochemistry, and episodic or chronic mobilization of particulates, dissolved radionuclides, heavy metals, and radon gas.

Airborne transfer from abandoned uranium mine wastes operates through two mechanistic modes: gaseous emanation and particulate entrainment. ^{222}Rn is produced in situ by the decay of ^{226}Ra in tailings and waste rock and diffuses through pore spaces as the physical matrix becomes exposed or more permeable [12, 13]. Where radon accumulates in enclosed spaces or near-surface dwellings, it contributes a significant inhalation dose and elevated lung-cancer risk. Concurrently, desiccated tailings and finely milled residues behave as a source of respirable dust: wind shear over exposed surfaces entrains silt- and clay-sized particles that carry adsorbed radionuclides (e.g., 230th, ^{226}Ra, and ^{210}Po) and associated heavy metals, enabling long-range transport and deposition on soils, vegetation, and built environments. Deposition of these particulates creates secondary exposure pathways through inhalation, incidental ingestion, and contamination of local food resources.

Hydrological transfer encompasses both surface and subsurface processes that mobilize dissolved and suspended contaminants. Precipitation, snowmelt, and runoff physically erode tailings and waste-rock piles, entraining solids into overland flow and fluvial systems; dissolved species released by leaching are carried in runoff and can sorb to suspended sediments or remain in solution to travel downstream. Infiltration of meteoric water into unconsolidated tailings produces porewaters that equilibrate with oxidized mineral surfaces; oxidative weathering, particularly of sulfide-bearing materials, generates acidic conditions that increase the solubility of uranium, radium, and many toxic metals, leading to acid mine drainage. This acidic runoff can severely impact aquatic ecosystems, harm water quality, and increase the bioavailability of toxic metals. Unlined or degraded containment allows seepage to reach shallow aquifers; once contaminants enter groundwater, they may form persistent plumes that serve as long-term sources to springs, baseflow, and drinking-water supplies or migrate to deeper regional aquifers depending on hydraulic gradients and geologic heterogeneity.

The food-chain route links environmental deposition and aqueous uptake to biotic accumulation. Contaminants deposited on soils or present in irrigation or surface waters can be taken up by plants or adhere to forage, leading to bioaccumulation in livestock and wildlife; human exposure then occurs through consumption of contaminated produce, meat, or traditional foods, a pathway of particular concern for communities relying on local subsistence resources [14, 15]. The partitioning behaviour of each element, especially its speciation, sorption affinity, and bioavailability, controls the degree of uptake and trophic transfer, with some radionuclides and metals exhibiting strong soil binding. In contrast, others remain mobile and more readily assimilated by organisms.

Direct contact and external exposure complete the suite of routes. Gamma radiation emitted from radionuclide-bearing materials penetrates beyond the waste mass and can deliver external doses to people living near or using contaminated materials; historical reuse of tailings and waste rock in construction has produced chronic, localized exposures and indoor radon problems. Extreme events like floods, dam breaches, and seismic activity can abruptly amplify all these transfer routes by redistributing large volumes of contaminated solids and water, overwhelming natural attenuation and remediation systems, and rapidly expanding the spatial footprint of contamination.

8.3.3 Factors Facilitating Hazard Transfer

Hazard transfer from abandoned uranium mines is driven by the interaction of hydrological and atmospheric transport processes with the altered physicochemistry of mine wastes, compounded by historical management failures; these factors convert tailings and waste rock into persistent sources of airborne particulates, dissolved contaminants, and radon that migrate into human and ecological exposure pathways.

The mobility of contaminants from abandoned uranium mining and processing sites is fundamentally controlled by environmental transport vectors (i.e., surface water, groundwater, and air) and by the intrinsic geochemical reactivity of the waste materials. Surface water processes mobilize both particulate and dissolved species: intense precipitation, snowmelt, and episodic runoff physically erode exposed tailings and waste-rock, entraining fine, radionuclide-bearing particles into overland flow and adjacent streams. Once entrained, particulates can settle in downstream sediments while dissolved uranium, radium, and associated metals remain in solution and are transported with the flow, linking tailings impoundments to fluvial systems and aquatic receptors.

Subsurface hydrology is an equally important pathway. Unlined or poorly contained tailings facilities permit infiltration and seepage, allowing porewaters to equilibrate with oxidized, weathered mineral surfaces and to carry soluble radionuclides into shallow and deeper aquifers. Changes in groundwater gradients, whether from mine dewatering, seasonal recharge, or anthropogenic pumping, can redirect contaminant plumes, extend their reach, and alter residence times, thereby converting a localized source into a regional groundwater problem. Catastrophic hydrological events, such as tailings dam failures or extreme floods, can produce large, rapid discharges of both solids and contaminated water, overwhelming natural attenuation and dispersing contaminants over broad areas [16].

Atmospheric processes facilitate long-range and near-field transfer through wind erosion and radon emanation. Dried tailings with fine grain sizes behave like fine sand and silt; wind entrainment generates respirable aerosols that carry adsorbed radionuclides and heavy metals, creating inhalation and secondary ingestion exposure routes when deposited on soils, crops, or built environments. Concurrently, ^{222}Rn produced by decay of ^{226}Ra in tailings diffuses through pore spaces and is emitted to the atmosphere; in poorly ventilated structures or subsurface voids, radon can accumulate to concentrations that significantly increase lung cancer risk for occupants.

Geochemical drivers amplify physical transport. Oxidative weathering of sulfide minerals in waste rock produces acid mine drainage (AMD), lowering pH and mobilizing uranium, radium, and a suite of toxic metals, including arsenic, lead, and copper. The chemical speciation of each element (i.e., its oxidation state, complexation with carbonate or sulfate, and sorption affinity for mineral surfaces) determines whether it remains particle-bound or enters the aqueous phase; for example, thorium tends to remain relatively insoluble at near-neutral pH, whereas uranium mobility increases under oxidizing, circumneutral to alkaline conditions where uranyl complexes form.

Human and management factors determine the initial and evolving vulnerability of sites. Many historic operations left the bulk of radioactivity (the long-lived decay products) in mill tailings, often without engineered liners or durable covers; installed engineered barriers can degrade through desiccation, root penetration, freeze-thaw, or mechanical disturbance, re-exposing wastes to air and water and re-initiating contaminant release. Consequently, the interplay of inadequate containment, evolving physical stability, and reactive geochemistry sustains long-term hazard transfer from abandoned uranium mine sites unless active remediation and long-term stewardship are implemented.

8.4 Potential Human Health Effects

8.4.1 Exposure Routes

People can be exposed to uranium through three principal routes: inhalation of aerosols, ingestion of contaminated water and food, and dermal contact, with an additional radiological component from external irradiation when enriched or short-lived progeny are present. Occupational and environmental releases produce respirable particles and vapours, which are the primary concern for inhalation exposure; ingestion becomes important when groundwater or foodstuffs are contaminated. Dermal uptake of uranium is possible for some soluble salts but is generally insignificant compared with inhalation and ingestion. For a concise summary of likely sources and pathways, see Table 8.1, which lists possible uranium-pathway sources and possible exposure pathways and complements the discussion below.

Table 8.1 Potential uranium sources and exposure pathways in the human body

Possible sources	Possible exposure pathway
1. Air	• Only about 0.76–5% of the uranium a person breathes will enter the bloodstream through the respiratory tract (i.e., nose, mouth, throat, and lungs) • Some uranium compounds are slowly cleared from the lungs
2. Food and water	• Only about 0.1–6% of the uranium a person ingests will get into the bloodstream through the gastrointestinal tract (i.e., mouth, stomach, and intestines) • Uranium compounds that dissolve in water enter the bloodstream more quickly than uranium compounds that are poorly soluble in water
3. Dermal contact	• A minimal amount of uranium can be absorbed through the skin; water-soluble uranium compounds are the most easily absorbed

8.4.1.1 Mechanisms After Inhalation

When uranium-containing particles are inhaled, their aerodynamic diameter determines where they deposit in the respiratory tract and how quickly they clear. Insoluble oxides such as UO_2 and U_3O_8 tend to deposit in the deep lung. They can be retained for long periods, increasing the potential for local radiological dose from uranium and its short-lived progeny and for chemical toxicity from decay products (e.g., lead isotopes in the decay chain). Soluble uranium compounds dissolve in lung fluids and are transferred to blood; for some soluble forms, >20% of inhaled material may be absorbed into the blood, but the absorbed fraction depends strongly on solubility and particle-size distribution [17–19].

8.4.1.2 Mechanisms After Ingestion

Ingested uranium behaves differently depending on chemical speciation. Insoluble uranium compounds are poorly absorbed from the gastrointestinal tract and are excreted mainly in the faeces, resulting in low systemic retention. Soluble uranium species (e.g., uranyl ion, UO_2^{2+}, and complexes) are more readily taken up: typical gut absorption rates reported for environmental exposures are ≈2% for soluble forms and ≈0.2% for insoluble forms, meaning roughly 98% of ingested uranium is not absorbed under normal dietary conditions [6, 20].

8.4.1.3 Systemic Distribution and Excretion

Once in the blood, uranium is rapidly filtered and excreted by the kidneys; about 70% of blood-borne uranium is excreted in urine within 24 h, with urinary excretion rising to higher fractions over several days (commonly cited values approach 90% within 48–96 h for the rapidly cleared fraction). Long-term retention occurs mainly in mineralized tissues: the skeleton is the principal long-term reservoir, with substantial fractions also transiently present in liver, kidneys, and other soft tissues. Typical whole-body burdens from routine environmental intake are on the order of tens to hundreds of micrograms (a commonly cited mean of ~90 μg), with the skeleton containing the majority of retained uranium [21–23].

8.4.1.4 Implications

Chemical nephrotoxicity, or kidney damage, is the primary health concern associated with soluble uranium that enters systemic circulation. Therefore, the most effective strategies for reducing exposure include controlling contamination of water and food sources as well as limiting inhalation of soluble aerosols. Conversely, exposure to insoluble particulates poses long-term radiological and local chemical risks to the lungs. When assessing the associated risks, it is essential to consider factors, such as particle size, solubility, and decay-chain composition [24, 25].

8.4.2 Uranium Toxicology

Uranium toxicity combines primarily chemical nephrotoxicity from soluble uranyl species and a secondary radiological cancer risk from long-term exposure; soluble uranyl (UO_2^{2+}) forms that complex with carbonate and serum proteins are the most biologically active and drive kidney and cellular damage through metal-ion substitution, oxidative stress, and phosphate-backbone hydrolysis [7]. Table 8.2 summarizes the principal health effects of uranium by exposure route, highlighting that inhalation poses the most significant combined radiological and chemical risk, while oral and dermal routes pose primarily chemical risks and are generally less severe at environmental levels.

Table 8.2 The principal health effects of uranium by exposure route

	Exposure route			
Effects category	Inhalation	Oral	Dermal	Other routes
Systemic effects	Lung inflammation Systemic distribution to the kidney and bone	Gastrointestinal absorption of soluble uranyl, leading to renal accumulation	Minimal absorption for insoluble forms	Direct systemic entry increases kidney and bone burden
Immunological and lymphoreticular	Altered immune markers	Limited evidence of immune modulation	Local irritation possible	Local and systemic immune responses are possible
Neurological	Possible cognitive and peripheral effects	Low-level effects uncommon	Unlikely except with significant systemic uptake	Depends on systemic dose
Reproductive	Reduced fertility markers in some studies	Evidence at environmental doses	Unlikely at typical exposures	Risk increases with systemic dose
Developmental	Foetal growth and developmental delays	Developmental toxicity at high doses in animals	Unlikely	Risk increases with systemic dose
Cancer	Increased lung cancer risk from retained particulates and radon progeny exposure	No substantial evidence of oral-route radiogenic cancer	Negligible from the dermal route alone	Retained insoluble particles in tissues can raise long-term cancer risk

Note: Other routes of exposure include wounds, embedded fragments, and accidental intravenous

8.4.2.1 Chemical Mechanisms and Systemic Distribution

Soluble uranium in oxidizing environments exists predominantly as the uranyl cation, written as UO_2^{2+}. In blood, UO_2^{2+} partitions between carbonate complexes, notably the bis- and tris-carbonate species $UO_2(CO_3)_2^{2-}$ and $UO_2(CO_3)_3^{4-}$, and protein-bound forms (e.g., albumin, transferrin, ceruloplasmin, and hemopexin) with largely 1:1 stoichiometry for many serum proteins. The carbonate complexes are more diffusible into soft tissues and bone, whereas protein-bound uranyl is more readily cleared by glomerular filtration and tubular handling in the kidney [25–30].

8.4.2.2 Nephrotoxicity: Targets and Molecular Actions

The kidney is the primary target of chemical uranium toxicity. Proximal tubular epithelial cells accumulate uranyl species, and injury manifests as cellular dysfunction and tubular necrosis. UO_2^{2+} acts as a hard Lewis acid that can displace essential divalent cations (e.g., Mg^{2+}) from metalloenzymes; displacement at the hexokinase active site impairs the first glycolytic step (glucose → glucose-6-phosphate), reducing cellular ATP generation and contributing to tubular cell injury. This displacement can be represented conceptually as:

$$Mg^{2+} - \text{enzyme} + UO_2^{2+} \rightleftharpoons UO_2^{2+} - \text{enzyme} + Mg^{2+}$$

Additional membrane-level effects include disruption of transporters (e.g., para-aminobenzoic acid transport) and peritubular membrane integrity, amplifying nephrotoxic outcomes [31, 32].

8.4.2.3 Genotoxicity and Oxidative Stress

Uranyl ions interact with nucleic acids and proteins. Coordination of UO_2^{2+} to phosphate groups in DNA increases susceptibility to phosphodiester hydrolysis. It can promote single-strand breaks and cross-links, producing genomic instability in exposed lymphocytes and other cells. Concurrently, uranyl exposure induces oxidative stress in pulmonary and renal cells, depleting antioxidants, such as reduced glutathione, and impairing enzymes, such as superoxide dismutase, thereby exacerbating macromolecular damage and altering the cellular proteome [33–35].

8.4.2.4 Radiological Considerations and Exposure Limits

Although chemical toxicity of uranium predominates at typical environmental levels, its radioactivity contributes to long-term cancer risk when internalized in particulate or insoluble forms that lodge in the lung or bone. Alpha emissions from uranium isotopes have low penetration but high linear energy transfer, increasing lung cancer risk after inhalation of aerosols and increasing bone-sarcoma risk after incorporation in bone; radon progeny from mine sites further elevate lung cancer risk in nearby communities. International guidance for public exposure recommends an effective dose limit of 1 mSv per year, with short-term allowances up to 5 mSv in exceptional circumstances, provided the five-year average remains ≤1 mSv/year; an equivalent dose to the skin is limited to 50 mSv/year [30, 36, 37].

8.4.3 The Uranium Toxicokinetics

Uranium entering the body from mine-derived dust, soil, or water follows the classical toxicokinetic sequence of absorption, distribution, (bio)transformation, and excretion. The chemical form and oxidation state control each step: hexavalent uranium (U(VI)) predominates in oxidizing environments as the uranyl ion, UO_2^{2+}, which forms soluble carbonate complexes such as $UO_2(CO_3)_2^{2-}$ and $UO_2(CO_3)_3^{4-}$ and readily enters the bloodstream; tetravalent uranium (U(IV)) tends to form insoluble oxides and particulates that remain localized in the respiratory tract after inhalation [38].

Absorption varies strongly by route and solubility. Inhalation of fine aerosols is the most consequential route at mine sites because a fraction of inhaled uranium dissolves and is absorbed into blood (reported ranges ~0.76 to 5%), while insoluble particles can deposit in alveoli or lymph nodes and persist, delivering a local chemical and radiological dose. Oral absorption from contaminated water or soil is generally low (approximately 0.1–6% of ingested uranium, higher for soluble salts), so most ingested uranium is excreted in faeces; dermal uptake is minimal except for highly water-soluble forms or damaged skin.

Once absorbed, uranium distributes preferentially to the kidneys, bone, liver, and spleen. The kidneys receive a high early burden because glomerular filtration and tubular reabsorption concentrate uranyl species in proximal tubular cells, where chemical interactions (e.g., uranyl acting as a hard Lewis acid and displacing essential cations) can cause nephrotoxicity. Bone acts as a long-term reservoir: a significant fraction of body uranium is sequestered in bone mineral, and reported biological half-lives in bone range from ~70 to 200 days, whereas most non-bone uranium is eliminated within 1–2 weeks [39].

Excretion is dominated by urinary elimination of absorbed uranium; the rate depends on ongoing dissolution of retained particles and renal function. Insoluble particles retained in the lung or embedded in tissue may slowly solubilize, sustaining systemic exposure and, for enriched materials, a persistent internal radiological hazard. Because natural and depleted uranium exposures at environmental levels are typically dominated by chemical heavy-metal effects (notably proximal tubular kidney injury), risk assessment must nevertheless account for the combined chemical and radiological potential of enriched or retained particulate uranium, where alpha emissions can add stochastic cancer risk to the chemical burden [40].

8.4.4 The Uranium Toxicodynamics

Uranium in mine wastes exists in multiple chemical forms, and their speciation controls absorption, transport, and toxicity. Weathering of ore and tailings alters pH and redox conditions, converting relatively insoluble U(IV) phases to more mobile U(VI) uranyl complexes that dissolve into groundwater and surface water as carbonate, sulfate, or phosphate complexes. These aqueous uranyl species are the primary bioavailable forms that drive uptake by organisms and subsequent toxicodynamics [40].

Absorption routes differ by exposure pathway: inhalation of dust and aerosols delivers particulate uranium to the respiratory tract, where soluble uranyl salts are absorbed across alveolar epithelium, while insoluble particles can persist and slowly dissolve, prolonging exposure. Oral ingestion of contaminated water or food results in variable gastrointestinal uptake, which is strongly dependent on compound solubility, dietary factors, and gut chemistry. Typically, only a small fraction of ingested uranium is absorbed, but chronic ingestion of soluble uranyl complexes can lead to cumulative systemic burdens [7, 41].

Once absorbed, uranium is deposited preferentially in the kidney and bone. The uranyl ion binds to serum proteins and small anions, is filtered at the glomerulus, and reabsorbed in the proximal tubule, where it concentrates and exerts direct nephrotoxic effects. In bone, uranium substitutes for calcium in hydroxyapatite, leading to long-term retention and a persistent internal source of both chemical and radiological insult. Biokinetic models show that tissue half-lives vary widely: rapid clearance from blood contrasts with prolonged residence in bone and kidney, which underlies chronic toxicity risk even after external exposures decline [42].

At the cellular level, the chemical toxicity of the uranium arises from direct interactions with macromolecules and from oxidative stress. The uranyl ion forms strong complexes with phosphate and carboxylate groups on proteins and nucleic acids, disrupting enzymatic activity and DNA integrity. Uranyl exposure induces mitochondrial dysfunction, generation of reactive oxygen species, lipid peroxidation, and activation of cell death pathways; in renal proximal tubular cells, these events culminate in tubular necrosis and impaired renal function, which are the hallmark outcomes of significant uranium exposure [3].

Metabolic handling and excretion are dominated by renal processes; urinary excretion is the principal elimination route for soluble uranium, but renal reabsorption and intracellular accumulation in tubular cells create a feedback loop that concentrates toxicity in the kidney. Chelation and complexation chemistry influence elimination: carbonate and citrate complexes increase mobility and urinary clearance, whereas adsorption to particulates or incorporation into bone reduces excretion and prolongs internal dose [3, 43].

Ecotoxicologically, the toxicodynamics of uranium extend to aquatic and terrestrial biota through bioavailability, which is governed by water chemistry and trophic transfer. In freshwaters, complexation with carbonate or organic ligands alters uptake by algae and invertebrates; chronic sublethal exposures impair growth, reproduction, and enzymatic systems in aquatic organisms, while accumulation in plants and herbivores creates food-web pathways that sustain ecological exposure long after mining ceases [7, 43].

Taken together, the toxicodynamics of uranium from abandoned sites are defined by chemical form, environmental transformation, and organ-specific biokinetics, with kidney, lung, and bone as primary targets, and ecological impacts mediated by altered speciation and trophic transfer; these mechanistic insights guide monitoring priorities and remediation strategies [7].

8.5 Potential Ecotoxicological Effects

8.5.1 Compartmental Risks

Abandoned uranium mines threaten four tightly linked ecological compartments: surface water, groundwater, soil/sediment, air, and the living organisms that connect them. Contaminants move between these compartments by leaching, runoff, wind dispersion, radon emanation, and biological uptake, producing both acute and long-term ecological and human health risks. This subsection describes the compartments, the physical and geochemical mechanisms that mobilize contaminants, and how those processes translate into exposure for biota and people [44].

Surface waters receive the first and most visible flux of contaminants when precipitation and surface runoff interact with exposed waste rock and tailings. Rainwater percolating through tailings dissolves uranium, radium, and associated trace metals and metalloids (e.g., arsenic, molybdenum, selenium, and nickel), and acidic conditions produced by sulfide oxidation (acid mine drainage) increase metal solubility and mobility. Under low pH, metal cations desorb from mineral surfaces and complex with sulfate or carbonate ligands, enhancing transport into streams and lakes; conversely, neutral to alkaline conditions can promote uranium complexation with carbonate and formation of mobile uranyl-carbonate species. Once in the water column, dissolved metals and radionuclides can be toxic to fish and benthic invertebrates and may be transported downstream or sequestered in sediments [45].

Groundwater contamination occurs when seepage from unlined tailings impoundments, open pits, or mine workings infiltrates the subsurface. Percolating leachate carries dissolved radionuclides and metals into aquifers, where redox conditions, mineralogy, and groundwater flow control fate and persistence. In reducing zones, uranium (VI) (i.e., soluble uranyl) can be reduced to uranium(IV) and precipitate as uraninite or adsorb to organic matter and iron oxides, temporarily immobilizing it; changes in redox or chemistry can remobilize these stores, making groundwater contamination a long-term legacy that can affect drinking water supplies for decades [44, 45].

Soils and sediments act as both sinks and secondary sources. Windblown dust from dry tailings deposits radioactive particulates onto surrounding soils, while fluvial transport deposits contaminated particulates into river and lake sediments, where metals accumulate. Sorption to clay minerals, organic matter, and iron oxy-hydroxides governs short-term retention. However, physical erosion, bioturbation, or chemical changes (e.g., acidification and reductive dissolution of iron oxides) can release contaminants back into the water column or render them bioavailable to benthic organisms. Sediment-bound contaminants thus pose chronic exposure risks to organisms that live in or feed on sediments [46].

Airborne pathways include particulate dust and gaseous radon. Radon gas, produced by decay of radium in tailings, diffuses and emanates from porous waste piles into the atmosphere, where it disperses; inhalation of radon and attached particulates is a primary radiological exposure route. Fine dust can travel substantial distances, be inhaled by animals and humans, or deposit on vegetation and water surfaces, creating additional ingestion and dermal exposure pathways [42, 46].

Biota integrate exposures across media through ingestion, inhalation, and dermal contact. Plants can take up uranium from contaminated soils, with uptake influenced by soil pH, speciation, and plant physiology; herbivores and livestock then ingest contaminated vegetation or lichen, and predators bioaccumulate contaminants through trophic transfer. Aquatic organisms accumulate metals from water and sediments, and humans are exposed via drinking water, inhalation of radon or dust, and consumption of contaminated fish, meat, or crops. The combined ecological and human health risk depends on contaminant speciation, environmental persistence, and food-web dynamics [47].

8.5.2 Environmental Exposure Pathways

Abandoned uranium mines create distinct ecological exposure pathways, namely surface water, groundwater, soils and sediments, air, and biota, through which uranium and its decay products move, transform, and become available to organisms, producing both acute and chronic ecological effects.

Surface waters receive direct contaminant loads when precipitation and runoff mobilize soluble uranium, radium, and co-occurring metals from tailings and waste rock; acid mine drainage from sulfide oxidation can lower pH and markedly increase metal solubility and transport, while neutral to alkaline waters favour formation of mobile uranyl-carbonate complexes that enhance downstream dispersion. Once in the water column, contaminants partition between dissolved phases, suspended particulates, and eventual deposition in sediments, determining exposure routes for plankton, benthic invertebrates, and fish [45].

Subsurface pathways link surface releases to groundwater systems when leachate from unlined impoundments or flooded workings infiltrates aquifers. Redox gradients in the subsurface control uranium speciation and mobility: oxidizing conditions favour soluble uranium (VI) species, whereas reducing conditions promote reduction to uranium(IV) and precipitation or strong sorption to mineral and organic phases, creating heterogeneous zones of immobilization that can later be remobilized by changes in hydrology or chemistry and supply contaminated porewater to wetlands and riparian habitats [48].

Soils and sediments act as both sinks and secondary sources. Windblown dust from dry tailings deposits radionuclide-bearing particulates on terrestrial surfaces, while fluvial transport concentrates contaminated particulates in river and lake sediments, where sorption to clays, organic matter, and iron oxyhydroxides governs retention. Physical disturbance, bioturbation, or geochemical shifts (for example, reductive dissolution of iron oxides) can release bound contaminants back into porewaters or the overlying water column, making sediments a long-term reservoir that controls chronic exposure of benthic communities [46].

Airborne transport redistributes particulate contaminants and gaseous decay products across the landscape. ^{222}Rn emanating from radium in tailings diffuses into the atmosphere, and its short-lived progeny attach to aerosols; wind erosion of tailings generates fine particulates that deposit on soils and vegetation, creating inhalation and deposition pathways that make contaminants available to terrestrial and aerial organisms and seed contamination of remote habitats via long-range transport.

Biota integrate exposures from all abiotic compartments through uptake, ingestion, and contact. Plant uptake of uranium depends on soil chemistry and speciation, with foliar deposition from dust and root absorption both contributing to tissue burdens; soil invertebrates and benthic organisms ingest contaminated particulates and organic matter, enabling trophic transfer to herbivores and predators. Patterns of bioaccumulation and biomagnification vary by element and organism physiology. However, the net effect is propagation of contaminants through food webs, with potential impacts on growth, reproduction, and community composition. Microbial processes in soils and sediments further mediate speciation and bioavailability, catalysing redox transformations that either immobilize or remobilize uranium depending on local conditions [49].

8.5.3 Biological Receptors in the Aquatic Food Chain

Uranium released from abandoned mines is first taken up by microbes and primary producers, then transferred through macrophytes and invertebrates to fish, resulting in measurable bioaccumulation and physiological effects across aquatic food webs. Across the aquatic food web, bioaccumulation and trophic transfer create pathways by which uranium moves from dissolved pools into biomass and up to higher trophic levels; emergent aquatic insects provide a notable vector for transferring aquatic contamination to terrestrial consumers. The net ecological outcome at a contaminated site depends on uranium speciation, water chemistry (pH, redox, and carbonate), sediment binding phases, and the feeding ecology of resident species, all of which determine bioavailability, assimilation, and ecological risk [50].

8.5.3.1 Algae, Cyanobacteria, and Aquatic Microorganisms

Algae, cyanobacteria, and heterotrophic microorganisms form the immediate biological interface with dissolved uranium and strongly influence its chemical speciation and fate. Uranium binds rapidly to cell surfaces via adsorption to charged functional groups and to extracellular polymeric substances (EPS), and certain microbes can internalize uranium through ligand-mediated transport or complexation with siderophores. EPS matrices concentrate metals, promote nucleation of particulate phases, and create microenvironments where localized redox reactions can reduce U(VI) to less soluble U(IV), altering mobility and bioavailability. These microbial processes both sequester uranium from the water column and make it available to grazers that consume biofilms, thereby initiating trophic transfer [51, 52].

8.5.3.2 Aquatic Vascular Plants

Aquatic vascular plants and bryophytes accumulate uranium from water and sediments, typically partitioning a significant fraction of the burden to roots, rhizomes, and basal tissues. Uptake mechanisms include adsorption to root cell walls, complexation with root exudates, and limited translocation to shoots, depending on species and water chemistry; carbonate and pH conditions strongly influence the fraction available for uptake. Submerged mosses and macrophytes can act as effective phytofilters, lowering dissolved concentrations locally while concentrating uranium in biomass and associated sediments. This creates concentrated biological reservoirs accessible to herbivores and detritivores, which can influence sediment geochemistry through organic matter inputs and redox alteration [53].

8.5.3.3 Aquatic Invertebrates

Aquatic invertebrates accumulate uranium through direct exposure to contaminated water and by ingesting contaminated food, such as biofilm, detritus, and plant material. Uptake occurs across gill or integument surfaces and via dietary assimilation; assimilation efficiency depends on particle chemistry, speciation, and gut physiology. Benthic taxa (chironomid larvae, amphipods, and isopods) often show elevated tissue burdens because they ingest sediment-bound particulates and biofilm. Emergent aquatic insects can export uranium to terrestrial systems upon emergence, creating cross-ecosystem transfer. Ecotoxicological responses in invertebrates include reduced growth, developmental delays, and physiological stress at environmentally relevant concentrations, making them key vectors for trophic transfer and community-level effects [52].

8.5.3.4 Fish

Fish integrate exposure from water, diet, and sediments and tend to accumulate uranium in organs with high metabolic or mineral affinity, notably liver, kidney, and bone. Uranium exposure in fish can induce oxidative stress, genotoxic effects in erythrocytes, and neurophysiological changes, such as altered acetylcholinesterase activity; these effects reflect both the chemical toxicity of the uranium and the radiological contributions of associated radionuclides. Bioaccumulation in fish is influenced by trophic position, feeding habits, and habitat use, and sublethal impacts on growth, reproduction, and behaviour can propagate through predator-prey interactions, altering population dynamics and ecosystem function [50, 54].

8.5.4 Biological Receptors in the Terrestrial Food Chain

Uranium released from abandoned mines enters terrestrial food webs through soil microbes, plants, and invertebrates and is transferred to higher vertebrates, producing combined chemical and radiological stress that is mediated by speciation, soil chemistry, and trophic interactions.

The biological receptors of uranium in the terrestrial food chain can be summed up as:

8.5.4.1 Soil Microorganisms

Soil bacteria, fungi, and protists are the first biological receptors for uranium in contaminated terrestrial systems and strongly influence its speciation, mobility, and bioavailability. Actinobacteria, Firmicutes, Pseudomonas, and metal-tolerant fungi interact with uranium through cell-wall adsorption, extracellular polymer production, and intracellular sequestration; microbial metabolism and biosynthesis pathways (including altered amino- and fatty-acid production and DNA/cell-wall stress responses) can confer tolerance and enable reduction of soluble U(VI) to less soluble U(IV) phases, thereby immobilizing uranium or transforming it into particulate forms that enter food webs as particulate-bound material [55].

8.5.4.2 Terrestrial Nonvascular and Vascular Plants

Plants take up uranium primarily through roots, with uptake and translocation governed by soil pH, carbonate concentration, and root exudates that complex uranium. Some species, such as mustards and ryegrass, are efficient accumulators and have been studied for phytoextraction. In contrast, species like *Carlina corymbosa* show low uptake, concentrating most of the uranium in roots and rhizosphere soils rather than in shoots. Root adsorption to cell walls, complexation with organic ligands, and limited xylem translocation determine whether uranium becomes available to herbivores or remains sequestered in belowground biomass and rhizosphere sediments [8].

8.5.4.3 Terrestrial Invertebrates

Terrestrial invertebrates, litter dwellers, herbivorous insects, and predators, such as spiders, acquire uranium by ingesting contaminated plant material, soil organic matter, and associated microbes. Dietary assimilation and gut physiology control uptake efficiency, and detritivores that process litter can concentrate uranium in tissues, creating a pathway for transfer to insectivorous vertebrates. Although biomagnification is not universally strong for uranium, invertebrate accumulation can produce localized hotspots of exposure and facilitate movement of uranium through trophic links and into burrow systems where small mammals forage [56].

8.5.4.4 Amphibians and Reptiles

Amphibians and reptiles in contaminated habitats are exposed through dermal contact with soils and porewaters, ingestion of contaminated invertebrates, and use of contaminated breeding pools. Kidney toxicity from soluble uranium species is a primary chemical hazard for these taxa, while chronic exposure to radionuclides can cause cellular and genetic damage. Life-history traits, such as permeable skin in amphibians and site fidelity, increase exposure duration and ecological risk.

8.5.4.5 Birds and Mammals

Birds and mammals ingest uranium through contaminated vegetation, invertebrate prey, soil ingestion, and drinking water, and they also receive exposure from dust deposition on forage. Small mammals and larger herbivores can accumulate uranium in kidney and liver tissues, where chemical nephrotoxicity is the dominant acute effect, and radiological effects (i.e., DNA damage and reproductive impairment) are of concern under chronic exposure. The degree of trophic transfer and population-level impact depends on feeding ecology, home-range overlap with contaminated substrates, and the balance between dietary intake and physiological elimination [57].

8.5.5 *Bioconcentration Factors and Trophic Transfers*

Uranium released from abandoned mines and mills is taken up and transferred through food webs, processes controlled by chemical speciation, environmental chemistry, and organism physiology. Bioconcentration and trophic transfer are therefore site-specific but follow predictable mechanistic pathways that determine ecological and human exposure.

8.5.5.1 Aquatic Ecosystem Exposure

In aquatic systems, uranium bioavailability and bioconcentration are governed primarily by aqueous speciation, which is controlled by pH, carbonate concentration, calcium hardness, and dissolved organic matter. Uranyl UO_2^{2+} forms carbonate and calcium-uranyl complexes that alter free ion activity and thus uptake by primary producers and invertebrates. Phytoplankton and periphyton take up dissolved uranyl species directly across cell membranes or via adsorption to cell surfaces, producing the first step in trophic transfer; measured bioconcentration factors (BCFs) for algae and macrophytes vary widely with water chemistry, often increasing under low-hardness, low-carbonate conditions where free uranyl predominates. Benthic invertebrates accumulate uranium both from dissolved phases and from ingestion of contaminated sediments and biofilms; the biota-sediment accumulation factor (BSAF) can be substantial where fine-grained, organic-rich sediments concentrate uranium, creating a persistent benthic reservoir that sustains exposure to higher trophic levels [55, 58].

Fish accumulate uranium through gill uptake of dissolved species and by dietary intake of contaminated prey; gill uptake is sensitive to competition with calcium and to complexation by carbonate and natural organic ligands, which reduce free uranyl activity and lower uptake rates. Empirical studies and modelling indicate that trophic magnification of uranium is generally limited compared with highly lipophilic organic contaminants, because uranium is hydrophilic and binds to mineral phases and proteins rather than partitioning into lipid tissues; nevertheless, trophic transfer occurs and can lead to elevated concentrations in piscivorous species where dietary exposure and sediment-linked pathways dominate. Site-specific factors, such as episodic acidification from acid mine drainage or redox changes that mobilize particulate-bound uranium, can abruptly increase dissolved, bioavailable fractions and thus raise BCFs and trophic transfer rates, making temporal monitoring essential for risk assessment.

8.5.5.2 Terrestrial Ecosystem Exposure

In terrestrial systems, soil chemistry and plant physiology control uranium uptake and transfer into food webs. Uranium in soils is often bound to mineral surfaces, iron oxides, and organic matter, limiting root uptake under neutral to alkaline conditions; however, acidic soils, chelating organic ligands, and phosphate competition can increase uranium solubility and plant availability, raising plant-soil accumulation factors (BASF) and enabling entry into herbivore diets. Root uptake mechanisms include apoplastic adsorption to cell walls and symplastic transport, where uranyl complexes mimic essential cations; mycorrhizal associations and rhizosphere microbial activity can either immobilize uranium through precipitation and sorption or mobilize it via ligand production, thereby modulating plant uptake and subsequent trophic transfer.

Herbivores ingest uranium via contaminated forage and soil ingestion, and bone deposition is a key long-term sink for vertebrates because uranium substitutes for calcium in hydroxyapatite, producing prolonged internal exposure that can be transferred to predators and scavengers through consumption of tissues and bone fragments. While biomagnification factors for uranium across terrestrial food chains are typically low relative to persistent organic pollutants, bioaccumulation in specific compartments (e.g., bone and kidney) and repeated dietary exposure can produce significant internal doses in wildlife and humans relying on local resources, necessitating integrated measurements of soil, plant, invertebrate, and vertebrate tissues to characterize exposure pathways and to parameterize bioconcentration and trophic-magnification models for remediation planning.

8.6 Other Prevalent Chemical Hazards

Abandoned uranium mines and processing sites pose a suite of interlinked chemical hazards beyond uranium itself, driven by exposed ore minerals, long-lived decay products, and legacy processing reagents. These hazards are persistent, mobile under changing geochemical conditions, and present both human toxicological and broader ecological risks. Most chemical hazards at abandoned uranium mines, particularly metals and metalloids, processing reagents, and many radionuclide decay products, pose chemical and ecotoxicological risks that frequently exceed the direct toxicological risks from uranium itself [59].

8.6.1 Toxic Metals and Metalloids

Toxic heavy metals and metalloids liberated from uranium-bearing rock include arsenic, lead, molybdenum, selenium, cadmium, nickel, and copper. These elements differ in speciation and mobility; for instance, arsenic commonly forms oxyanions that become mobile under reducing conditions, lead and cadmium can persist in soils and bioaccumulate, and molybdenum and selenium exert endocrine and hepatic effects at elevated doses. Human toxicology ranges from neurodevelopmental impairment (e.g., lead) to carcinogenesis (e.g., arsenic) and renal or hepatic injury (e.g., cadmium and selenium), while ecotoxicological outcomes include reduced microbial diversity, impaired invertebrate reproduction, and trophic transfer through plant and animal uptake that sustains chronic exposure in ecosystems.

8.6.2 Chemical Hazards of Radionuclides

Abandoned uranium mines present a complex matrix of chemical and radiological hazards in which radionuclides act as both sources of ionizing radiation and chemically active contaminants that interact with biota, soils, and waters, prolonging ecological degradation and human exposure. Weathering of ore and tailings alters redox and pH conditions, changing speciation and mobility. This geochemical dynamism means that chemical toxicity can be as persistent and biologically relevant as radiotoxicity [59].

8.6.2.1 Thallium

Thallium is frequently associated with uranium-bearing ores and becomes mobilized during mining and milling. Chemically analogous to potassium, thallium readily enters cells via potassium transport pathways, disrupting mitochondrial function and ionic homeostasis. Clinically, this manifests as neurological and cardiovascular dysfunction, alopecia, and multi-organ failure after chronic exposure. In aquatic systems, thallium salts are highly soluble and bioavailable, producing acute toxicity to fish and invertebrates and creating rapid trophic transfer risks in contaminated streams and ponds [60].

8.6.2.2 Thorium

Thorium is a naturally occurring radionuclide often found alongside uranium in ore deposits. While its radiotoxicity is relatively low due to its long half-life, thorium's chemical toxicity arises from its strong binding to proteins and nucleic acids. Once internalized, thorium accumulates in bones and liver tissues, where it can interfere with enzymatic functions and cellular replication. In soils, thorium exhibits low mobility, but its persistence contributes to long-term contamination, especially in areas with acidic conditions that enhance its solubility [61].

8.6.2.3 Bismuth

Bismuth, particularly in its radioactive isotopes such as bismuth-214, is produced by the decay of uranium and radium. Although elemental bismuth is considered relatively non-toxic, its radioactive isotopes can pose significant chemical hazards. Bismuth compounds can disrupt renal function and have been implicated in gastrointestinal toxicity. In mining environments, bismuth may co-precipitate with other heavy metals, complicating remediation efforts and increasing the cumulative toxic burden.

8.6.2.4 Radium

Radium is a highly radiotoxic and chemically hazardous element that was historically extracted from uranium ores for medical and industrial uses. Chemically, radium behaves similarly to calcium, leading to its incorporation into bones, where it can cause severe damage to bone marrow and increase the risk of cancers. In aquatic systems, radium can be mobilized under reducing conditions, contaminating groundwater and bioaccumulating in aquatic organisms. Its long-lived isotopes contribute to persistent environmental toxicity.

8.6.2.5 Radon

Radon is a noble gas produced from the decay of radium and is notorious for its radiological hazards due to its inhalation risk. However, its chemical inertness belies its indirect chemical toxicity. Radon decay products, such as polonium and bismuth isotopes, attach to airborne particulates and can be inhaled, depositing toxic metals in lung tissues. In abandoned mines, radon accumulation in enclosed spaces poses a significant hazard to both humans and wildlife, especially in poorly ventilated shafts and tunnels.

8.6.2.6 Polonium

Polonium, particularly polonium-210, is one of the most chemically toxic radionuclides known. It is a decay product of radon and is highly bioavailable once internalized. It exemplifies the convergence of chemical and radiological toxicity: highly bioavailable once internalized, polonium binds to sulfhydryl groups in proteins, disrupts cellular respiration, and causes rapid, systemic tissue damage even at trace doses, while its volatility and particulate association enable entry into food chains and localized ecological poisoning. Even trace amounts can be lethal, underscoring the need for stringent monitoring and containment [42].

8.6.3 Processing Reagents and Chemical Residuals

Processing reagents and residual industrial chemicals amplify metal mobility and ecological harm. Sulfuric acid used in leaching produces acid mine drainage, lowering pH and accelerating metal dissolution. Cyanide-derived salts, alkali washes, and residual organic solvents (e.g., kerosene and tertiary amines) persist in soils as reactive or non-aqueous phase liquids, posing acute toxicity to aquatic organisms and chronic contamination of groundwater. These reagents also alter metal speciation and complexation, often increasing bioavailability and complicating remediation strategies.

8.6.4 Other Industrial Contaminants

Other industrial contaminants that compound human and ecological risk include silicon dust, nitrogen and sulfur oxides. Silicon is related to uranium mining and processing in several key ways: as a naturally occurring element in ore deposits; as a component in engineered materials for uranium extraction and remediation; and as an impurity that affects the properties of uranium alloys. Silica dust generated by historic crushing and tailings handling is a respirable hazard that increases the risk of silicosis and other chronic lung diseases among site visitors and remediation workers.

Nitrogen oxides (NO_x) and sulfur oxides (SO_x) are of the chemicals used during the extraction and milling of uranium ore. They are not naturally occurring emissions from the ore itself. Emissions of nitrogen and sulfur oxides from historical milling and ore roasting can have legacy deposition effects, acidifying soils and waters and mobilizing metals already present in tailings. The combined impact of mixed contaminants is synergistic: acidification, organic complexation, and altered redox conditions alter metal speciation and bioavailability, making remediation technically challenging and prolonging ecological recovery timescales [1, 59].

8.7 Perspectives

Uranium is significant not only as the heaviest naturally occurring element present in weighable amounts and as a nuclear fuel, but also for its chemical and radiological implications for human health and the environment. Exposure to environmental uranium has been associated with kidney and liver damage, cancer, congenital disabilities, and endocrine disruption that can increase risks of fertility problems and reproductive cancers [4–6, 38]. For instance, the predicted no-effect concentration (PNEC) is listed in Table 8.3, showing toxicity across different compartments and organisms.

8.7.1 Environmental Mobility and Geochemistry

Uranium's environmental behaviour is strongly controlled by its geochemical form. In oxidizing conditions, uranium predominates as hexavalent U(VI). It is present as the uranyl ion, which forms readily soluble species across a wide range of pH and redox

Table 8.3 Selected predicted no-effect concentration (PNEC) of uranium for the types of non-human biota in different ecosystem compartments

Compartment	PNEC	
1	Terrestrial plants	250 mg U kg^{-1} dry soil
2	Soil biota (from microbes to animals)	100 mg kg^{-1} dry soil
3	(a) Freshwater plants	0.005 mg L^{-1}
	(b) Freshwater invertebrates	0.005 mg L^{-1}
	(c) Freshwater benthos	100 mg kg^{-1} dry sediment
	(d) Freshwater fish at the water	(i) 0.4 mg L^{-1} in very soft water (<10 mg $CaCO_3$ L^{-1})
		(ii) 2.8 mg L^{-1} in soft water (10–100 mg $CaCO_3$ L^{-1})
		(iii) 23 mg L^{-1} in hard water (>100 mg $CaCO_3$ L^{-1})
4	Mammals	0.1 mg kg^{-1} body weight d^{-1}

Source: Sheppard et al. [62]

conditions and therefore moves easily with percolating water into surface and groundwater. The water pathway is the primary exposure route for uranium, although airborne dust can also transport particulate uranium. Geochemical factors such as carbonate concentration, pH, redox potential, and competing cations (e.g., Ca^{2+}, Mg^{2+}, and Fe) strongly influence uranium speciation, solubility, and mobility [9, 63, 64].

8.7.2 Sources, Pathways, and Co-Contaminants

Small amounts of uranium occur naturally in soil, rock, and water, so trace dissolved uranium is normal in many natural waters. Concern arises when concentrations exceed background or guideline levels. Anthropogenic sources dominate the most alarming contaminations and include mining, milling, processing of uranium ores, phosphate fertilizer production, coal combustion residues, military uses, and especially the washing of waste and tailings at abandoned ore mining and processing sites. Releases from tailings and waste heaps are frequently non-point and temporally variable, ranging from short, concentrated pulses to chronic low-level seepage, and are commonly accompanied by other toxic metals and metalloids, such as cadmium, lead, radium, and arsenic.

8.7.3 Human Health Considerations

Because uranium is both chemically toxic and radioactive, health assessments must consider both pathways. Chemical toxicity, particularly nephrotoxicity, is often the dominant near-term concern where uranium is soluble and bioavailable. Chronic exposures through contaminated drinking water can produce systemic effects, and the endocrine-disrupting properties of uranium raise additional concerns for reproductive health. Airborne dust from disturbed tailings can provide a secondary exposure route, especially for nearby communities and workers.

8.7.4 Ecotoxicological Effects Across Compartments

Ecotoxicological risks from uranium are generally driven by its chemical toxicity and vary across ecosystem compartments, organism types, and assessment endpoints. Soil bacteria often tolerate uranium better than higher plants, and plant communities colonizing natural orebodies show visible toxicity at rhizosphere concentrations around 50 ppm, with acute symptoms at roughly ten times that level. Organismal responses are frequently nonlinear: many taxa exhibit hormetic or acclimatory responses at low chronic exposures, and toxicity occurs only under acute conditions, complicating the interpretation of ecological harm.

Effective assessment, therefore, depends on selecting appropriate endpoints vis molecular biomarkers of metabolic disruption, cellular-level oxidative stress indicators, population-level growth and reproduction metrics, and ecosystem-level community composition measures. Long-term observations are essential because ecosystems may show gradual, modified responses following initial acute effects [65, 66].

8.7.5 Uptake, Retention, and Trophic Transfer

Uranium uptake and retention in biota are controlled by site chemistry and organism physiology. In terrestrial plants, clay and organic matter content, acidity, and the presence of competing cations influence root uptake; uranium commonly localizes in roots and seeds rather than leaves. In aquatic systems, carbonate complexation enhances uranium solubility and bioavailability. Evidence for biomagnification through food webs is limited and site-specific, so trophic transfer cannot be generalized across ecosystems without targeted study [67, 68].

8.7.6 Remediation Approaches and the Role of Bioremediation

Classical remediation at decommissioned uranium sites has relied on civil and chemical engineering measures, capping tailings, applying neutralizing or stabilizing materials, and treating contaminated waters with chemical or electrochemical methods. These approaches can be effective but are often expensive, may simply transfer contamination, and can be impractical for the many legacy sites in resource-limited settings. Because discharges from tailings are frequently diffuse and unpredictable, there is growing interest in alternative, lower-cost, and longer-lasting approaches that harness natural processes. Bioremediation, which uses biological processes to immobilize or transform uranium (for example, via microbial reduction or biomineralization), is advocated as a promising option, though field-scale validation, long-term performance data, and careful site-specific design remain critical [6, 69].

8.7.7 Uncertainties and Research Needs

Key uncertainties persist in assessing uranium ecotoxicity and long-term risk at abandoned sites. Data gaps include the ecological significance of hormetic responses, the long-term stability of biologically immobilized uranium, the potential for remobilization under changing environmental conditions, and the extent of trophic transfer in different ecosystems. Addressing these gaps requires coordinated long-term monitoring, standardized endpoints across studies, and comparative field trials of remediation technologies.

8.8 Concluding Perspectives

This chapter explores the core fate and effects of uranium, focusing on toxicology and ecotoxicology at abandoned mining and processing sites. Toxicology and ecotoxicology are broad scientific fields that provide the frameworks, methodologies, and benchmarks necessary to assess harm. However, their application to uranium is context-dependent. The chemical mobility of uranium, particularly in the form of soluble uranyl species, creates distinct exposure pathways and potential hazards, especially through contaminated water and dust.

Real-world impacts are influenced by a complex interplay of local geochemistry, ecological community structure, exposure patterns, and human factors. In practice, toxicological responses to uranium can vary among individuals based on exposure routes and doses. These responses are further influenced by lifestyle, nutritional status, pre-existing health conditions, and genetic predispositions. Ecotoxicological outcomes also differ across ecosystems, as factors such as soil texture, organic matter, pH, carbonate content, redox dynamics, and the presence of competing cations and co-contaminants affect uranium speciation, bioavailability, and biological uptake.

Given that many abandoned sites release uranium alongside other metals and metalloids, and that these discharges can occur either episodically or chronically, risk assessments and remediation strategies must be site-specific, interdisciplinary, and adaptable. Standard engineering controls can mitigate or reduce immediate exposures but may be costly or temporary. Conversely, biologically based approaches present promising, lower-cost alternatives. However, these require rigorous field validation and long-term monitoring to ensure stability amid changing environmental conditions.

Ultimately, responsible management depends on integrating general toxicological principles with detailed local data, continuous observation, and community-sensitive decision-making. This approach ensures that interventions are guided by scientific understanding while accounting for the diverse vulnerabilities of individuals and ecosystems affected by legacy uranium contamination.

References

1. Mkandawire, M., Pompe, W., & Dudel, E. G. *der Deutschen Gesellschaft für Limnologie (DGL) und Societas Internationalis Limnologie (SIL)* (pp. 124–125). der Deutschen Gesellschaft für Limnologie (DGL).
2. NEA. (2002). *Environmental remediation of uranium production facilities*. (OECD.
3. Mehra, R., & Kaur, S. (2020). In D. K. Gupta & C. Walther (Eds.), *Uranium in plants and the environment* (pp. 217–241). Springer International.
4. Schuttmann, W. (1993). Schneeberg lung disease and uranium mining in the Saxon Ore Mountains (Erzgebirge). *American Journal of Industrial Medicine, 23*, 355–368.
5. Sram, R. J., et al. (1993). Monitoring genotoxic exposure in uranium miners. *Environmental Health Perspectives, 99*, 303–305.
6. WHO. (2001). *Depleted uranium sources, exposure and health effects*. Department of Protection of the Human Environment, World Health Organization.
7. Guéguen, Y., & Frerejacques, M. (2022). Review of knowledge of uranium-induced kidney toxicity for the development of an adverse outcome pathway to renal impairment. *International Journal of Molecular Sciences, 23*. https://doi.org/10.3390/ijms23084397
8. Cui, Q., et al. (2023). A critical review of uranium in the soil-plant system: Distribution, bioavailability, toxicity, and bioremediation strategies. *Critical Reviews in Environmental Science and Technology, 53*, 340–365. https://doi.org/10.1080/10643389.2022.2054246
9. Kalin, M., Wheeler, W. N., & Meinrath, G. (2005). The removal of uranium from mining waste water using algal/microbial biomass. *Journal of Environmental Radioactivity, 78*, 151–177.
10. Landa, E. R. (2004). Uranium mill tailings: nuclear waste and natural laboratory for geochemical and radioecological investigations. *Journal of Environmental Radioactivity, 77*, 1–27.
11. IAEA. *An International Workshop on Environmental Contamination from uranium production facilities and their remediation* (Vol. 272). International Atomic Energy Agency.
12. Biswas, P., & Wu, C. Y. (1998). Control of toxic metal emissions from combustors using sorbents: A review. *Journal of the Air & Waste Management Association, 48*, 113–127.
13. Sims, R., Lawless, T. A., Alexander, J. L., Bennett, D. G., & Read, D. (1996). Uranium migration through intact sandstone: Effect of pollutant concentration and the reversibility of uptake. *Journal of Contaminant Hydrology, 21*, 215–228.
14. Riley, S. J. (1995). Geomorphic estimates of the stability of a uranium mill tailings containment cover: Nabarlek, Northern Territory, Australia. *Land Degradation and Rehabilitation, 6*, 1–16.
15. Roussel, C., Bril, H., & Fernandez, A. (2000). Arsenic speciation: Involvement in evaluation of environmental impact caused by mine wastes. *Journal of Environmental Quality, 29*, 182–188.
16. IAEA. (2004). The long term stabilization of uranium mill tailings: Final report of a co-ordinated research project 2000–2004. In *Report No. IAEA-TECDOC-1403, 311 International Atomic Energy Agency, Vienna, Austria.*

17. Hoffmann, W., Kranefeld, A., & Schmitz-Feuerhake, I. (1993). Radium-226-contaminated drinking water: hypothesis on an exposure pathway in a population with elevated childhood leukemia. *Environmental Health Perspectives, 101*, 113–115.
18. Kusiak, R. A., Ritchie, A. C., Muller, J., & Springer, J. (1993). Mortality from lung cancer in Ontario uranium miners. *British Journal of Industrial Medicine, 50*, 920–928.
19. Simon, O., & Garnier-Laplace, J. (2005). Laboratory and field assessment of uranium trophic transfer efficiency in the crayfish Orconectes limosus fed the bivalve C. fluminea. *Aquatic Toxicology, 74*, 372–383.
20. Veiga, L. H. S., Amaral, E. C. S., & Fernandes, H. M. (1998). Human health risk screening of radioactive and nonradioactive contaminants due to uranium industry operation. *Journal of Environmental Radioactivity, 39*, 69–85.
21. Piao, C. Q., & Hei, T. K. (1993). The biological effectiveness of radon daughter alpha particles. I. Radon, cigarette smoke and oncogenic transformation. *Carcinogenesis, 14*, 497–501.
22. Morris, S., & Meinhold, A. F. (1995). Probabilistic risk assessment of nephrotoxic effect of uranium in drinking water. *Health Physics, 69*, 897–908.
23. Zhang, L., et al. (2022). Health effects of particulate uranium exposure. *Toxics, 10*, 575.
24. Henge Napoli, M. H., Archimbaud, M., Ansoborlo, E., Metivier, H., & Gourmelon, P. (1995). Efficacy of 3,4,3-LIHOPO for reducing the retention of uranium in rat after acute administration. *International Journal of Radiation Biology, 68*, 389–393.
25. Russell, J. J., Kathren, R. L., & Dietert, S. E. (1996). A histological kidney study of uranium and non-uranium workers. *Health Physics, 70*, 466–472.
26. Dang, H. S., Pullat, V. R., & Sharma, R. C. (1995). Distribution of uranium in human organs of an urban Indian population and its relationship with clearance half-lives. *Health Physics, 68*, 328–331.
27. Periyakaruppan, A., Kumar, F., Sarkar, S., Sharma, C., & Ramesh, G. (2007). Uranium induces oxidative stress in lung epithelial cells. *Archives of Toxicology, 81*, 389–395.
28. Rybicka, K. K. (1996). Glycosomes—The organelles of glycogen metabolism. *Tissue and Cell, 28*, 253–265.
29. Miller, A. C., et al. (1998). Transformation of human osteoblast cells to the tumorigenic phenotype by depleted uranium uranyl chloride. *Environmental Health Perspectives, 106*, 465–471.
30. Van Horn, J. D., & Huang, H. (2006). Uranium(VI) bio-coordination chemistry from biochemical, solution and protein structural data. *Coordination Chemistry Reviews, 250*, 765–775.
31. De Castro, L., et al. (2023). Renal toxicity and biokinetics models after repeated uranium instillation. *Scientific Reports, 13*, 4111. https://doi.org/10.1038/s41598-023-31073-1
32. Miller, R. C., et al. (1995). The biological effectiveness of radon-progeny alpha particles: II. Oncogenic transformation as a function of linear energy transfer. *Radiation Research, 142*, 54–60.
33. Stearns, D. M., et al. (2005). Uranyl acetate induces hprt mutations and uranium-DNA adducts in Chinese hamster ovary EM9 cells. *Mutagenesis, 20*, 417–423.
34. Simon, O., & Garnier-Laplace, J. (2004). Kinetic analysis of uranium accumulation in the bivalve Corbicula fluminea: effect of pH and direct exposure levels. *Aquatic Toxicology, 68*, 95–108.
35. Yazzie, M., Gamble, S. L., Civitello, E. R., & Stearns, D. M. (2003). Uranyl acetate causes DNA single strand breaks in vitro in the presence of ascorbate (Vitamin C). *Chemical Research in Toxicology, 16*, 524–530.
36. Tsuruta, T. (2006). Bioaccumulation of uranium and thorium from the solution containing both elements using various microorganisms. *Journal of Alloys and Compounds, 408–412*, 1312–1315.
37. Bird, G. A. (2012). In R. A. Meyers (Ed.), *Encyclopedia of sustainability science and technology* (pp. 11220–11262). Springer.
38. ATSDR. (2013). In Agency for Toxic Substances and Disease Registry (Ed.), *Vol. CAS#: 7440-61-1*. U.S. Department of Health and Human Services, Public Health Service. Retrieved from https://wwwn.cdc.gov/TSP/ToxProfiles/ToxProfiles.aspx?id=440&tid=77

39. Yusuf, I. (2024). *Uranium biokinetics and internal dosimetry*. Linköping University Electronic Press.
40. Yue, Y.-C., Li, M.-H., Wang, H.-B., Zhang, B.-L., & He, W. (2018). The toxicological mechanisms and detoxification of depleted uranium exposure. *Environmental Health and Preventive Medicine, 23*, 18. https://doi.org/10.1186/s12199-018-0706-3
41. Gao, N., Huang, Z., Liu, H., Hou, J., & Liu, X. (2019). Advances on the toxicity of uranium to different organisms. *Chemosphere, 237*, 124548. https://doi.org/10.1016/j.chemosphere.2019.124548
42. IAEA. (2023). In International Atomic Energy Agency (Ed.), *Vol. STI/DOC/010/488* (p. 361). International Atomic Energy Agency.
43. Tasat, D. R., Orona, N. S. S., Bozal, C. B. B., Ubios, A., & Cabrini, R. (2012). In P. Bubulya (Ed.), *Cell metabolism—Cell homeostasis and stress response*. IntechOpen.
44. Antunes, I. M. H. R., Carvalho, P. C. S., Albuquerque, M. T. D., & Santos, A. C. T.P. Stanley, C. Wolkersdorfer, & K. Wolkersdorfer (Eds.). *IMWA 2021—"Mine Water Management for Future Generations"*. International Mine Water Association.
45. EPA. (2007). *Investigation of potential health, geographic, and environmental issues of abandoned uranium mines* (p. 131). U.S. Environmental Protection Agency, Office of Radiation and Indoor Air Radiation Protection Division (6608J).
46. Stone, J. J., & Stetler, L. D.R. I. Barnhisel (Ed.). *2009 National Meeting of the American Society of Mining and Reclamation: Revitalizing the environment: proven solutions and innovative approaches* (pp. 1394–1411). America Society of Mining and Reclamation.
47. Kalin, M. A. (1983). Long-term ecological behaviour of abandoned uranium mill tailings: 1. Synoptic survey and identification of invading biota. In *Report No. EPS 4-ES-83-1*. Institute for Environmental Studies, University of Toronto.
48. Lin, Y., Hoover, J., Beene, D., Erdei, E., & Liu, Z. (2020). Environmental risk mapping of potential abandoned uranium mine contamination on the Navajo Nation, USA, using a GIS-based multi-criteria decision analysis approach. *Environmental Science and Pollution Research, 27*, 30542–30557. https://doi.org/10.1007/s11356-020-09257-3
49. Carvalho, F. P., et al. (2023). *The Environmental Behaviour of Uranium. Report No. 488*. International Atomic Energy Agency.
50. Croteau, M.-N., Fuller, C. C., Cain, D. J., Campbell, K. M., & Aiken, G. (2016). Biogeochemical controls of uranium bioavailability from the dissolved phase in natural freshwaters. *Environmental Science & Technology, 50*, 8120–8127. https://doi.org/10.1021/acs.est.6b02406
51. Croteau, M.-N., Fuller, C. C., Cain, D. J., & Campbell, K. M. (2025). Dietary bioavailability of uranium to a model freshwater invertebrate. *Environmental Science & Technology, 59*, 16641–16651. https://doi.org/10.1021/acs.est.5c05140
52. Bergmann, M., & Graça, M. A. S. (2020). Bioaccumulation and dispersion of uranium by freshwater organisms. *Archives of Environmental Contamination and Toxicology, 78*, 254–266. https://doi.org/10.1007/s00244-019-00677-y
53. Zerbini, M., et al. (2024). Exploring uranium bioaccumulation in the brown alga *Ascophyllum nodosum*: Insights from multi-scale spectroscopy and imaging. *Scientific Reports, 14*, 1021. https://doi.org/10.1038/s41598-023-49293-w
54. Rogiers, T., et al. (2022). Molecular mechanisms underlying bacterial uranium resistance. *Frontiers in Microbiology, 13*. https://doi.org/10.3389/fmicb.2022.822197
55. Kumar, A., et al. (2023). A comprehensive review of Uranium in the terrestrial and aquatic environment: Bioavailability, immobilization, tolerance and remediation approaches. *Plant and Soil, 490*, 31–65. https://doi.org/10.1007/s11104-023-06101-8
56. Hinck, J., et al. (2010). Biological pathways of exposure and ecotoxicity values for uranium and associated radionuclides: Chapter D in Hydrological, geological, and biological site characterization of breccia pipe uranium deposits in Northern Arizona. In *Report No. 2010-5025D* (pp. 283–351).

57. Chen, H., et al. (2025). Uranium contamination mediating soil and ore microbial community assembly at four mining sites, South China. *Frontiers in Microbiology, 16*. https://doi.org/10.3389/fmicb.2025.1553072
58. Meinrath, G. (1998). Aquatic chemistry of uranium. A review focusing on aspects of environmental chemistry. *Freiberg On-line Geoscience (FOG), 1*, 1–101.
59. Randhawa, J. S., Robin, Kaur, P., & Meehnian, H. (2024). A comprehensive review on health and environmental hazards of uranium: analytical techniques, mitigation strategies and its toxicity treatments. *Journal of Radioanalytical and Nuclear Chemistry, 333*, 3693–3711. https://doi.org/10.1007/s10967-024-09527-1
60. Chang, Y., & Chiang, C.-K. (2024). The impact of thallium exposure in public health and molecular toxicology: A comprehensive review. *International Journal of Molecular Sciences, 25*, 4750.
61. Rump, A., Hermann, C., Lamkowski, A., Popp, T., & Port, M. (2023). A comparison of the chemo- and radiotoxicity of thorium and uranium at different enrichment grades. *Archives of Toxicology, 97*, 1577–1598. https://doi.org/10.1007/s00204-023-03484-6
62. Sheppard, S. C., Sheppard, M. I., Gallerand, M.-O., & Sanipelli, B. (2005). Derivation of ecotoxicity thresholds for uranium. *Journal of Environmental Radioactivity, 79*, 55–83.
63. Al Hashimi, A., Evans, G. J., & Cox, B. (1996). Aspects of the permanent storage of uranium tailings. *Water, Air, & Soil Pollution, 88*, 83–92.
64. Trontelj, A., & Ponikvar Zorko, P. (1998). Influence of a uranium mine on the macrozoobenthic communities of the streams in the nearest environs, Slovenia. *Water Science and Technology, 37*, 235–241.
65. Carvalho, I. G., et al. (2005). Environmental impact of uranium mining and ore processing in the Lagoa Real District, Bahia, Brazil. *Environmental Science & Technology, 39*, 8646–8652.
66. Mkandawire, M. (2005). *Fate and effect of uranium and arsenic in surface waters of abandoned uranium mines: Studies with Lemna gibba L.* Shaker.
67. Bowden, R. A., & Shaw, R. P. (2007). The Kayelekera uranium deposit, Northern Malawi: Past exploration activities, economic geology and decay series disequilibrium. *Applied Earth Science, 116*, 55–67. https://doi.org/10.1179/174327507x167082
68. Mkandawire, M., Taubert, B., & Dudel, E. G. (2004). Capacity of *Lemna gibba* L. (Duckweed) for uranium and arsenic phytoremediation in mine tailing waters. *International Journal of Phytoremediation, 6*, 347–362.
69. Giannardi, C., & Dominici, D. (2003). Depleted uranium, the longterm effects of the US bombing of Iraq. *Ponte, 59*, 227–238.

Chapter 9
Radioecology of Abandoned Uranium Mines

Abstract The radioecology of abandoned uranium mines and processing facilities is examined through the interactions between radionuclide sources and complex geochemical, hydrological, and ecological systems. Field observations, long-term monitoring, conceptual models, and numerical simulations from multiple regions form the basis of a systematic framework for evaluating exposure pathways and ecological impacts. The text highlights how site chemistry, groundwater and surface-water dynamics, soil-plant interactions, and food-web processes shape uranium and radionuclide behaviour and bioavailability. It does not distinguish among mining methods, since the intrinsic radioactivity of mine wastes is broadly similar. Instead, it stresses that radionuclide transport and exposure outcomes depend on site-specific factors, including geochemistry, hydrology, climate, and background geology. These factors guide monitoring strategies, risk assessment, and remediation planning.

Keywords Radiotoxicological hazard · Dose-effect relationship · Radionuclide transport · Ecological timescales · Exposure pathways · Physicochemical partitioning · Biological assimilation

9.1 Introduction

The radioecology of abandoned uranium mines elucidates the mechanisms by which radioactive elements disseminate into the surrounding environment, thereby posing significant health risks through contamination of soil, water, and air [1]. Central to this discourse are the long-lived radionuclides, specifically radium and thorium, which are frequently concentrated within mill tailings and waste rock. The decay of radium not only generates radon gas but also heightens the risk of lung cancer. Exposure to uranium and other radionuclides can occur through contaminated drinking water and the inhalation of radioactive dust. The enduring legacy of these mining operations is often characterized by elevated radiation levels in adjacent communities and ecosystems, engendering substantial public health and

M. Mkandawire, *Fate and Effects of Uranium in Abandoned Uranium Mines*,
https://doi.org/10.1007/978-3-032-10447-2_9

environmental challenges that necessitate protracted, complex, and costly remediation efforts [2–4].

In this context, radioecology extends beyond the mere documentation of contamination; it seeks to elucidate the intricate processes governing the migration, transformation, and interaction of radionuclides with biotic components. Within abandoned mine and processing sites, various factors collectively shape the pathways through which contaminants are distributed and their subsequent bioavailability [5–7]. The factors include the weathering of tailings, episodic runoff, groundwater flow, aeolian transport of dust, and biological activity. Other factors, such as chemical speciation, redox conditions, sorption dynamics with minerals and organic matter, and microbial transformations, play pivotal roles in determining whether uranium and its decay products remain immobilized or become mobile, thereby entering food webs [8–11].

At the landscape scale, mill sites and waste rock piles create heterogeneous source zones that facilitate the spread of contamination into adjacent soils, sediments, waterways, and the atmosphere. Radon emanation from mill tailings poses risks of chronic exposure, both indoors and outdoors. At the same time, fine particulate matter generated by eroding waste can transport adsorbed radionuclides over extensive distances. Furthermore, seasonal hydrology and climate variability significantly modulate these contaminant pathways, influencing the dynamics among dilution, sequestration, and remobilization, complicating the forecasting of future exposure scenarios [12–14].

The interconnection between human and ecological receptors is evidenced through shared exposure pathways, such as drinking water sourced from contaminated aquifers, agricultural products and wild foods cultivated or foraged from affected lands, and the inhalation of dust or radon-contaminated air. Wildlife can bioaccumulate radionuclides within their tissues and transfer these through trophic interactions, resulting in population-level effects that parallel human health concerns [15]. Consequently, effective site characterization, risk assessment, remediation, and long-term stewardship rely on a comprehensive mechanistic understanding of radioecological processes rather than on point measurements of radioactivity alone.

9.2 Radiation Concerns and Risks

Abandoned uranium mines and processing sites remain radiologically mainly active because of the decay products of uranium rather than uranium metal or uranyl salts themselves. Most operational extraction removes a significant fraction of uranium as the commercial commodity. The residual uranium in tailings and waste rock is often present at lower specific activity and, critically, has a very long half-life, producing daughter nuclides with greater radiotoxicity and mobility on human and ecological timescales. Uranium isotopes such as ^{238}U and ^{235}U have half-lives on the order of geological time ($t_{1/2}{}^{238}\mathrm{U} \approx 4.47 \times 10^9 years$, $t_{1/2}{}^{235}\mathrm{U} \approx 7.04 \times 10^8$ $years$), so their specific activity per unit mass is relatively low compared with many of their short-lived daughters that transiently accumulate in environmental reservoirs.

The most consequential hazard arises from the uranium decay series. For the ^{238}U series, the simplified chain of principal concern is

$$^{238}\mathrm{U} \rightarrow {}^{234}\mathrm{U} \rightarrow {}^{230}\mathrm{Th} \rightarrow {}^{226}\mathrm{Ra} \rightarrow {}^{222}\mathrm{Rn} \rightarrow \ldots \rightarrow {}^{210}\mathrm{Pb} \rightarrow \ldots \rightarrow {}^{206}\mathrm{Pb}(\text{stable})$$

Key radiotoxic and mobile members of this chain are radium-226 (^{226}Ra), radon-222 (^{222}Rn), lead-210 (^{210}Pb), and polonium-210 (^{210}Po). Radium has a half-life. $t_{1/2}$(^{226}Ra $\approx$ 1600 *years*) and emits high-energy alpha radiation upon decay to radon; radon is a noble gas (^{222}Rn, $t_{1/2} \approx$ 3.82 *days*) and readily emanates from tailings and fractured bedrock, creating inhalation exposures remote from the source. Lead-210 ($t_{1/2} \approx$ 22.3 *years*) and polonium-210 ($t_{1/2} \approx$ 138 *days*) are chemically prone to association with particulates and biota and contribute to internal doses through ingestion and incorporation into food webs.

Geochemical controls govern which daughter radionuclides become mobile and bioavailable. Under oxidizing, acidic conditions commonly produced in exposed tailings, hexavalent uranium as the uranyl ion, UO_2^{2+}, forms soluble complexes (e.g., uranyl-carbonate complexes $UO_2(CO_3)_3^{4-}$) that can be transported with groundwater. Conversely, under reducing conditions, uranium is often immobilized as a tetravalent species (e.g., UO_2 precipitates) or strongly sorbed to mineral surfaces; the half-reaction describes this redox sensitivity

$$UO_2^{2+} + 4H^+ + 2e^- \rightarrow U^{4+} + 2H_2O$$

So small changes in Eh and pH can shift uranium between mobile and immobile forms. Radium behaves differently: Ra^{2+} is an alkaline-earth cation chemically analogous to Ba^{2+} and Ca^{2+}, exhibiting limited redox chemistry but pronounced affinity for ion-exchange sites in sediments and for co-precipitation with sulfate or carbonate minerals. As a result, radium mobility often correlates with sulfate content, ion-exchange capacity, and competitive cation concentrations rather than redox state.

Radon emanation is governed by emanation coefficients and diffusion transport: radon produced by the alpha decay of ^{226}Ra either remains trapped in mineral grains, is released into pore gas and diffuses through tailings, or dissolves into porewater and is transported hydraulically. The radon flux J from a porous medium can be approximated by a diffusion-advection expression:

$$J = -D_{\mathrm{eff}} \frac{\partial C}{\partial x} + vC$$

where D_{eff} is the effective diffusion coefficient in the porous medium, C is the radon concentration, and v is any advective pore-gas velocity; coarse, dry, and desiccated tailings typically increase D_{eff} and therefore radon exhalation and dust-borne alpha/beta activity.

Particle-bound radionuclides represent another major vector. Fine particulate matter produced by wind erosion or mechanical disturbance carries sorbed lead, polonium, and uranium daughters. The mass activity (A_m) of such particles depends on specific activity and particulate mass m:

$$A_m = a_s \cdot m$$

where a_s is the activity per mass; small particles have higher surface area-to-mass ratios and therefore a greater fraction of sorbed daughter products per unit mass, increasing inhalation dose per unit mass of dust inhaled.

Ecological transfer is driven by physicochemical partitioning and biological assimilation. Soil-plant transfer factors (TF, unitless) for radionuclides describe the ratio of activity concentration in plant tissue to that in soil and are strongly nuclide- and species-dependent:

$$TF = \frac{C_{\text{plant}}}{C_{\text{soil}}}$$

Values for ^{226}Ra and ^{210}Pb are generally lower than for some soluble forms of uranium under oxidizing conditions, yet ^{210}Po exhibits high biological affinity and can concentrate in invertebrates and grazing animals, producing elevated doses in specific food web nodes. Trophic transfer factors and metabolic assimilation lead to internal doses from alpha emitters, which have high linear energy transfer and strong biological effectiveness when lodged in the lungs, bone, or soft tissues.

Dose metrics must account for radiation type and energy. Alpha emitters such as radium and polonium deliver high localized dose despite low penetration depth; the committed effective dose from ingestion or inhalation of alpha-emitting species can therefore exceed that from comparable beta/gamma emitters. Quantitatively, the activity A (Bq) of a radionuclide with decay constant λ relates to the number of nuclei N by

$$A = \lambda N$$

and the decay constant to half-life by

$$\lambda = \frac{\ln 2}{t_{1/2}}$$

Shorter half-life daughters produce higher activity per nucleus and greater instantaneous dose rates than their long-lived uranium parent for the same atom count, which explains why daughters often drive hazard even when uranium mass dominates the inventory.

Temporal evolution of hazards at legacy sites reflects in situ ingrowth and outgrowth of daughters. Immediately after ore processing, daughter-product activities can be out of secular equilibrium with the parent uranium. Over time, the chain tends towards secular equilibrium when the production and decay rates of daughters balance, with daughter activities approaching the parent activity for daughters with half-lives short relative to the parent. The secular equilibrium condition for a daughter D produced by parent P occurs when

$$A_D \oplus A_P \quad \text{if}\, t_{1/2,D} \ll t_{1/2,P}$$

Because many daughter isotopes have half-lives many orders of magnitude shorter than uranium, they can dominate site activity for centuries to millennia until long-lived daughters such as ^{226}Ra decay away.

Human exposure scenarios that dominate radiological risk at abandoned sites are inhalation of radon and dust carrying alpha-emitting daughters, ingestion of contaminated water and foodstuffs containing bioaccumulated daughters, and external gamma exposure from sediments and tailings with elevated concentrations of gamma-emitting daughters, such as ^{214}Bi or ^{214}Pb. The relative contribution of each pathway depends on site form and use: open tailings and eroding waste rock favour airborne and external exposure; contaminated surface water and springs amplify ingestion pathways; sediments and flooded workings concentrate particulate-affiliated daughters and sustain benthic food web exposure.

The persistent radiological risk at abandoned uranium and processing sites is principally a consequence of the chemical and radiological behaviour of uranium daughter products, like radium, radon, lead, and polonium, whose shorter half-lives, higher specific activities, and distinct geochemical affinities render them more mobile, bioavailable, and radiotoxic than the largely depleted uranium parent. Understanding redox controls on uranyl speciation, ion exchange, and co-precipitation behaviour of radium; radon emanation dynamics; particulate transport mechanisms; and biological transfer factors is essential for accurate characterization of environmental exposure and dose from legacy uranium sites.

9.3 Radiation Transfer into Ecosystem Compartments

Abandoned uranium mining and processing sites remain persistent sources of radioactive uranium and its associated radionuclides to air, soil, surface water, groundwater, vegetation, and biota for decades to centuries after closure. The radiologically and chemically important nuclides derive principally from the ^{238}U decay series, notably ^{238}U, ^{234}U, ^{230}Th, ^{226}Ra, ^{210}Pb, and ^{210}Po, and, where ore mineralogy or processing introduces thorium-rich ores, contributions from the ^{232}Ra series are also relevant. Radioactive uranium in ore, waste rock, and tailings exists in multiple mineral and aqueous forms: reduced uraninite (UO_2), coffinite ($USiO_4$), oxidized uranyl species UO_2^{2+}, uranyl-carbonate complexes such as $UO_2(CO_3)_3^{4-}$, and uranyl-organic complexes ($UO_2 - DOC$). Oxidation state, coordination chemistry, and aqueous complexation determine mobility, bioavailability, and, therefore, transfer into ecosystem compartments. Radiological assessment and mechanistic modelling must, hence, represent both near-equilibrium metrics for long-term stewardship and time-dependent parameters that capture routine discharges, episodic releases, and kinetic leaching from solid wastes. Secular disequilibrium within decay chains is common in weathered tailings and must be explicitly accounted for in predictive frameworks.

9.3.1 Influence of Radionuclide Speciation

Radionuclide speciation determines the chemical identity and phase of radionuclides present in an abandoned uranium mine setting, thereby controlling their mobility, bioavailability, and the dominant pathways for ecosystem transfer. In mine-impacted waters and porewaters, speciation is typically dominated by aqueous complexes and low-molecular-mass (LMM) species when oxidizing, carbonate-rich conditions prevail; hexavalent uranium as the uranyl ion, UO_2^{2+} , forms strong, highly soluble complexes with carbonate and calcium that remain in the dissolved phase and are rapidly transported in groundwater and surface runoff. Under reducing, anoxic conditions found in saturated sediments or organic-rich soils, uranium is converted to tetravalent U(IV) phases that precipitate as poorly soluble oxide, hydroxide, or uraninite-like solids. This dramatically lowers dissolved concentrations but creates long-lived insoluble reservoirs that may be remobilized if redox conditions change.

Solid phases in mine tailings, waste rock, and sediments host a mixture of high-molecular-mass (HMM) species, particle-associated radionuclides, and true radioactive particles. Radionuclides sorbed to mineral surfaces, occluded within secondary authigenic phases (e.g., Fe/Mn oxides and phosphate minerals), or encapsulated in colloidal organic-mineral aggregates behave differently from dissolved species: they are less immediately bioavailable but represent a persistent source that can release mobile forms through mineral weathering, changes in ionic strength, bioturbation, or physical erosion. Colloid-facilitated transport is a critical consideration because radionuclide-laden colloids can remain suspended and migrate beyond the range predicted from dissolved-phase chemistry alone. Radionuclide-laden colloids include iron oxyhydroxide nanoparticles, organic colloids, and clay fragments [16]. Under turbulent flows or during resuspension events, these colloids increase the effective transport distance of otherwise immobile isotopes.

Oxidation state controls for actinides such as neptunium and plutonium introduce additional complexity. Neptunium in oxidizing mine waters commonly exists as Np(V) as the neptunyl ion, which is relatively mobile and forms stable aqueous complexes; reduction to Np(IV) leads to insoluble hydroxides and markedly lower mobility [17–19]. Plutonium can partition among Pu(III), Pu(IV), and Pu(V/VI) states. Even at low dissolved concentrations, it often associates strongly with colloids or organic matter, exhibiting pseudo-conservative transport under some conditions. Radium, as a group II analogue of calcium, tends to be present as an aqueous

Ra^{2+} ion in porewaters and surface waters but exhibits strong removal in sediments by co-precipitation with barite or adsorption to sulfate-rich mineral phases; its behaviour is sensitive to sulfate concentration, competing cations, and the carbonate chemistry that influences exchange with solid matrices.

The sediment and soil compartment acts as the principal sink for many radionuclides, but the form in which they are sequestered determines long-term availability. Sorption onto clay minerals and surface complexation to Fe- and Mn-oxide coatings immobilize many radionuclides, while incorporation into authigenic sulfide or phosphate phases during diagenesis produces geochemically stable hosts. Organic matter exerts dual influences: dissolved organic carbon (DOC) can complex actinides, increasing their aqueous mobility, while particulate organic matter and humic-rich colloids enhance particulate-bound transport and may facilitate uptake by benthic organisms. Therefore, pH and redox potential (Eh) act as master variables that minor shifts in pH or transitions from oxic to suboxic/anoxic conditions can change dominant complexes, alter partition coefficients (K_d), and switch a radionuclide from a labile dissolved form to a fixed solid phase or vice versa.

Biological compartments record the integrated outcome of these chemical controls. Plant uptake is governed by root zone speciation: soluble LMM species and small organometallic complexes are more readily absorbed, whereas particle-associated radionuclides are largely excluded from direct root uptake but can contribute via foliar deposition or ingestion of contaminated particulates by grazers. Aquatic organisms accumulate radionuclides via direct uptake of dissolved species across gill surfaces and indirectly through trophic transfer. Seasonal changes in water chemistry, temperature, and primary productivity modulate bioaccumulation factors and concentration ratios, producing temporal variability in organismal burdens that must be resolved when assessing exposure to wildlife and humans.

For risk assessment and remediation planning, the distinction between total activity and speciated forms is fundamental. Speciation data enable prediction of which compartments will act as ongoing sources or sinks, inform the selection of appropriate K_d and concentration ratio values for models, and guide engineered interventions: oxidant or reductant amendments, reactive barriers targeting specific phases (e.g., sulfate-reducing systems for metal sequestration), or amendments that promote formation of low-solubility mineral hosts. Accurate conceptual site models, therefore, require coupled geochemical and hydrological characterization: field and laboratory speciation measurements (sequential extractions, in situ voltammetry, and filtered/unfiltered activity coupled with colloid size-fractionation), redox and pH profiling, and kinetic information on particle weathering and microbial-mediated transformation to capture both equilibrium and rate-limited processes that determine long-term radionuclide behaviour in abandoned uranium mine ecosystems.

9.3.2 Transfer Routes into Ecosystem Compartments

Radionuclides liberated from abandoned uranium mining and processing sites are redistributed across the landscape by a small set of powerful, interacting transport mechanisms. These mechanisms ultimately determine which ecosystem compartments become contaminated and how exposure risks propagate. Waterborne processes are often dominant where mine wastes and tailings are exposed to precipitation and surface flow. Dissolution and leaching mobilize soluble radionuclide species from tailings, waste rock, and contaminated soils into percolating water, carrying contaminants into shallow groundwater and through the soil profile. Episodic or intense rainfall converts contaminated surfaces into active sources through surface runoff that entrains particles and dissolved species and routes them to streams, wetlands, and lakes. Legacy mine waters and discharges, frequently acidic and metal-rich, provide continuous or intermittent point-source inputs to receiving waters, where radionuclides partition between dissolved phases and settling particulates. In the receiving environment, river and lake sediments serve as long-term sinks, as particulate-bound radionuclides accumulate in bed and floodplain deposits, creating secondary contamination zones that can remobilize under changing hydrodynamic or geochemical conditions.

Airborne transport moves radioactivity beyond immediate mine footprints by aerosolizing dust and releasing radon gas. Wind erosion of tailings, exposed waste piles, and dried sediments produces contaminated dust plumes that deposit radionuclides over adjacent soils, vegetation, and water bodies, forming a broad, diffuse pattern of contamination. Radon, generated during the decay of radium in uranium-bearing materials, escapes as a gas and creates an inhalation exposure pathway that is spatially and temporally independent of particulate transport. Both dust deposition and radon release link the site directly to human and ecological inhalation and dermal exposure routes, and they seed terrestrial surfaces where subsequent waterborne and biological processes can act on the deposited material.

Biological pathways convert environmental contamination into ecological and human health risk through uptake and trophic transfer. Plants take up dissolved radionuclides from the root zone, and some species exhibit preferential accumulation of specific radionuclides, concentrating activity in leaves, stems, or root tissues. Foliar interception of dust further increases radionuclide burdens on vegetation, creating combined root and surface exposure. Herbivores ingest contaminated forage, soil, or sediments, incorporating radionuclides into tissues and organs. Predators then accumulate these radionuclides through dietary intake, leading to biomagnification patterns in both terrestrial and aquatic food webs. In aquatic systems, plankton and benthic organisms concentrate radionuclides from water and sediments and pass them upward through invertebrates, fish, and piscivorous birds and mammals. In terrestrial food chains, small mammals, birds, and larger predators acquire activity via foraging and soil ingestion. These food chain transfers extend the impact of a site far beyond the original footprint and connect environmental contamination directly to human exposure through consumption of contaminated water, fish, crops, livestock, and wild game.

Direct exposure pathways operate concurrently with transport and bioaccumulation. Gamma-emitting radionuclides in contaminated soils, tailings, and sediments deliver external dose to organisms and people present at or near contaminated surfaces, independent of ingestion or inhalation. Inhalation of airborne particulates and radon constitutes an efficient internal exposure route with particular relevance to mine workers, nearby residents, and wildlife occupying burrows or shelters within contaminated zones. Ingestion of contaminated water, foodstuffs, or soil is a direct route for internalizing radionuclides, with dose and health consequences governed by the type of radionuclide, its chemical form, and the duration and frequency of exposure.

The primary ecosystem compartments that receive and retain radionuclides are soil, sediments, water, vegetation, and biota. Soil serves as the first long-term repository for deposited dust and the locus of exchange between solid and aqueous phases. Sediments in rivers, lakes, and wetlands accumulate particle-bound radionuclides and act as secondary sources that can prolong ecological exposure for decades. Surface water and groundwater convey dissolved radionuclides through the landscape and supply contaminated drinking and irrigation water. Vegetation reflects both soil-derived uptake and atmospheric deposition, serving as the immediate food resource that links contamination to herbivores. Aquatic and terrestrial fauna concentrate radionuclides through ingestion and inhalation, and their movement links contamination across habitats and to human consumers.

Understanding the relative importance of each transfer route at a particular site requires integrating geochemistry, hydrology, meteorology, and ecology. Site-specific factors such as tailings composition, waste management history, local climate and precipitation patterns, landscape drainage, vegetation cover, and the presence of wetlands or lakes will dictate whether leaching, runoff, dust mobilization, radon emanation, or biological uptake dominates. Seasonal cycles and episodic events such as storms, droughts, or fire can shift dominant pathways by altering erosion rates, water table levels, or vegetation cover. Effective risk assessment and remediation planning, therefore, rest on characterizing these pathways, identifying contaminated compartments and secondary sinks, and prioritizing actions that interrupt mobility and reduce routes by which radionuclides enter food chains and human exposure pathways.

9.3.3 Transfer to Farmland and Adjacent Ecosystems

Farmland downgradient or adjacent to tailings and process areas is exposed to contaminated irrigation, plume seepage, atmospheric dry and wet deposition, particulate resuspension, and direct contact with contaminated soils. Deposition on crops occurs through two main pathways: surface-adhered particulates that can be removed, and a soluble fraction that may be absorbed through the leaves. For uranium-series nuclides, foliar uptake of soluble uranyl species (UO_2^{2+}, $UO_2\left(CO_3\right)_n^{2-2n}$). It is generally limited by cuticular barriers, and translocation to the grain is usually small relative to root uptake. The weathering half-life, T_w, of canopy-intercepted material depends on rainfall intensity, leaf morphology, and particle composition; fine soluble particulates wash to soil rapidly, whereas refractory grains can persist on foliage and be incorporated into harvestable biomass. Root uptake is controlled by the chemical activity of uranium in root-zone porewater, not by bulk soil activity; uranyl-carbonate complexation and complexation with DOC substantially increase aqueous mobility at moderate to high bicarbonate and DOC levels, so porewater measurements with suction lysimeters or rhizons are essential to constrain site-relevant F_v values.

In the soil matrix, radionuclide interactions are governed by sorption to mineral surfaces, co-precipitation with Fe-Mn oxides or sulfides under reducing conditions, complexation with organic matter, and physical sequestration in fine pores. Operational K_d values must be reported concurrently with soil metrics (pH, redox potential (Eh), OC, CEC, and grain size). The retardation factor R, $R = 1 + (\rho_b \cdot K_d)/\theta$, where ρ_b is the bulk density and θ the volumetric water content, converts solute to groundwater velocities and is central to advective transport modelling. Kinetic sorption/desorption, diffusion into intraparticle sites, and slow mineral dissolution often invalidate the simple equilibrium K_d for transient tailings leaching. Column leaching tests and two-site kinetic sorption models are required to parameterize non-equilibrium release and breakthrough behaviour.

On farmland, plant type and management influence transfer markedly: leafy vegetables and shallow-rooted crops generally exhibit higher uranium Fv than cereals because of topsoil contamination and interception. Flooded rice paddies are a special case in which reducing anoxic conditions can convert U(VI) to U(IV) via precipitation as UO_2(s), reducing porewater dissolved U while creating particulate sources in sediments. Tropical and subtropical farm soils, frequently acidic and highly weathered, tend to form uranyl-organic complexes and may exhibit higher mobility and plant uptake relative to temperate soils. However, local DOC, pH, and mineralogy control the magnitude.

Animal exposure on farmland occurs primarily via ingestion of contaminated forage, soil, and water. The gastrointestinal absorption fraction f1 varies with chemical form: soluble uranyl species generally yield higher f1 than refractory particulates. Radium behaves chemically like calcium, with a relatively high skeletal affinity and long biological residence time. Uranium concentrates preferentially in the kidney and liver, with lower levels in muscle. Transfer coefficients to milk and meat depend strongly on species, lactation stage, diet composition, and mineral nutrition; controlled feeding trials that reflect local breeds and husbandry practices are therefore required for reliable Fa and Fm values.

Forested land intercepts airborne contaminants effectively; evergreen canopies retain deposition longer than deciduous stands, transferring activity slowly into litter and organic horizons. Fungal mycelia and fruiting bodies often concentrate daughters such as ^{210}Pb and ^{210}Po and may also bioaccumulate actinide daughters, producing elevated aggregated transfer factors for mushrooms and game. Disturbances, such as wildfire, logging, or insect outbreaks, can remobilize stored activity into the atmosphere and runoff, creating episodic exposure pathways that can exceed steady-state fluxes.

Climatic extremes alter transfer pathways: arid systems minimize leaching and sustain long-term surficial reservoirs of resuspendable dust, increasing inhalation risks; carbonate crusts may immobilize some actinides at the surface, while wind erosion mobilizes fine, contaminated particles. Arctic and alpine ecosystems cycle radionuclides slowly; lichens and mosses efficiently trap atmospherically deposited radionuclides, and grazers such as reindeer concentrate them in edible tissues, producing strong seasonal peaks. Seasonal sampling, including post-snowmelt, monsoon peaks and dry-season lows, is therefore essential.

Surface waters and groundwater require explicit partitioning among dissolved, colloidal, and particulate phases. In oxic surface waters, uranium commonly exists as uranyl-carbonate complexes (e.g., $UO_2\left(CO_3\right)_3^{4-}$). In high DOC waters, uranyl-organic complexes dominate and can behave as truly dissolved species or as stable colloids capable of long-range transport. Colloid-facilitated transport, therefore, often controls the apparent mobility of particle-reactive daughters. Measurement protocols should include sequential filtration and ultrafiltration, size-fractionated suspended solids analysis, and porewater sampling. Sediment core profiles dated with ^{210}Pb provide a temporally integrated record of historical discharges and are invaluable for constraining episodic release timelines.

9.3.4 Radiation Uptake and Accumulation in Aquatic Biota

Uranium and its daughters enter aquatic ecosystems as dissolved ions, complexes, colloids, and particulates; organismal accumulation depends on exposure pathway (filter feeding, grazing, sediment ingestion, and uptake across permeable epithelia), speciation, and particle reactivity. Bacteria and microbial biofilms sorb and sometimes intracellularly immobilize radionuclides via surface complexation to carboxyl and phosphoryl groups, EPS entrapment, and enzymatic reduction of U(VI) to insoluble U(IV). Reported concentration ratios (CR; $Bq \cdot kg^{-1}$ wet organism per $Bq \cdot L^{-1}$ water) for microbial mats and biofilms vary widely: values of 104 are reported for particle-reactive radionuclides under high particulate loading, while truly dissolved uranyl complexes produce lower CRs ($10 - 10^2$). Pertinent microcosm experiments that control redox and DOC are necessary to derive reliable, site-relevant microbial CRs.

Phytoplankton accumulate radionuclides via surface adsorption and cellular uptake, with CRs for uranium typically on the order of thousands and can be larger for daughters that strongly adsorb to organic or iron oxide particles. Feeding mode, gut retention time, and trophic ecology determine the organism-specific CR.

Fish generally show lower muscle CRs than invertebrates and plants; typical fish muscle CRs for uranium are often 10^{-2}–1 $Bq \cdot kg^{-1}$ per $Bq \cdot L^{-1}$, with higher concentrations in kidney and liver, where CRs may be 10–100 times greater than muscle. Benthic feeding fish and omnivores that ingest sediment or contaminated prey will show greater burdens than pelagic species. Egg and gonad concentrations can be significant for reproductive and human exposure assessments and require targeted sampling. Measurement practice must partition water into dissolved (<0.45 or <0.02 μm) and colloidal fractions, size-fractionate particulates, and pair organism sampling with concurrent sediment and porewater analysis. CRs used in models should be explicitly referenced to the whole organism, edible fraction, or specified organs and presented with GM/GSD.

9.3.5 Radiation Uptake and Accumulation in Soil Microorganisms

Soil microorganisms, including bacteria, archaea, and fungi, mediate uranium redox transformations, sorption, and biomineralization, thereby directly influence mobility and bioavailability. Microbial cell walls and EPS provide high-affinity ligands for uranyl sorption; microbial reduction of U(VI) to U(IV) yields localized precipitation of UO_2(s) that reduces aqueous concentrations but creates particulate pools that may remain bioavailable to detritivores. Quantitative microbial accumulation expressed as microbial biomass CR (Bq·kg^{-1} biomass per Bq·kg^{-1} soil) spans wide ranges (10^{-2} – 10^{1}) depending on OC, moisture, and proximity to particulate contamination, with elevated accumulation where microbes colonize contaminated particles or create reductive microzones.

Fungal hyphae and mycelia may translocate radionuclides along hyphal networks, redistributing complexed uranium from particle surfaces into rhizosphere zones and plant roots. Mycorrhizal associations can either immobilize radionuclides in fungal tissues or facilitate host plant uptake, depending on the species and chemical properties of the radionuclides. Microbial mediation of uranium can therefore either attenuate or enhance plant and higher trophic-level exposure. For modelling, include microbially mediated rate processes, such as Monod-type kinetics for enzymatic U(VI) reduction, and parameterize rates using microcosm experiments that measure reduction velocity, sorption to biomass, and desorption under varying DOC, pH, and electron acceptor conditions. Separating extracellularly sorbed radionuclide from intracellular biomass is analytically essential and is achieved by gentle rinse/extraction protocols combined with biomass proxies (microbial C or N) to normalize burdens.

9.3.6 Radiation Uptake and Accumulation in Terrestrial Animals

Terrestrial animals, domestic livestock and wild mammals, birds, reptiles, and invertebrates, accumulate radionuclides primarily through ingestion of contaminated forage, direct soil ingestion, and drinking water. Chemical form governs bioavailability: soluble uranyl species generally have higher gastrointestinal absorption than refractory particulates. Radium is assimilated similarly to calcium and preferentially partitions to bone, while uranium concentrates in the kidney and liver. Typical magnitudes place uranium transfer to milk Fm in the 10^{-3}–10^{-2} range and radium Fm in the 10^{-2}–10^{-1} range, although these are highly species- and diet-dependent. Organ concentrations (kidney, liver, and bone) are often orders of magnitude higher than in muscle, so tissue-specific sampling is required where organ consumption may occur.

Wild herbivores grazing on contaminated vegetation or lichens (e.g., reindeer) can accumulate high tissue burdens due to dietary reliance on radionuclide-rich resources and soil/lichen ingestion. Predators and scavengers receive radionuclides via contaminated prey. For many actinides and their daughters, trophic magnification is limited compared to organic contaminants. However, specific daughters with a strong affinity for particular tissues may be transferred through the food web. Age, reproductive state, dietary mineral balance (calcium and phosphorus), and husbandry management (housing, supplemental feed, and grazing rotation) significantly influence animal uptake and can be managed to mitigate transfer to foodstuffs. Controlled feeding studies, tracer trials, and paired faecal/tissue sampling are the basis for robust transfer coefficient derivation; compartmental kinetic models (gut, blood, liver, kidney, bone, and muscle) parameterized to empirical clearance rates yield time-resolved predictions of tissue burdens.

9.4 Radiation Effects on Human Health

9.4.1 Drivers of the Radiotoxicological Hazard

Abandoned uranium mines pose a complex radiotoxicological hazard driven by the presence and mobility of radionuclides, including uranium isotopes (primarily ^{238}U and its decay-series progeny, including ^{226}Ra, ^{210}Pb, and ^{210}Po), radium, and short-lived decay products that emit alpha, beta, and gamma radiation. Uptake pathways to humans from mine sites include the inhalation of dust and radon progeny, the ingestion of contaminated water, food (wild plants, livestock, and fish), or soil, and, less commonly, direct dermal contact when contaminated particulates or dissolved radionuclides transfer across broken skin or wounds. The relative importance of each pathway depends on site characteristics: radon gas and attached particulates dominate inhalation exposure in underground workings and sealed tunnels; waterborne uranium and radium control ingestion exposure where groundwater or surface waters are contaminated; and food chain uptake concentrates certain radionuclides in biota, with root uptake and bioaccumulation determining doses to consumers.

9.4.2 Mechanisms of Uptake and Internal Distribution

Mechanisms of uptake and internal distribution reflect the chemistry and particle size of the source material, together with physiological transport processes. Soluble uranium species (hexavalent U) are more readily absorbed from the gut and transported in blood, where they partition to the kidney and bone, producing both chemical nephrotoxicity and radiological dose to radiosensitive tissues. Radium mimics calcium and preferentially deposits in bone, delivering localized alpha and beta doses to marrow and bone surfaces, thereby increasing the long-term risk of bone sarcomas and leukaemia. Lead-210 and polonium-210, as decay products of radon and uranium, accumulate in soft tissues and the marrow, delivering alpha and beta doses with high linear energy transfer in critical microenvironments. Inhaled particulate-bound radionuclides deposit in the respiratory tract, where mucociliary clearance, translocation to lymphatics, or dissolution controls the fraction that becomes systemic; attached radon progeny deliver highly localized alpha doses to bronchial epithelium, a principal mechanism for increased lung cancer risk following prolonged exposure.

9.4.3 The Dose-Effect Relationship

Radiotoxicity arises from the energy deposited in biological material and the specific biological interactions that follow. When ionizing radiation penetrates tissue, it deposits energy quantified as the absorbed dose, measured in grays (Gy), where 1 Gy is 1 joule per kilogram. Different radiation types produce different biological damage per unit absorbed energy; to capture this, the equivalent dose H_T is calculated by weighting absorbed doses by radiation-specific factors W_R using the relation

$$H_T = \sum_R W_R * D_{T,R}$$

where $D_{T,R}$ is the absorbed dose in tissue T from radiation R and H_T is expressed in sieverts (Sv). For example, an absorbed dose of 1 Gy delivered by alpha particles corresponds to an equivalent dose of 20 Sv when a regulatory weighting factor of 20 is used, whereas the same absorbed dose of gamma radiation uses a weighting factor of 1. To reflect differential tissue sensitivity, the effective dose E aggregates equivalent doses across tissues using tissue weighting factors W_T, yielding

$$E = \sum_T W_T * H_T = \sum_T W_{T \sum_R W_R} * \overline{D_{T,R}}$$

The effective dose, also expressed in sieverts, provides a single metric correlated with stochastic health risk for whole-body or partial-body exposures.

The dose-effect relationship for ionizing radiation distinguishes deterministic (tissue-reaction) effects from stochastic effects. Deterministic effects, observed at relatively high absorbed doses, have thresholds and increasing severity with dose; examples include radiation-induced skin injury, acute radiation syndrome, and organ failure when critical tissues receive large doses over short periods. Stochastic effects, principally cancer and heritable genetic damage, are probabilistic: the probability of occurrence increases with dose, while the severity of an individual case is independent of dose. Epidemiological data robustly establish elevated cancer rates after high acute or protracted exposures, but quantifying risk at long-term low levels (commonly below about 200 mSv effective dose) is inherently uncertain because background rates of cancer and confounding exposures obscure small excess risks. Consequently, low measured radiation levels at abandoned uranium mines do not automatically equate to negligible risk; the actual hazard depends on exposure duration, the mix and bioavailability of radionuclides, and the pathways that deliver dose to sensitive tissues.

9.4.4 *Risk Management*

Risk management, therefore, relies on conservative application of dose metrics, pathway analysis, and exposure-time considerations. The effective dose provides a practicable way to compare exposures from different scenarios and to prioritize interventions, recognizing that alpha-emitting contaminants bound to dust or ingested materials can deliver disproportionately large equivalent doses to localized tissues compared with the same absorbed energy from penetrating gamma radiation. In practice, mitigation at abandoned mine sites focuses on reducing inhalation of radon and particulates, preventing contaminated water and food chain uptake, and lowering occupancy or use of contaminated areas so that cumulative effective dose to individuals remains as low as reasonably achievable.

9.5 Radioecological Prediction and Modelling

Radiation measurement and predictive modelling at abandoned uranium mines must be conceived as a single, traceable field-to-model workflow in which sampling geometry, detector response, laboratory protocols, and environmental heterogeneity are explicitly linked to state variables and model operators. Measurement campaigns should combine long-term fixed-station gamma spectrometry, repeated spatial transects, and stratified grab and composite sampling of solids and porewaters so that the measurement support volume is known and reproducible; laboratory analyses must use matrix-matched standards, chemical recovery spikes, and high-resolution detectors (HPGe for chain-resolved gamma, alpha spectrometry and LSC for separated actinides) while recording complete metadata for instrument, calibration, geometry, moisture, grain size, and extraction procedure to enable unambiguous scale transformation between observed activity and model compartments.

Quality assurance and uncertainty characterization are integral to model credibility: replicate samples, field and laboratory blanks, inter-laboratory comparisons, and full analytical error models must be propagated into transport and chain models. Propagate measurement, spatial sampling, temporal variability, and parameter uncertainties using Monte Carlo or Bayesian hierarchical frameworks that preserve covariance among correlated radionuclides and pathways; retain posterior parameter distributions for scenario projections and decision thresholds rather than single best-fit values and weight objective functions by observation uncertainty and support-volume mismatch during calibration [20–25].

Key model parameters are defined and used consistently: porosity n, bulk density ρ_b, porewater velocity v, hydrodynamic dispersion D, distribution coefficient K_d (or kinetic sorption constants k_s, k_d), decay constants λ_i with $t_{1/2,i} = \ln 2/\lambda_i$, biota concentration factors CF or kinetic uptake/loss rates k_u, k_l, sediment porosity n_s, sediment bulk density $\rho_{b,s}$, sediment diffusion D_s, and domain descriptors, such as support volume V_s and characteristic length L. Priors for these parameters may be informed by carefully chosen analogues selected for lithology, climate, hydrology, and weathering state, with analogue-derived hierarchical variance explicitly included to quantify transferability.

Transport, sorption, decay, and chain kinetics must be coupled mathematically. For aqueous activity concentration $C_i(x,t)$ with instantaneous linear sorption, use the retardation representation

$$R_i = 1 + \frac{\rho_b K_{d,i}}{\theta}$$

and the advective-dispersive decay equation

$$R_i \frac{\partial C_i}{\partial t} = -v\frac{\partial C_i}{\partial x} + D\frac{\partial^2 C_i}{\partial x^2} - \lambda_i C_i + \frac{S_i}{\theta}$$

When sorption kinetics matter, represent aqueous and solid-phase activities with coupled equations

$$\theta\frac{\partial C_i}{\partial t} + \rho_b\frac{\partial S_i}{\partial t} = -v\frac{\partial C_i}{\partial x} + D\frac{\partial^2 C_i}{\partial x^2} - \lambda_i\theta C_i + S_i^{(ext)}$$

$$\frac{\partial S_i}{\partial t} = k_s C_i - k_d S_i - \lambda_i S_i$$

and evaluate the sorption Damköhler number Da_s = \dfrack$_s$Lv to decide whether equilibrium or kinetic formulations are appropriate.

Decay chains require explicit, coupled production-decay-transport matrices. In vector form for a chain of n nuclides,

$$R\frac{\partial C}{\partial t} = -v\frac{\partial C}{\partial x} + D\frac{\partial^2 C}{\partial x^2} + AC + S(x,t)$$

where $R =$ diag (R_i) and A has diagonal entries $-\lambda_i$ and off-diagonals representing in-growth fractions from parents. Parent-daughter pairs in open, transport-active systems follow:

$$\frac{\mathrm{d}C_P}{\mathrm{d}t} = -\lambda_P C_P - k_{\mathrm{loss},P} C_P$$

$$\frac{\mathrm{d}C_D}{\mathrm{d}t} = \lambda_P C_P - \lambda_D C_D - k_{loss,D} C_D$$

with Bateman solutions recovered in closed systems. Specific activity a_i (Bq per unit mass of element) is useful when the element mass M is tracked using $A_i = a_i M$, but secular equilibrium assumptions must be tested: in mine-impacted environments, preferential mobility of daughters (e.g., radium relative to uranium) frequently breaks equilibrium and introduces a significant bias if ignored [26].

Sorption and sediment exchange should be represented using multi-reservoir formulations that allow for labile and refractory pools, ageing, burial, and diffusive exchange. A practical three-reservoir sediment-water representation uses porewater, C_w, labile solid S_l, and refractory solid S_r with explicit adsorption, desorption, ageing, and burial rate constants so that long-term availability and loss to burial are simulated rather than assumed removed. Biota uptake can be treated as equilibrium concentration factors for screening, $C_{biota} = BAF_i C_{water}$, or as first-order kinetics [27–30]

$$\frac{\mathrm{d}C_{biota}}{\mathrm{d}t} = k_u C_{water} - k_l C_{biota} - \lambda_{bio} C_{biota}$$

and trophic transfer must be resolved using ingestion rates, assimilation efficiencies, and diet-composition matrices, with food web exposures being decision-relevant.

Parameter estimation and model evaluation require joint inversion of tracer, solute, and radionuclide observations, with objective functions that account for observation uncertainty and support volume. Bayesian hierarchical models allow analogue priors while enabling site-level updates and preserving posterior covariances that influence exceedance probabilities. Compute exceedance probabilities for regulatory thresholds via:

$$P\left(C_i\left(x_r,t\right) > C_{crit}\right) = \int_{\Theta} 1\left\{C_i\left(x_r,;t,;\theta\right) > C_{crit}\right\}\pi\left(\theta \mathrm{data}\right)\mathrm{d}$$

and use variance-based sensitivity metrics such as Sobol indices to prioritize measurements that will most reduce predictive uncertainty.

Practical model deployment requires reporting probabilistic forecasts of metrics relevant to exposure and remediation, explicit statements of equilibrium-versus-kinetic sorption assumptions, inclusion or exclusion of full decay-chain in-growth, and transparent use of analogue-derived priors. Validation should include retrospective comparisons with historical observations and prospective testing against new monitoring data collected with predefined support volumes and QA protocols. These combined measurement, modelling, and uncertainty practices ensure that radioecological predictions for abandoned uranium mines are technically defensible, site-specific, and actionable for experts managing legacy mine risks [31].

9.6 Perspectives

Radioecology at abandoned uranium sites is a discipline that examines how radioactive materials released by historic mining and milling interact with soils, sediments, water, biota, and human communities. Perceived risks and hazards centre on external gamma exposure from ore and tailings, inhalation of radon and radon progeny, ingestion of contaminated water and foodstuffs, and the long-term mobilization of radionuclides through weathering and hydrological pathways. These hazards are not uniform; their magnitude depends on the radionuclide inventory, physical disposition of wastes, and the accessibility of contaminated media to people and ecological receptors. The combination of radiological and non-radiological hazards at many legacy uranium sites means that radioecological assessment and management must address both acute access/safety concerns and chronic exposure pathways.

9.6.1 Variation with Site-Specific Environment

Radioecological behaviour at any abandoned mine is controlled first by the chemical and physical form of the uranium and associated decay-series radionuclides. Co-located mine wastes, which contain finely divided tailings, dust-prone rock piles, or crushed mill residues, present elevated short-term inhalation and dust-redistribution risks compared with intact, bedrock-hosted residues. Geochemical conditions, such as pH, redox potential, and the presence of complexing ligands, determine whether uranium and its daughter products remain relatively immobile or are transported into groundwater and surface waters. Microbial activity and organic matter can further alter redox conditions and mediate uranium mobility, producing spatially heterogeneous contaminant patterns within a site.

Hydrology and geomorphology set the dominant transport routes. Sites with shallow groundwater, permeable alluvial deposits, or active stream networks facilitate radionuclide migration off-site and increase exposure potential for downstream aquatic ecosystems and human water users. Conversely, sites on low-permeability bedrock or in endorheic basins may concentrate radionuclides locally but reduce off-site dispersal. Physical erosion, seasonal freeze-thaw cycles, and extreme precipitation events can episodically redistribute contaminated sediments and dust, temporarily amplifying exposures in ways that conventional steady-state sampling may miss.

Vegetation, soil type, and land use modulate ecological uptake and human contact. Sandy, low-organic soils tend to support greater leaching and downward migration of soluble radionuclides, while clay-rich soils can retard vertical movement but can also store contaminants near the surface, where root uptake and surface grazing by animals occur. Plant species differ markedly in radionuclide uptake: some grasses and bryophytes accumulate uranium daughters more readily than woody species, shaping food web transfer and the profile of ecological risk. Traditional and subsistence land uses, such as hunting, grazing, and the consumption of locally harvested plants and fish, create culturally specific exposure pathways that standard exposure models can underrepresent.

9.6.2 Geographical, Climatic, and Geological Factors

Regional climate exerts a robust, often underappreciated control on radioecology. Arid and semi-arid regions amplify wind-driven dust transport and favour radon emanation from exposed tailings, elevating inhalation and surface-deposition risks over large areas. Humid temperate regions promote chemical weathering and fluvial transport, increasing the potential for radionuclides to enter aquatic food webs and affect fisheries and irrigation supplies. Cold climates with permafrost or long periods of snow cover can temporarily immobilize contaminants but may also create pulse-release dynamics during thawing that mobilize large contaminant loads in short periods.

Geological context defines both the background radionuclide signature and the mechanical behaviour of mine wastes. Uranium deposits hosted in carbonate rocks interact differently with groundwater chemistry than those in silicate matrices, producing distinct speciation and mobility patterns. Fractured crystalline bedrock may transmit contaminants rapidly through preferential pathways, while thick, cohesive regoliths can act as sinks that slowly release contaminants over decades. The broader regional geology also determines the natural background of certain radionuclides, complicating the attribution of observed concentrations to mining legacy versus geogenic sources [31, 32].

Geography and socioecological setting determine receptor presence and vulnerability. Remote sites with sparse populations may still pose ecological hazards to migratory species, or long-range radiological dispersion may affect communities far from the mine. Sites within Indigenous territories or areas of subsistence harvesting require radioecological assessments that integrate land use, cultural practices, and treaty rights into exposure and risk appraisal.

9.6.3 Radioecological Management of Abandoned Mines

Effective management of abandoned uranium sites requires policies that combine precautionary containment with adaptive, site-specific stewardship. First, adopt a tiered assessment framework that rapidly screens legacy sites for acute hazards, then progressively applies more detailed radioecological characterization when screening indicates potential human or ecological exposure. Screening should include standardized metrics for surface gamma, radon flux, particulate dust potential, and basic hydrochemical indicators, allowing for transparent prioritization of thousands of legacy sites.

Second, require that remediation and closure objectives be explicitly tied to land use and receptor scenarios developed in partnership with local and Indigenous communities. Risk-based standards must therefore be informed by the realistic uses and cultural values of affected lands rather than by generic, distant reference scenarios. Management plans should document locally relevant exposure pathways, set measurable remediation endpoints, and identify post-closure institutional controls for residual contamination.

Third, prioritize engineered containment and source control for sites with mobile wastes or those in hydrologically connected landscapes. Stabilizing tailings, regrading and vegetating erodible piles, and installing low-maintenance covers reduce dust and radon generation and limit surface water infiltration that mobilizes radionuclides [33]. Wherever groundwater contamination exists or is likely, policy should mandate hydrological monitoring networks and remediation approaches designed to limit contaminant discharge to surface waters.

Fourth, institutionalize long-term monitoring and adaptive management funded through the polluter-pays principle or, where responsible parties are absent, dedicated public programs. Monitoring must be multi-media and multi-scale, combining continuous radon and gamma monitoring at high-priority sites with periodic sediment and biota sampling designed to detect ecological accumulation. Data management practices should ensure that monitoring results are publicly available and interpretable, enabling communities and regulators to detect trends and trigger adaptive responses.

Fifth, incorporate climate resilience into closure design and long-term stewardship. Design standards should anticipate increased precipitation intensity, altered freeze-thaw cycles, and other climate-driven processes that can change erosion and transport regimes. Robust remedial designs that withstand a range of plausible climate scenarios will reduce the risk that legacy containment will fail as regional conditions shift.

Finally, strengthen governance through multiagency coordination, transparent prioritization, and meaningful engagement with affected communities. Integrated programs that bring together public health, environmental protection, land management, and Indigenous governance reduce duplication, align remediation objectives, and ensure that decisions reflect both scientific evidence and local values. Where extensive inventories of orphaned sites exist, a strategic plan that maps sites, ranks risks, and sequences interventions allows resources to be applied where they yield the largest reduction in human and ecological harm.

These policy approaches, when combined with robust technical assessment and sustained community partnership, shift the management of abandoned uranium mines away from reactive patchwork towards a durable, equitable strategy that acknowledges ecological complexity, respects local land uses, and safeguards people and ecosystems across the long-time scales that radioecology demands.

9.7 Concluding Remarks

The radioecology of abandoned uranium mines encompasses interlinked physical, chemical, biological, and social processes that together determine long-term environmental and human health outcomes. Residual radiation from mine wastes and altered landscapes produces persistent concerns shaped by radionuclide speciation, decay properties, and geochemical mobility. Speciation controls solubility, sorption, redox behaviour, and bioavailability, thereby dictating the dominant transfer routes into air, surface and ground water, soils, and biota. Windborne dust, surface runoff, groundwater flow, and biological vectors move contaminants between compartments, concentrating them in sediments, floodplain soils, and agricultural fields, where food chain integration amplifies exposure potential. Farmland and adjacent ecosystems face particular risk where irrigation, grazing, and soil amendment create direct pathways into crops and livestock. Aquatic systems exhibit pronounced accumulation driven by sediment interactions and trophic transfer, resulting in measurable uptake by fish, invertebrates, and aquatic plants. Soil microorganisms mediate transformation and retention processes that alter radionuclide speciation and long-term release dynamics. Terrestrial animals bioaccumulate radionuclides through foraging, linking environmental contamination to human dietary exposure [34–36].

Human health effects are governed by exposure pathways, cumulative dose, and radionuclide-specific radiobiology; dose-response relationships remain the quantitative basis for risk assessment and remediation targets. Effective management requires integrated, site-specific monitoring, paired with mechanistic and empirical radioecological models that integrate hydrology, geochemistry, ecology, and land use. Predictive modelling supports scenario analysis, prioritizes interventions, and informs adaptive remediation strategies. Together, targeted monitoring, speciation analysis, ecological transfer studies, dose-based health evaluation, and robust modelling form the foundation for stewardship of post-mining landscapes. Prioritizing these elements will protect ecosystems, secure food resources, and reduce human health risks while guiding phased remediation and long-term land-use decisions.

References

1. Pettersson, H. B. L., Hassstadius, L., & Hedvall, R. (1988). Radioecology in the vicinity of prospected uranium mining sites in a subarctic environment. *Journal of Environmental Radioactivity, 6*, 25–40.
2. Ishimori, Y., Lange, K., Martin, P., Mayya, Y. S., & Phaneuf, M. (n.d.). In International Atomic Energy Agency (Ed.), *IAEA* (Technical Reports Series No. 474). International Atomic Energy Agency.
3. Landa, E. R. (1992). *Field studies of radon in rocks, soils, and water* (p. 10). CRC Press.
4. Gil-Pacheco, E., et al. (2020). A radiological index for evaluating the impact of an abandoned uranium mining area in Salamanca, Western Spain. *Environmental Pollution, 258*, 113825. https://doi.org/10.1016/j.envpol.2019.113825
5. Bone, S. E., Dynes, J. J., Cliff, J., & Bargar, J. R. (2017). Uranium(IV) adsorption by natural organic matter in anoxic sediments. *Proceedings of the National Academy of Sciences, 114*, 711–716. https://doi.org/10.1073/pnas.1611918114
6. Denison, F. (2007). *Uranium(VI) speciation: Modelling, uncertainty and relevance to bioavailability models. Application to uranium uptake by the gills of a freshwater bivalve*. Université d'Aix-Marseille.
7. Francis, A. J. (1998). Biotransformation of uranium and other actinides in radioactive wastes. *Journal of Alloys and Compounds, 271–273*, 78–84.
8. Anderson, R. S., Sørensen, M., & Willetts, B. B. (n.d.). (pp. 1–19). Springer Vienna.
9. Cumberland, S. A., Douglas, G., Grice, K., & Moreau, J. W. (2016). Uranium mobility in organic matter-rich sediments: A review of geological and geochemical processes. *Earth-Science Reviews, 159*, 160–185. https://doi.org/10.1016/j.earscirev.2016.05.010
10. Mkandawire, M., Taubert, B., & Dudel, E. G. (n.d.). In N. Kalogerakis (Ed.), *Second European Bioremediation Conference* (pp. 357–360) (Technical University of Crete, and European Community (High-level Scientific Conference Programme)).
11. Mkandawire, M., Taubert, B., & Dudel, E. G. (n.d.). In N. Kalogerakis (Ed.), *Second European Bioremediation Conference* (pp. 405–408) (Technical University of Crete, and European Community (High-level Scientific Conference Programme)).
12. Deb, D., & Chakma, S. (2023). Colloid and colloid-facilitated contaminant transport in subsurface ecosystem, a concise review. *International Journal of Environmental Science and Technology, 20*, 6955–6988. https://doi.org/10.1007/s13762-022-04201-z
13. Hirose, K. (1995). The relationship between particulate uranium and thorium-complexing capacity of oceanic particulate matter. *Science of the Total Environment, 173–174*, 195–201.
14. Markich, S. J. (2002). Uranium speciation and bioavailability in aquatic systems: An overview. *Scientific World Journal, 2*, 707–729.
15. Thomas, P. A., & Gates, T. E. (1999). Radionuclides in the lichen-caribou-human food chain near uranium mining operations in northern Saskatchewan, Canada. *Environmental Health Perspectives, 107*, 527–537.
16. Dreissig, I., Weiss, S., Hennig, C., Bernhard, G., & Zänker, H. (2011). Formation of uranium(IV)-silica colloids at near-neutral pH. *Geochimica et Cosmochimica Acta, 75*, 352–367. https://doi.org/10.1016/j.gca.2010.10.011
17. Newsome, L., Morris, K., & Lloyd, J. R. (2014). The biogeochemistry and bioremediation of uranium and other priority radionuclides. *Chemical Geology, 363*, 164–184. https://doi.org/10.1016/j.chemgeo.2013.10.034
18. Songkasiri, W., Reed, D. T., & Rittmann, B. E. (2002). Bio-sorption of neptunium(V) by Pseudomonas fluorescens. *Radiochimica Acta, 90*, 785–789.
19. Rapantova, N., Licbinska, M., Babka, O., Grmela, A., & Pospisil, P. (2013). Impact of uranium mines closure and abandonment on groundwater quality. *Environmental Science and Pollution Research International, 20*, 7590–7602. https://doi.org/10.1007/s11356-012-1340-z

20. Conlon, E. M., Postier, B. L., Methe, B. A., Nevin, K. P., & Lovley, D. R. (2009). Hierarchical Bayesian meta-analysis models for cross-platform microarray studies. *Journal of Applied Statistics, 36*, 1067–1085. https://doi.org/10.1080/02664760802562480
21. Iglesias, C. P., & Claxton, K. (2006). Comprehensive decision-analytic model and Bayesian value-of-information analysis: Pentoxifylline in the treatment of chronic venous leg ulcers. *PharmacoEconomics, 24*, 465–478. https://doi.org/10.2165/00019053-200624050-00005
22. Qian, S. S., King, R. S., & Richardson, C. J. (2003). Two statistical methods for the detection of environmental thresholds. *Ecological Modelling, 166*, 87–97.
23. Denison, F. H., & Garnier-Laplace, J. (2005). The effects of database parameter uncertainty on uranium(VI) equilibrium calculations. *Geochimica et Cosmochimica Acta, 69*, 2183–2191.
24. Laloue, N., Laroche, C., Jobic, H., & Methivier, A. (2009). Kinetic Monte Carlo modelling to study diffusion in zeolite understanding the impact of dual site isotherm on the loading dependence of n-hexane and n-heptane diffusivities in MFI zeolite, as revealed by QENS experiments. *Oil & Gas Science and Technology, 64*, 773–793. https://doi.org/10.2516/Ogst/2009065
25. Sun, H. B., & Furbish, D. J. (1995). Moisture content effect on radon emanation in porous media. *Journal of Contaminant Hydrology, 18*, 239–255.
26. Wiramanaden, C. I. E., Orr, P. L., & Russel, C. K. (2015). Assessment of radium-226 bioavailability and bioaccumulation downstream of decommissioned uranium operations, using the caged oligochaete (Lumbriculus variegatus). *Environmental Toxicology and Chemistry, 34*, 507–517. https://doi.org/10.1002/etc.2852
27. Abramets, A., & Rovdan, E. (2002). *Plant nutrition* (pp. 484–485).
28. Fahselt, D., Wu, T. W., & Mott, B. (1995). Trace element patterns in lichens following uranium mine closures. *Bryologist, 98*, 228–234.
29. Fresquez, P. R., Armstrong, D. R., Mullen, M. A., & Naranjo, L. (1998). The uptake of radionuclides by beans, squash, and corn growing in contaminated alluvial soils at Los Alamos National Laboratory. *Journal of Environmental Science and Health - Part B: Pesticides, Food Contaminants, and Agricultural Wastes, 33*, 99–121.
30. Gadd, G. M. (2010). Metals, minerals and microbes: Geomicrobiology and bioremediation. *Microbiology, 156*, 609–643. https://doi.org/10.1099/mic.0.037143-0
31. Blanchette, M. L., & Lund, M. A. (2016). Pit lakes are a global legacy of mining: An integrated approach to achieving sustainable ecosystems and value for communities. *Current Opinion in Environmental Sustainability, 23*, 28–34. https://doi.org/10.1016/j.cosust.2016.11.012
32. Diehl, P. (2003). *Uranium mining in Eastern Germany: The WISMUT Legacy*.
33. LaRosa, J. J., et al. (1992). Radiochemical methods used by the IAEA's laboratories at Seibersdorf for the determination of ^{90}Sr, ^{144}Ce and Pu radionuclides in environmental samples collected for the international Chernobyl project. *Journal of Environmental Radioactivity, 17*, 183–209.
34. DeForest, D. K., Brix, K. V., & Adams, W. J. (2007). Assessing metal bioaccumulation in aquatic environments: The inverse relationship between bioaccumulation factors, trophic transfer factors and exposure concentration. *Aquatic Toxicology, 84*, 236–246.
35. Morris, S., & Meinhold, A. F. (1995). Probabilistic risk assessment of nephrotoxic effect of uranium in drinking water. *Health Physics, **69***, 897–908.
36. Tan, X., Yim, S.-Y., Uppu, P., & Kleinow, K. (2010). Enhanced bioaccumulation of dietary contaminants in catfish with exposure to the waterborne surfactant linear alkylbenzene sulfonate. *Aquatic Toxicology, 99*, 300–308. https://doi.org:citeulike-article-id:7271176

Chapter 10
Uranium Biogeochemical Behaviour in Abandoned Mines

Abstract This chapter synthesizes current understanding of the biogeochemistry of natural uranium and co-occurring radionuclides and metals in abandoned uranium mine environments, focusing on how microbial, plant, and faunal interactions mediate speciation, mobility, bioavailability, and ecological effects. It examines redox-driven transformations, sorption and complexation processes, biotic uptake and trophic transfer, and the feedbacks between biological activity and geochemical gradients that control contaminant fate and risk. Parts of this chapter were first published as a review in Environmental Science & Technology in 2013 and have been updated and reproduced with permission from Springer Nature Publishing GmbH.

Keywords Radon emanation · Borehole mining · Pyrite minerals · Bioremediation · Radioactive isotopes · Co-precipitation · Biosystem compartments

10.1 Introduction

Discharges of uranium, associated radionuclides, heavy metals, and metalloids from waste and tailing dumps at abandoned uranium mining and processing sites pose contamination risks to the surface and groundwater. As a result, efforts have been made to rehabilitate the mining sites, initially using classical remediation techniques based on high chemical and civil engineering technology.

Recently, bioremediation technology has been advocated as an alternative to the classical approach because of, among other things, (a) the high demand for sites requiring remediation, (b) the economic implication of running and maintaining the facilities due to high energy and workforce demand, and (c) the pattern and characteristics of most contamination discharge and composition from the former uranium mining and processing sites. Effective bioremediation requires appropriate means to harness uranium's interactions with the entire ecosystem, which can be achieved only with precise knowledge of the biogeochemical processes that determine its mobility and fate in abandoned uranium mines.

M. Mkandawire, *Fate and Effects of Uranium in Abandoned Uranium Mines*,
https://doi.org/10.1007/978-3-032-10447-2_10

Thus, the discussion in this chapter focuses on, but is not limited to, the water pathway. To give readers a perspective, the chapter starts by recapping what constitutes an abandoned uranium mining site, the uranium mining techniques, and the geochemical properties of uranium that favour its interaction with the biological system, enabling bioremediation.

10.2 Abandoned Uranium Mining Site

Not every decommissioned uranium mine is an abandoned mine. However, only sites where advanced exploration, mining, or mine production has ceased without comprehensive rehabilitation or restoration are classified as abandoned mines, also called orphaned mines [1–3]. All abandoned uranium mining sites share identical environmental problems, ranging from disturbed physical landscapes to large piles of rocks, mining waste, and tailings to environmental contamination with radionuclides and heavy metals [4, 5]. Consequently, the primary concerns associated with abandoned uranium mines are public health and safety, environmental safety, and aesthetic concerns. However, the mining and processing procedure determine the characteristics of an abandoned mining site.

10.2.1 Mining and Processing Sites

Chapter 2 thoroughly discusses abandoned uranium mines, clarifying that not all decommissioned mines are abandoned. The uranium ore mining and processes have been discussed in Chap. 2. Briefly, several techniques are used to extract uranium from its ore deposits, but the most commonly used are open-pit mining, underground mining, in situ leaching (ISL), and borehole mining. In open-pit and underground mining, the first step is to crush the ore-bearing rock into a fine powder. Then, vast amounts of water, leaching agents, and thickeners are added to the pulverized ore to oxidize and dissolve the uranium minerals. The uranium minerals are usually available as UO_2 is oxidized to UO_3, followed by the formation of either a uranyl sulfate ($UO_2(SO_4)_3^{4-}$) or a uranyl-carbonate (predominantly $UO_2(CO_3)_3^{4-}$) in acid leaching or carbonate leaching, respectively. There is no crushing in the ISL and borehole mining procedures. After the leaching step, uranium is concentrated and purified using either ion exchange or solvent extraction. Then, the extract is subsequently dewatered and dried to produce hydrated uranium peroxide ($UO_4{\cdot}2H_2O$), which is centrifuged and precipitated to the yellowcake (U_3O_8) containing at least 75% uranium oxides [6, 7].

10.2.2 Characteristics of the Sites

10.2.2.1 Mining Wastes and Tailing

Vast piles of mine waste and tailings are the first characteristic of abandoned open-pit and underground mines, as the ores' uranium content is often between 0.1% and 0.7% [4, 5, 8]. Thus, many rocks are excavated, and large amounts of tailings are generated during uranium extraction. A mine can generate up to 40 tonnes of waste rock for every tonne of uranium ore produced, while underground mines produce about one tonne of waste rock per tonne of ore [2, 9–11]. The landscape is disturbed by mountains of waste and debris and by lakes formed by flooding of open pits. Apart from creating undesired physical features, waste debris and tailings create massive stockpiles of radioactive and toxic waste rock and sand-like tailings, which become a serious source of radioactive, toxic metals, and radioactive air pollutants. With most of the uranium removed, the mine wastes and tailings remain highly radioactive because almost 85% of the radioactive elements and uranium decay progenies in the original uranium ore end up in the tailings [2, 11].

10.2.2.2 Mine Water Quality

Water contamination is a common characteristic of abandoned uranium mining sites. The tailings and wastes from the uranium milling process consist of ground rock particles, non-extracted radioactive and heavy metals, residual uranium, and leach and mill chemicals. As a result, chemically hazardous substances in waste and tailings, such as lead, cadmium, arsenic, and mercury, are more readily discharged into the environment through leaching and blown into the water and air pathway [12, 13]. In addition, the excavated and ground rock particles have a high surface area, which increases their contact with air and water, naturally favouring geochemical processes, especially redox and acidification. Hence, precipitation leaches out radionuclides (i.e., progeny of uranium and debris), heavy metals, and metalloids, which may contaminate surface and groundwater. Depending on the type of rock, these wastes may be acid-generating. For instance, surface mine water discharged from a few abandoned mining sites in eastern Germany (which these authors have extensively studied) has pH values between 2.7 and 3.4. The dissolved arsenic concentrations range from 0.8 to 700 mg L^{-1}, iron from 750 to 3000 mg L^{-1}, and sulfate from 2,000 to 8000 mg L^{-1}. However, 20–60% of the dissolved arsenic is immobilized by co-precipitation with Fe (III) [14–16]. Nevertheless, this results in high groundwater arsenic concentrations and sediment enrichment, which can be as much as 500% above background levels [17, 18].

10.2.2.3 Radon Gas Emanation and Air Quality

Another characteristic of abandoned uranium mining is air pollution, manifested in odours from water, dust, or both. However, the most dangerous air pollution associated with abandoned uranium mining is odourless radon (^{222}Rn) emanation. The source of ^{222}Rn gas is the natural radioactive decay of ^{238}U, which produces its daughter product, radium-226 (^{226}Ra). Uranium-rich minerals such as granite, phosphate, shale, and pitchblende host natural high concentrations of ^{222}Rn. Therefore, the rock piles and tailings in abandoned uranium mines continuously release Rn into the environment. Besides significant Ra gas sources, winds may also blow dust containing radionuclides, heavy metals, and particulate matter from the rock debris and tailing piles. Furthermore, waste rock and tailings dumps of abandoned uranium mines release nitrogen oxides (NO_x), volatile organic compounds (VOCs), and carbon dioxide (CO_2). In addition, acid plants producing acid for milling operations release large amounts of sulfur dioxide (SO_2)—a major contributor to acid rain.

10.2.2.4 Acid Mine Drainage

Acid mine drainage (AMD) is common at most abandoned uranium mining and processing sites because uranium ores are often associated with pyrite (FeS_2). Thus, pyrite minerals are abundant in the mine waste rock [17]. When the pyrite-rich waste rocks and tailings are exposed to water and air, they break down, leading to sulfuric acid (H_2SO_4) and $FeSO_4$. Then, the $FeSO_4$ undergoes further reactions, eventually forming ferrous hydroxide ($Fe(OH)_3$). Fe(OH)3 is responsible for the reddish or orange colour in seepage water from mine waste and tailings, indicating that acid drainage is occurring. The whole breakdown process can be summarized in the following chemical reactions:

$$2FeS_2 + 7O_2 + 2H_2O \rightarrow 2FeSO_4 + 2H_2SO_4 \tag{10.1}$$

$$2Fe^{2+} + \frac{1}{2}O_2 + 2H^+ \rightarrow 2Fe^{3+} + H_2O \tag{10.2}$$

$$Fe^{3+} + 3H_2O \rightarrow Fe(OH)_3 + 3H^+ \tag{10.3}$$

$$FeS_2(s) + \frac{15}{4}O_2 + \frac{7}{2}H_2O \rightarrow 4H + +2SO_4^{2-} + Fe(OH)_3(s) \tag{10.4}$$

Logically, acid mine drainage alters the pH of receiving waters due to the high levels of Fe^{2+} and H^+ ions. However, pH measurements may not detect heavy acid mine drainage in some streams with high dissolved carbonates, as pH only indicates the concentration of H ions. In addition, acid mine drainage depletes water's buffering capacity by neutralizing carbonate and bicarbonate ions, forming carbonic acid

(H_2CO_3), which readily dissociates into water and carbon dioxide (see Reactions 10.5 and 10.6).

$$H^+ + CO_3^{2-} \rightleftharpoons HCO_3^- + H^+ \rightleftharpoons H_2CO_3 \quad (10.5)$$

$$H_2CO_3 \rightarrow H_2O + CO_2 \quad (10.6)$$

10.2.2.5 Aquifer Contamination from In Situ Leaching

Although there is little surface disturbance and no generation of waste rock and tailings, the abandoned ISL mines have severe groundwater contamination that is difficult to remediate. During ISL mining, bacterial contamination is common, sometimes facilitating leaching. When the mines are abandoned, bacterial activity affects the leaching of minerals into the aquifers. Under prevailing unsaturated conditions, pyrite oxidizes, generating significant quantities of sulfuric acid, further mobilizing heavy metals, uranium, and radionuclides trapped in rock pores, and adding to the contaminant load in groundwater. The trapped liquids are characterized by elevated levels of heavy metals and metalloids, which can be 100–400% higher in Cd, As, Ni, and U than before the ISL began [17, 19]. In acid ISL mining, an important characteristic of sulfuric acid leaching is the deterioration of the ore zone's permeability due to chemical and gaseous plugging and rising salinity levels. In the alkaline ISL, large quantities of solutions escape beyond the mining zone due to gaseous oxygen plugs. However, alkaline agents tend to increase groundwater salinity only slightly. Nevertheless, the contamination potential of surrounding waters with U, Ra, SO_4^{2-}, Fe, and other heavy metals remains very high.

10.3 Geochemistry

10.3.1 Physicochemical Condition

Uranium, the heaviest naturally occurring element, is a ubiquitous element with an average concentration of 3 mg kg^{-1} in the Earth's crust [20]. The concentrations of U in geological materials are highest in continental-type rocks, and abandoned mines can potentially have over 234 mineral species, depending on the bedrock and geological background of the site. Since elemental uranium hardly exists due to surface oxidation of its metallic state [21, 22], abandoned mines have almost exclusively oxides of U(IV) and U(VI) out of the six potential oxidation states of uranium. U(III) quickly oxidizes to U(IV) under most redox conditions, while U(V) is readily disproportionate to U(IV) and U(VI) in nature [23–25]. Due to uranium radioactivity, the geochemistry of abandoned mining sites is dominated by the chemistry of decay products, unextracted chemical elements (e.g., As, Fe, Ca, Mg,

V, P, and S) that accompany uranium in ore deposits, and other minerals, such as pyrite.

Uranium has 16 radioactive isotopes, but only three occur naturally: ^{238}U, ^{235}U, and ^{234}U [26]. The uranium isotopes emitted, mainly alpha and, to a lesser degree, beta particles, are accompanied by a small emission of gamma radiation, especially during the decay of ^{235}U [27, 28]. The emissions are all ionizing radiation, strong enough to break chemical bonds and thereby damage or destroy living cells [29–31]. In terms of radioactivity in nature, ^{235}U contributes about 2.2%, ^{238}U about 48.6%, and ^{234}U about 49.2%. The total activity level of uranium depends on the ratio of its isotopes. The ^{235}U and ^{234}U isotopes pose a greater radiological toxicity risk than ^{238}U because they have much shorter half-lives [32].

10.3.2 *Aquatic Chemistry of Uranium*

10.3.2.1 Behaviour in Surface Mine Waters

A linear dioxo cation governs the aquatic chemistry of uranium UO_2^{2+} ($[O=U=O]^{2+}$) [33, 34], which is mainly controlled by the pH, the redox potential, and the type of the available complexing ligand, such as carbonates, phosphates, vanadates, fluorides, sulfates, and silicates [35–37]. The U(III) is very easily oxidized, while the U(V) is prone to disproportionation [38, 39]. Figure 10.1 presents the speciation of uranium in aquatic system at atmospheric CO_2 exchange. In oxidized subsurface waters with redox potential (Eh) > 200 mV, dissolved UO_2^{2+} form carbonate complexes of varying stoichiometry as a function of the pH and the partial pressure of $CO_2(g)$ [33]. The uranyl-carbonate species prevail in the neutral to alkaline range, while the UO_2^{2+} hydrolyses at low pH values. Most hydrolysis products are oligomeric. Thus, oligomeric hydrolysis $(UO_2)_2(OH)_2^{2+}$ species are usually the most dominant. These are followed by other oligomeric and monomeric hydrolysis species, which include $[(UO_2)_3(OH)_5]^+$, $[(UO_2)_3(OH)_4]^{2+}$, $[(UO_2)_2OH_3]^+$, $[(UO_2)_3(OH)_7]^-$, $(UO_2)_4(OH)_7^+$, UO_2OH^+, $UO_2(OH)_2$, $UO_2(OH)_3^-$, and $UO_2(OH)_4^{2-}$ [25, 35, 40–43]. The monomeric carbonate species include $UO_2CO_3^0$, $UO_2(CO_3)_2^{2-}$, and $UO_2(CO_3)_3^{4-}$, while oligomeric carbonate compounds include $(UO_2)_3(CO_3)_6^{6-}$ [33, 35, 44]. The distribution and occurrence of the uranyl species in water is summarized in Error! Reference source not found.. Under the reduced conditions with redox potential (Eh) below zero volts, the dissolved U(VI) species are reduced to the U(IV) and precipitate as $UO_2 \bullet 2H_2O$ [34, 45, 46]. This occurs mostly in sediments and very deep reservoirs. The U(IV) has a strong tendency to bind to organic material [34]. Uranium in mine water is derived mainly from the oxidation of uraniferous bitumen or the dissolution of carbonates [33, 47, 48]. Consequently, the dissolved U is associated with the concentrations of NO_3^-, CO_3^{2-}, SO_4^{2-}, Cl^-, and not silica [49].

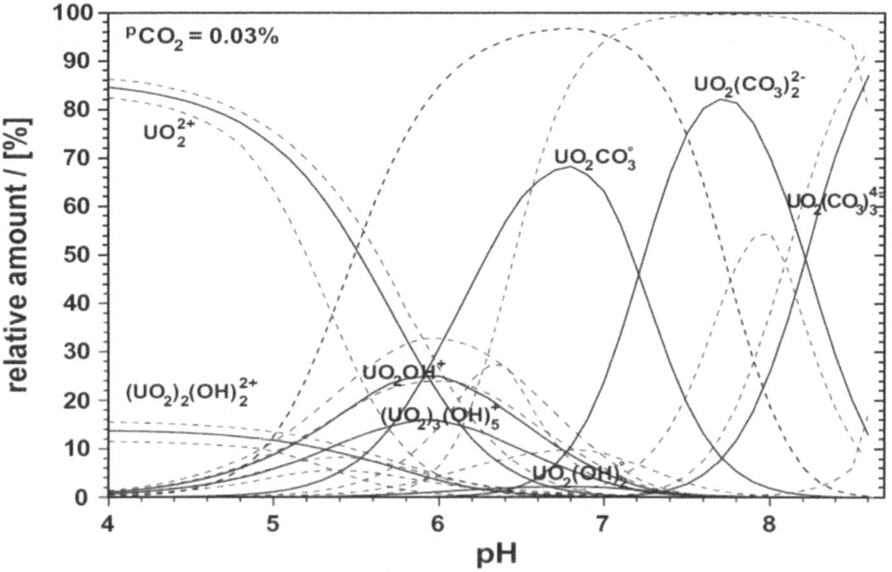

Fig. 10.1 Distribution of U(VI) aqueous species at 25 °C and ionic strength μ = 0.01 M NaCl as a function of pH at atmospheric CO_2 partial pressure. Dashed lines give 90% confidence limits. (The speciation diagram has been modified and used with courtesy from Meinrath [34])

10.3.2.2 Uranium Behaviour in Oxidizing Aquifers

The chemistry of uranium in the oxic aquifer is very similar to the chemistry in the surface waters. Under oxidizing conditions, U(VI) is the most stable form of uranium and exists mainly as the uranyl ion (UO_2^{2+}). It forms complexes readily with carbonate, phosphate, or sulfate ions. In low carbonate-free systems, the important species include UO_2^{2+}, $UO_2(OH)_2^0(aq)$, $UO_2(OH)_3^-$, $(UO_2)_3(OH)_5^+$, and $(UO_2)_2(OH)_2^{2+}$ [34, 41]. In predominantly carbonate-containing water, major uranyl species are carbonate anions with $UO_2(CO_3)_3^{4-}$ accounting for almost 60%, $UO_2(CO_3)_2^{2-}$ for 30%, and $(UO_2)_3(OH)_5^+$ [34]. However, the real distributions of the species in relation to other U(VI) species are a function of pH and the total concentration of dissolved U(VI). Concentrations of dissolved uranium are controlled by the aqueous speciation of U(VI) and adsorption of U(VI) species onto mineral and organic material [50–52]. The adsorption of U(VI) onto specific mineral phases primarily controls uranium mobility in oxic aquifers. The minerals prevalent in oxic conditions include carnotite [($K_2(UO_2)_2(VO_4)_2$], schoepite ($UO_3{\cdot}2H_2O$), rutherfordine (UO_2CO_3), tyuyamunite [$Ca(UO_2)_2(VO_4)_2$], autunite [$Ca(UO_2)_2(PO4)_2$], potassium autunite [$K_2(UO_2)_2(PO_4)_2$], and uranophane [$Ca(UO_2)_2(SiO_3OH)_2$] [26, 53–56]. However, adsorption onto colloidal organic matter facilitates uranium transport and dispersal in the oxic aquifers. Further, the solubility of U(VI)-bearing minerals controls the maximum dissolved uranium concentrations. In most abandoned uranium mines, uranium dissolution depends on the equilibrium between partially oxidized uraninite and the precipitation of coffinite [57]. Uranium co-precipitates with iron oxyhydroxides, which leads to the iron oxyhydroxides taking up nearly all the uranium dissolved in the mine waters before they reach the surface [58].

10.3.2.3 Uranium Behaviour in a Reducing Aquifer

Under the reducing conditions of negative Eh values, the dissolved U(VI) species are reduced to the less soluble U(IV) [59, 60]. Consequently, sparingly soluble U(IV) species or mixed U(IV)-U(VI) solids (e.g., U4O9) are present in reducing and anoxic environments, such as sediments and groundwater. The speciation of these sparingly dissolved U(IV) is dominated by hydrolytic species $U(OH)_3^+$ and $U(OH)_4^0(aq)$ in extremely low pH. The solid phase controlling uranium concentration in groundwater is partially oxidized pitchblende. Other important U(IV) minerals associated with reducing conditions include coffinite ($USiO_4$) and ningyoite ($CaU(PO_4)_2{\cdot}2H_2O$) [26]. There are usually accumulations of U(IV) precipitates in reducing aquifers because dissolved U(VI) species are, in most cases, transported through the permeable media, where they get reduced. This is how uranium roll-front deposits form in sandstone rocks. These orebodies occur when U(VI) carried by oxic groundwater contacts a zone containing a naturally occurring reductant, such as natural organic matter and H_2S (g) [59, 61]. The resulting U(IV)-containing minerals include uraninite and coffinite.

10.3.2.4 Uranium in Waste Rocks and Tailings Heaps

Generally, primary uranium ore in the waste rocks and tailings weathers into saprolites, which retain uranium as uranyl phosphates of the meta-autunite group, mainly as meta-uranocircite in saturated zones [26, 62, 63]. The uranium is retained in the abandoned mine wastes and tailings containing primary ore by two processes:

1. Incorporation into barium-strontium-calcium aluminium phosphate minerals of the crandallite group (mainly gorceixite)
2. Sorption of uranium with phosphorus onto iron oxides that coat the surfaces of other soil minerals

Thermodynamically, the meta-autunite group minerals are present in the saprolites in pH values below 4.5. Once exposed to the unsaturated environment, the meta-autunite reacts to form U(VI)-bearing aluminium phosphates and U(VI) surface complexes or nanoprecipitates associated with ferric oxides [26, 64, 183].

10.4 Uranium Interaction with Biosystem Compartments

10.4.1 Interaction with Microbiota

Unlike organic compounds, uranium and other metals do not break down into non-toxic components. Logically, biological activity can (1) only transform the metal from toxic to non-toxic species or (2) immobilize the metals from the pathway. Figure 10.2 illustrates uranium interactions with the biotic component and summarizes the mechanisms involved in metal bioremediation. Uptake, sorption, and precipitation by purified biopolymers derived from microbial cells provide alternative methods and additive processes for conventional physicochemical bioremediation [75]. Some plant and algal species can sequester uranium and other metals into their biomass, while others exude bioligands that trap uranium into bimetallic colloids. Some bacterial species can reduce mobile U(VI) to immobile U(IV) species [65].

Further, the algal-microbial biofilms can sequester uranium and other metals [66–68]. The bioreduction, sequestration, and trapping of uranium in bio-colloids can nucleate the biomineralization processes of uranium. However, practical applications of bioremediation techniques under natural conditions are still far from reality.

For instance, uranium can be immobilized through adsorption onto cell surfaces, extracellular polysaccharides, and low-molecular-weight organic acids, including biogenic colloids [69, 75]. Further, uranium can be internalized into cells via metabolic processes. In cells, uranium may be sequestered as inert compounds within organelles such as vacuoles, as well as in cell membranes and walls. Some species

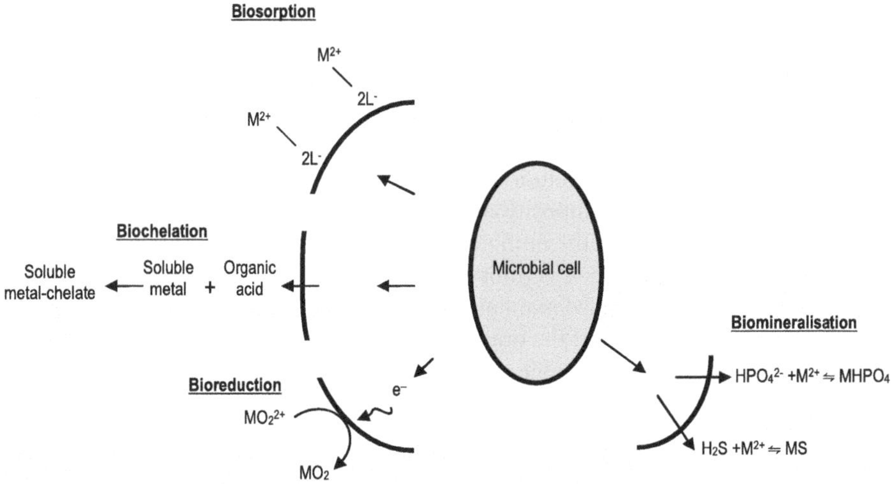

Fig. 10.2 The interaction of microorganisms with metals emphasizes the critical processes that lead to uranium immobilization in aquatic environments. (The structure is modified from an unknown source)

excrete the uranium compounds back into the water as precipitates or organometallic colloids. Under conducive conditions, the interaction may result in uranium mineralization. For instance, some marine and freshwater algae can incorporate uranium into aragonite ($CaCO_3$) mineral structures.

10.4.2 Biosorption

Uranium biosorption involves the deposition of metal ions onto the biosorbent via various sorption processes, including ion exchange, complexation, chelation, and microprecipitation [70–72]. The cell walls of most plant and algal or microbial materials are composed mainly of polysaccharides and carbohydrates (e.g., cellulose, xylan, and mannans) [73]. The backbone of polysaccharides includes ligands, such as amino, carboxyl, hydroxy, and sulfide groups, which eventually negatively influence the cell wall [74, 75]. The negatively charged groups attract and bind uranium cations [76, 77]. The chemistry at the cell surface involved can be simplified as follows:

$$R-COOH \rightleftharpoons R-COO^{-}+R^{+} \quad (10.7)$$

$$R-O-H \rightleftharpoons R-O^{-}+H^{+} \quad (10.8)$$

$$R-PO-H \rightleftharpoons R-COO^{-}+H^{+} \quad (10.9)$$

Deprotonation leaves negatively charged sites, convenient for uranyl adsorption

$$R-COOH \rightleftharpoons R-COO^{-}+UO_2^{2+} \quad (10.10)$$

The cell wall structure and the uranium ion's characteristics determine the capacity and efficiency of bioremediation of uranium water using sorption techniques. Uranium adsorption capacity onto the cell surface depends on the number of available ligand binding sites, uranium speciation, and the media's physicochemical properties. The chemical composition and cell size are reasonably constant for a given species, implying that the number of ligand groups on a surface is also fixed. Furthermore, the complexation is limited to a single layer [78–80].

Therefore, the cell's surface maintains a constant uranium concentration relative to the cell composition, even if the uranium ions exceed the concentration of affinity sites [80]. The number of available surface-binding sites can only increase as cells grow. Thus, the rate at which uranium ions are removed from water by biosorption depends on the growth rate of the biosorbent. A few studies have estimated the number of high-affinity surface-binding sites on algae and bacterial cells. For an instant, phytoplankton has about 10^8 uranium-binding sites per cell [80]. In contrast, the bacterium *Shewanella putrefaciens* has approximately 2.0×10^{19} carboxyl sites,

5.5×10^{18} phosphoryl sites, and 2.3×10^{19} amine sites per gram of bacteria, translating into a biosorption capacity of up to 100 mg of uranium [80–82].

There are many ligand types on a cell surface, which results in the selective binding of uranium cation species. Some ligand groups may carry a positive charge, thereby adsorbing anionic metal complexes [83, 84]. The mixture of positively and negatively charged groups on a cell surface determines the number of cations and anions that can be complexed. The uranium cation adsorbs to most cell surfaces, whereas binding anionic or complexed uranium species is more difficult because there are fewer positively charged ligands on the cell surface [25, 85, 86]. At a low pH value, the uranium is present in the solution mainly in free UO_2^{2+} . Furthermore, adsorb poorly to the cell surface due to competing with protons for the biomass's binding sites [82, 87]. At higher pH, the sorption is higher because of the ion exchange between the hydrolysed uranyl ions, UO_2OH^+, $(UO_2)_3(OH)_5^+$, $(UO_2)_2(OH)_2^{2+}$, and the protons [44, 80].

Living and dead cells behave similarly with respect to uranium adsorption, provided the cell wall structure remains intact [88–90]. Thus, the dead algae, microbes, and plant biomass still contain negatively charged ligands that can bind to uranium. With the help of invertebrate shredders, the biomass is reduced to particulate organic matter, to which more uranium is adsorbed because of the increase in the surface-to-volume ratio [80]. Furthermore, the dead biomass decays into humic and fulvic substances through microbial activities. The humic and fulvic substances are rich in dissolved ligands, probably because of the carboxyl and hydroxyl groups associated with lignin's breakdown. The carboxyl group has a dissociation constant (pKs) of about 5, facilitating complexation with uranium over a wide pH range [91–93].

10.4.3 Bioprecipitation

Bioprecipitation occurs when uranium species interact with the biogenic compound. The processes can take place either internally or externally in the cell. Bioprecipitation of uranium involves biosorption and active or metabolic internalization. Most plant and microbe populations release extracellular polysaccharides as a byproduct of metabolism [94–96].

Additionally, when resource stoichiometry is imbalanced, homeostatic regulation leads affected organisms to exude chelates, low-molecular-weight organic compounds, and polyphenols [80, 97]. These excreted byproducts, or exudates, interact with uranium, resulting in precipitate formation, but not necessarily on the cell surface. Furthermore, their introduction into the water alters the chemical speciation, thereby immobilizing uranium. For instance, microbes such as *Pseudomonas* (Gamma Proteobacteria) and *Citrobacter* (Omnibacteria) also produce large quantities of extracellular polysaccharides [66, 75], while some macrophytes, such as *Lemna* sp. and *Myriophyllum aquaticum*, exude oxalates and polyphenolic substances that complex uranium ions [2, 97–99]. There is also evidence that phosphate interacts with uranium, leading to the formation of very stable crystalline structures

[36, 50, 100]. For instance, homeostatic regulation of sudden uranium uptake by *Peltigera* sp. (a lichen) is the cellular formation of uranyl phosphate crystals stored in the cellular mycobiont [84, 101]. This formation of U(VI) phosphate crystals is widespread. It occurs mostly on the cell surface, as in bacteria such as *Citrobacter* sp. The mechanism involves the production of alkaline phosphatase, which causes cells to excrete phosphate to the cell surface. The phosphorus provides the nucleus for the precipitation of U(VI) phosphate [96]. Another homeostatic regulation of stoichiometry imbalance induced by uranium uptake is apparent in *Lemna gibba*. In *L. gibba*, uranyl oxalate crystals develop to counteract the presence of uranium in the cells [100].

10.4.4 Chelation

Chelates are organic molecules that trap or encapsulate highly reactive trace metal cations, preventing them from involvement in unwanted chemical reactions. Sometimes, the trapping between metals and chelates may form insoluble compounds. Thus, most plants and microorganisms use chelates to aid the uptake and transportation of essential metal nutrients. Due to unique binding properties, chelates are often used to stabilize metals, making them critical compounds in bioremediation [80]. Chelating agents produced by microorganisms are mostly siderophores, enhancing the dissolution of iron and increasing mobility and bioavailability. Siderophores are of low molecular weight and function primarily as iron transport compounds. Since there are similarities in the biological behaviour of ferric and uranyl ions, chelates can bind uranium like those used for iron [102]. Plants first release special chelates, known as phytosiderophores, that bind to metals in the rhizosphere and facilitate their absorption by the plant [103–105]. The function of chelates is stereoselective, depending on the recognition of the metal ion's coordination geometry.

As a result, some chelates, like EDTA, function poorly with uranium because of the linear structure of the uranyl cation [106]. Nonetheless, the benefits of the chelating agent on uranium remediation are enormous. Low-molecular-weight organic acids, such as citric and oxalic acids, exuded by most aquatic plants, facilitate the dissolution of uranium and its uptake by macrophytes and microorganisms [107]. Other chelates include metallothioneins and phytochelatins, which promote the uptake and transport of metals in organisms. The phytochelatins are heavy metal-complexing peptides composed of different amino acids (r-glutamic acid and cysteine) and n-glycine ($n = 3–7$), which are involved in the detoxification and homeostatic balance of heavy metals in plant cells [108–110]. Excess heavy metals are bound to the cell wall in a process called metathiolate formation through mercaptide complexes [111]. Consequently, chelating agents facilitate the bioreduction of uranium by altering its bioavailability.

10.4.5 Active Intracellular Uptake

The most widely recognized biological process for metal removal in wetlands is plant uptake. The main route of uranium uptake in aquatic plants is through roots in emergent and surface-floating plants. In contrast, in plants with either completely submerged leaves or both floating and submerged leaves, or free-floating plants, uptake occurs through leaves and roots. In the case of foliar uptake of uranium, a passive movement in the aqueous phase through cracks in the cuticle or the stomata to the cell wall and then the plasmalemma is involved. Submerged rooted plants have some potential for the extraction of metals from sediments as well as water, while rootless plants like *Wolffia* sp. or floating macrophytes like *Lemna* sp. can extract metals only from the water.

Previously, it has been thought that a few aquatic macrophytes and microorganisms, including algae and plants, actively pump metals across their cell membranes [112–114]. The energy source is usually coupled to photosynthesis and temperature. However, it is becoming increasingly clear that most macrophytes actively avoid the uptake of uranium ions into their systems. Uranium uptake occurs when external uranium concentrations exceed an arbitrary threshold of cell resistance. Consequently, the resistance gates are overwhelmed, allowing uranium to be taken up. Accidentally, uranium may be taken up by some cells when the uranium species mimics essential metals or interacts closely with nutrient resources like PO_4^{3-}. Once inside the cell, uranium may be concentrated and sometimes precipitated into vacuoles, as in some plants and algae, or sequestrated in the cell wall in a few uranium-resistant microphytes [115–117]. In some plant and algal species, uranium may also interact with internal inorganic phosphate to form uranium-phosphate crystals stored within the cellular mycobiont (e.g., in *Lemna* sp.) [116]. Some organisms change the redox status of uranium, bind it, and excrete it as non-mobile organometallic compounds upon recognizing its potential toxicity. Further, through the toxicity-avoidance mechanism, plant species can take up uranium and sequester it in roots, as in *Phragmites australis*, while other species may transfer it to aboveground parts, such as shoots. These different allocations of bioaccumulated uranium help design remediation procedures.

The uranium uptake rate by aquatic macrophytes and algae varies widely depending on plant species, growth rate, uranium speciation, and uranium concentration in plant tissue. The pH, organic matter context, and plant genotype have marked effects on uranium uptake. The UO_2^{2+}, uranyl-carbonate complexes, and $UO_2PO_4^-$ are the uranium species being preferentially taken up by the roots and transferred to the shoots [118]. Furthermore, when in mixtures, U(VI) competes with other metallic ions for uptake based on molecular size and shape and the ligand's configuration. For instance, lithium, Na, K, Fe, and NH_4 do not affect U accumulation. The transition group of metals and Al, Ca, Fe, and Zn ions compete with U(VI) by the biotic system, especially macrophytes and algae [80]. Uranium uptake is hindered by phosphate and carbonate ions but not affected by cations (Na, K, Mg, Ca, Mn, Co, Ni, and Zn), nitrates, sulfates, and thiosulfates. Nevertheless, the bioaccumulation procedure's contribution to uranium's overall immobilization in an aquatic system is small.

10.4.6 Biological-Induced Redox Reactions

Numerous natural-occurring metal dissimilatory and sulfate-reducing bacteria are capable of obtaining energy by coupling the oxidation of organic matter and H_2 to the reduction of soluble U(VI) to insoluble U(IV) (Fig. 10.3). Hydrocarbons are electron donors that provide electrons to reduce metals in redox reactions. For example, the half-reaction for U (VI) to U (IV) is:

$$UO_2^{2+} + 2e^- \rightleftharpoons UO_2 \tag{10.11}$$

When metabolized by the bacteria, the hydrocarbons provide the necessary electrons for the reaction. The remediation benefits from the process because the metals are less soluble and readily precipitate from solution once reduced. For instance, the reduction product, U(IV), is subsequently a non-toxic precipitate and a mineral, uraninite (UO_2) [26, 119]. However, many bacteria cannot use U(VI) as the sole terminal electron acceptor. Nevertheless, microbial reduction of soluble U(VI) to insoluble U(IV) is one strategy for immobilizing uranium from the water pathway

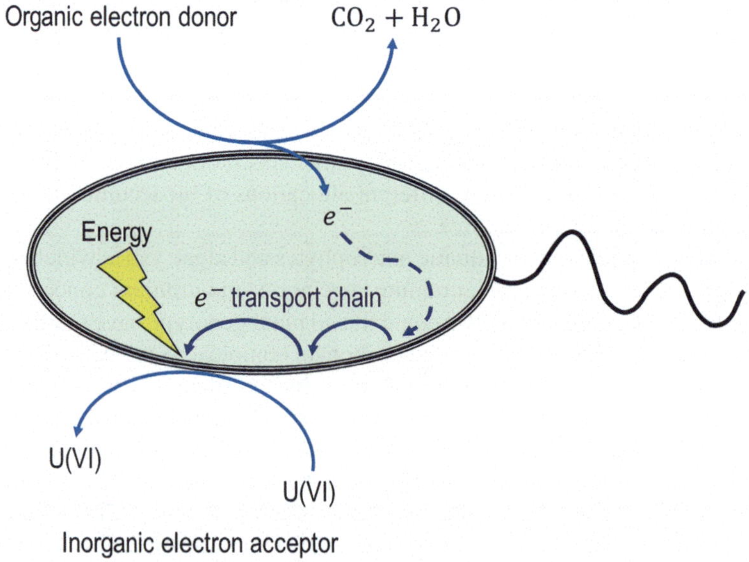

Fig. 10.3 Principle of bioreduction of U(VI) to immobile U(IV), which may result in uraninite minerals. (Modified from Ginder-Vogel et al. [121])

in abandoned uranium mines, as it promotes uranium mineralization under in situ environmental conditions [100, 120, 121].

Enzymatic Reduction Mechanism The enzymatic reduction pathway (VI) involves c-type cytochromes as reductases when hydrogen gas is supplied as the electron donor [122, 123]. In the H_2-utilizing bacteria, the electron is transferred by the tetraheme cytochromes c3, while the triheme periplasmic cytochrome c7 plays a role in transferring electrons from acetate to U(VI). There are suggestions that periplasmic and outer-membrane cytochromes also play an important role [96, 124]. The reduction of U(VI) to U(IV) requires two electrons. However, the microorganism can transfer only one electron. The one-electron reduction process initially generates U(V), for instance, the uranium reduction [45, 125]. Due to the instability of U(V) complexes, disproportionation to U(IV) occurs. However, the enzymatic reduction of uranium is hindered by the presence of some naturally occurring chemicals. For instance, bicarbonate ion concentrations reduce the enzymatic reduction of U(VI) because it lowers the half-cell potential of the U(IV)-U(VI) couple [126, 127]. Thus, the formation of ternary Ca-U(VI)-CO3 complexes makes it a less energetically favourable electron acceptor than U(VI)-carbonato complexes because of Ca′s direct interaction with the cells or electron donor. The presence of ferrihydrite can sometimes hinder U(VI) reduction [183].

Bioreduction Through Iron-Reducing Bacteria The majority of Fe(III)-reducing microorganisms can conserve energy by coupling the oxidation of hydrogen or organic compounds to the reduction of Fe(III), which can also reduce soluble U(VI) to insoluble U(IV) [45, 125, 128]. For example, *Geobacter* sp., which usually gains energy by reducing Fe (III) to Fe (II), can instead reduce U (VI) to U (IV). They can use U(VI) as an electron acceptor instead of Fe(III) [119]. Further, bacteria from both the delta and gamma subdivisions of the Proteobacteria, such as *Geobacter metallireducens, Geobacter sulfurreducens*, *Desulfuromonas palmitatis*, and *Desulfuromonas acetoxidans*, couple the complete oxidation of acetate to CO_2 via the citric acid cycle with Fe(III) as the electron acceptor [125, 129, 130]. In addition to acetate, *Geobacter* sp. can also oxidize ethanol, propionate, butyrate, valerate, pyruvate, propanol, and toluene, among others, to carbon dioxide, using Fe(III) as the electron acceptor [125, 129, 130]. The bacteria *Shewanella oneidensis*, *Shewanella alga*, and *Pseudomonas* sp. do not use organic electron donors such as acetate but mostly couple hydrogen oxidation to reduce Fe(III) [45, 130–132]. Given the wide range of Fe(III)-reducing bacteria, it is therefore not surprising that they exhibit a high degree of metabolic diversity, with various species able to reduce U(VI). Despite the apparent diversity of Fe(III)-reducing bacteria, Geobacter species are the dominant Fe(III)- reducing microorganisms [133–136].

Therefore, the capability to couple the oxidation of organic matter to the reduction of Fe(III) and, alternatively, U(VI) is through the following typical reactions:

$$[CH_3COCO_2]^- + 2Fe(III) + 2H_2O \rightleftharpoons [CH_3H3COO]^- + HCO_3^- + 2Fe(II) + 3H^+ \quad (10.12)$$

In addition, then:

$$[CH_3COO]^- + 8Fe(III) + 4H_2O \leftrightharpoons 2HCO_3^- + 8Fe(II) + 9H^+ \quad (10.13)$$

Concurrently or alternatively, the existence of Fe(III) with U(VI), the reactions include the following:

$$[CH_3COO]^- + 4U^{6+} + 4H_2O \rightleftharpoons 4U^{4+} + 2HCO_3^- + 9H^+ \quad (10.14)$$

$$H_2 + U^{6+} \rightleftharpoons U^{4+} + 2H^+ \quad (10.15)$$

The Fe(III) dissimilatory pathway involves the ferric iron reductase, which is probably located on the outer membrane in Gram-negative bacteria [126, 137, 138]. The ferric iron reductase acts as the terminal reductase of an electron transport chain, which is linked to the cytoplasmic membrane [136, 139]. Electrons are transferred down the transport chain to the ferric iron reductase, which transfers them onto insoluble extracellular Fe(III) oxides. The transfer of electrons is a mechanism for conserving energy by generating ATP [140, 141]. The Fe(III) is reduced by the ferric reductase before or after uptake into the cell, forming a weak Fe(II)-chelate complex from which the iron can be easily dissociated for use by the cell. Three possible mechanisms employed through which the Fe(III)-reducing bacteria reduce Fe(III) oxides, as well as U(VI), have been identified as follows:

(i) Direct contact, in which the microorganism contacts the insoluble iron oxide and transfers an electron from the cell to the Fe(III) oxide surface, with outer membrane-bound c-type cytochromes playing a role in this process [133, 135, 138, 142]. Some bacteria utilize pili in electron transfer from the cell to the mineral surface as highly conductive "nanowires" (a pilus is typically 6–7 nm in diameter).

(ii) Fe(III) chelators, where the bacteria produce siderophores to dissolve Fe(III) oxides which become more accessible to Fe(III) reductase than insoluble Fe(III) oxides [133, 135, 138, 142].

(iii) Extracellular electron-shuttling molecules, such as quinone-containing compounds or c-type cytochromes, transfer electrons from the cellular electron transport chain to the metal terminal electron acceptors [143–145]. Extracellular electron shuttles may already be present in the environment due to the decomposition of organic material, or produced by cellular processes and excreted into the environment [138]. Proteins never act as electron-shuttling compounds. At the same time, humic substances are excellent electron-shuttling compounds.

Bioreduction Through Sulfur-Reducing Bacteria
Sulfur-reducing bacteria can reduce U(VI) to U(IV), which can be salvaged for bioremediation in abandoned uranium mines. Like the Fe(III)-reducing bacteria, the chemical basis of SRB remediation involves microbial-mediated sulfate reduction coupled with organic matter oxidation. The reaction occurs as follows:

$$2CH_2O(aq) + SO_4^{2-} + H^+ \rightarrow H_2S + 2HCO_3 \rightarrow H_2S + H_2O + CO_2$$

where other organic substances can replace CH_2O, hydrogen sulfide (H_2S) then reacts with uranium to form a sulfide that precipitates out of the effluent:

$$H_2S + UO_2^{2+} \rightarrow UO_2S + 2H^+$$

Some bicarbonate (HCO_3^-) may be accompanied by CO_2 production in the sulfur redox reaction. This consumes acidity, raising the pH, which facilitates the precipitation of uranium and related heavy metals. Sulfate-reducing bacteria generally rely on simple carbon compounds, such as organic acids or alcohols, as electron donors for sulfate reduction, though some can use hydrogen [138, 146–148]. When organic matter is used as a carbon source, other heterotrophic bacteria degrade it into simple carbon compounds (Fig. 10.4). Similar to the mechanism in Fe(III)-reducing bacteria, sulfur-reducing bacteria transfer one electron to U(VI), reducing it to U(V). As mentioned earlier, U(V) is unstable and prone to disproportionate formation, forming U(VI) and U(IV).

Pre-conditions for Bioreduction Processes Generally, bacteria couple the oxidation of organic matter to the reduction of oxygen as the terminal electron acceptor in aerobic environments. Under anaerobic conditions, microbes can utilize NO_3^-, Mn(IV), Fe(III), SO_4^{2-}, and CO_2 as terminal electron acceptors to oxidize organic matter [130]. Thus, the amount of NO_3^-, Mn, and Ca (because of its relationship to CO_2 in water) influences the U(VI) reduction because of competition. Fortunately, Fe(III) reduction accounts for the oxidation of up to 65% of organic matter in some anaerobic sediment [149–151]. However, the reduction of Fe(III) to Fe(II) leads to other significant chemical changes, such as the release of trace metals and radionuclides into the water pathway and the inhibition of methane production [152, 153], as well as Fe(II) functioning as a catalyst in the reduction of U(VI) to U(IV).

Nevertheless, the bioremediation of uranium predominated in the mine water occurs much faster than abiotic reduction. For sulfur-reducing bacteria, anaerobic conditions with a redox potential below −200 mV are required, as oxygen, nitrate, manganese, and iron reduction all yield more energy per equivalent than sulfate reduction [146]. The redox conditions are also suitable for iron reduction to the ferrous (Fe^{2+}) state, which co-precipitates with sulfide. Furthermore, sometimes electron donors are not enough to reduce the metal. In this case, bioremediation strategies often add nutrients to provide an electron source. Conversely, a more suitable electron acceptor, such as oxygen, may be used in some situations. If this is the case, bioremediation strategies might focus on creating an anaerobic environment, such as a constructed wetland.

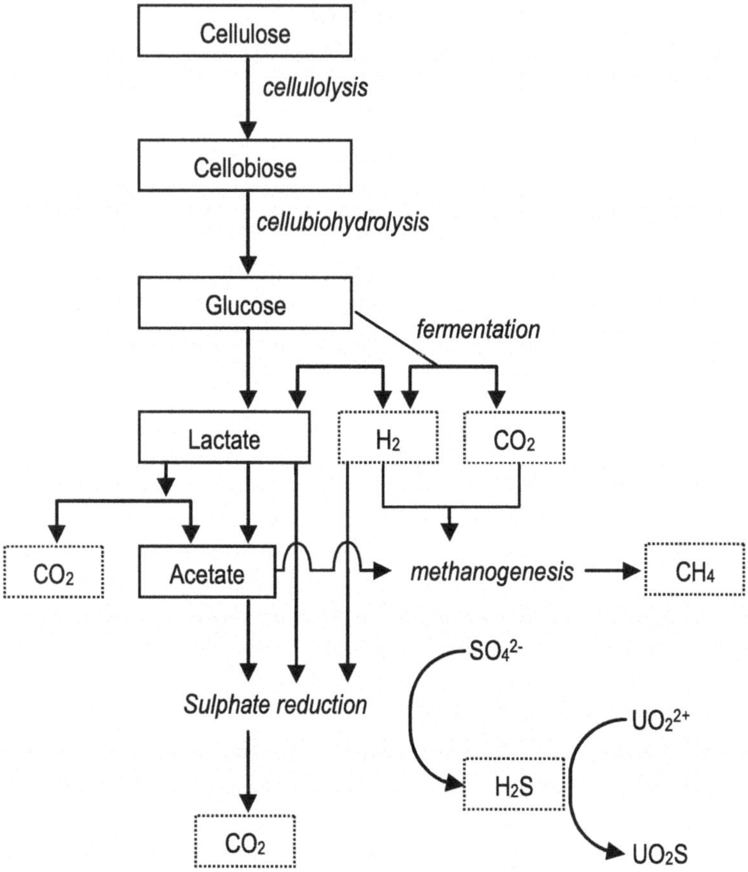

Fig. 10.4 Microbial processes involving oxidation of organic carbon substances coupled with sulfate reduction further reduce U(VI) to U(IV). (Modified from Doshi [146])

10.4.7 Phyto-interaction

Phytoremediation is the use of plants to clean up contaminated sites [154]. Metals can be phytoremediated by several different processes, including phytoextraction, rhizofiltration, and phytostabilization. Plants can also remove trace metals from water through biological uptake and surface adsorption [155, 156]. In phytoextraction, plants take up and concentrate the uranium and other metals in their tissues. Certain plant species, such as *Lemna* sp. and *Phragmites* sp., have evolved to extract uranium at exceptionally high levels. Once uranium and the other metals are taken up, they are either stored in the root or transported to the shoots through the xylem [157, 158].

Typically, uranium has difficulty migrating through the xylem due to its high cation exchange capacity. Most metals are chelated to facilitate the translocation, but this is not known for uranium. At the chute, it seems that uranium is stored in cell vacuoles and the cell wall. Inside the vacuole, uranium is bound to organic compounds, probably chelates, to prevent further damage to the plant.

Furthermore, uranium may account for crystals in the cell; for instance, uranium oxalate was observed in *Lemna gibba* [100]. Some plant species use their roots to filter uranium and other toxic metals from the water. With this technique, uranium is deposited in the macrophytes' rhizosphere either via precipitation, ion exchange, or chelation [80]. The processes can be direct, where the plant excretes or exudes organic substances that facilitate. For some plant species, uranium filtering occurs through the rhizospheric microflora that associates with the plants. Exudation may facilitate symbiotic and allelopathic associations.

10.4.8 Biogenic Facilitated Abiogenic Reduction of Uranium

Thermodynamics predicts the high potential of Fe (II) to reduce from U(VI) to U(IV). Many subsurface environments in abandoned uranium mines, especially sediments, are saturated with biogenic Fe(II) following long-term microbial reduction of Fe(III) [122, 159]. Therefore, the condition naturally immobilizes uranium, placing abiotic reduction above direct microbial reduction in uranium immobilization in wetland ponds of abandoned uranium mines [3, 160, 161]. Additionally, Fe(II) species adsorbed onto ferrihydrite, magnetite, and haematite surfaces can reduce U(VI) to U(IV). Reduced U(VI) is then present as a $UO_2(s)/Fe(OH)_3(S)$ precipitate in the water. Therefore, abiotic reductions of U(VI) by biogenic Fe(II) are likely the dominant reductive pathway. However, the Fe(II) does not significantly contribute to U(VI) reduction in the sediments. In most abandoned uranium mines, the reduction of U(VI) by Fe(II) is a minor process because U(VI) is complexed by carbonate.

10.5 Harnessing the Biogeo-Processes for Uranium Remediation

Various bioremediation strategies can be explored to reduce uranium and other contamination associated with abandoned uranium mining and processing sites. Here, three commonly applied strategies in abandoned uranium mining sites that offer some of the most promising bioremediation treatments are discussed. These strategies provide cost-effective alternatives to traditional metal decontamination methods, such as "pump and treat" soil-washing systems, excavation, and reburial of contaminated areas.

10.5.1 Constructed Wetlands

One strategy for bioremediating metals in abandoned uranium mines is the creation of artificial wetlands to treat contaminated surface water, groundwater, and acid drainage [162–165]. Groundwater is pumped into the constructed wetland for passive treatment. The wetland strategy relies on several basic physical, chemical, and biological processes to remove or immobilize uranium from the aquatic system. Thus, the only applicable strategy can involve all the bioremediation processes once well designed. The uranium remediation occurs in a wetland's three main compartments: sediment, hydrology, and vegetation.

The substrate harbours a microbial consortium responsible for uranyl reduction and transformation, nucleating uranium biomineralization. Wetlands with appropriate substrates promote the growth of sulfate-reducing bacteria in anaerobic conditions. Thus, in acid mine water rich in sulfates, these bacteria generate hydrogen sulfide, which reacts with metals, including insoluble metal sulfides [149, 166]. This is very important in controlling the generation of AMD from an abandoned uranium mine. The hydrology compartment contains heterogeneous polyligands like fulvic, humic, and tannic acids, amorphous metal oxyhydroxides, clay, bacterial surfaces, associated exocopolymers, and suspended macromolecules (e.g., polysaccharides and proteins) [37, 58, 167]. Anaerobic conditions also readily develop in the sediment's upper portion due to the hydric conditions that support hydrophytic vegetation. Uranyl ions can coordinate with most biological material, living or dead. It is also transported through the water column to the sediments. This is slow in well-oxygenated water and fast in anoxic water.

While wetland sediments serve as primary sinks for uranium, macrophytes absorb uranium through their roots and shoots. The wetland plants are also autotrophic organisms, producing biomass of reduced carbon compounds that provide nutrients to various micro- and macroscopic organisms [164]. The expansive rhizosphere of wetland plant species provides an enriched culture zone for microbes involved in uranium species biotransformation [168]. Physical removal processes of uranium in constructed wetlands include settling and sedimentation associated with particulate matter [25, 169, 170]. To some extent, mats of floating plants serve as sediment traps, while rooted and floating plants regulate the water flow rate by providing resistance. For particles, which are light or less dense than water, sedimentation is possible only after flocculation. Flocculates generally settle more rapidly in a wetland than do individual particles. Flocculation is enhanced by high pH, high suspended matter concentration, high ionic strength, and high algal concentration [80]. Before sedimentation, other chemical processes like precipitation, adsorption, and co-precipitation should occur first. More than 50% of the uranium is easily adsorbed onto the particulate matter in the wetland [15, 171, 172]. In sediments, uranium is adsorbed onto particles by either cation exchange or chemisorption, the uranium speciation changes over time as sediment conditions change.

10.5.2 Injection Wells

Well injection is another promising bioremediation strategy applicable in abandoned uranium mines [133]. This strategy is one of the major procedures for uranium bioremediation in groundwater, especially in abandoned in situ leaching mining operations and in leaching contaminated water and AMD from tailings and waste heaps. Mostly, bioremediation in the groundwater occurs through the bioreduction of uranium by iron and sulfur-reducing microorganisms. In most cases, underground environments are rich in microorganisms, but the conditions are usually not conducive to their growth or uranium reduction. The environments typically lack carbon sources and contain high levels of nitrates. Thus, biostimulation of microbial activity is necessary for uranium bioremediation. Thus, a common method to deliver carbon sources is through injection wells (Fig. 10.5). A series of wells is installed to control groundwater flow and allow the injection of solutes, such as ethanol, required to create a geochemical environment conducive to microbial growth and subsequent U(VI) reduction. For instance, ethanol is added to stimulate in situ denitrification of residual nitrate and reduction of U(VI) [173]. The addition of acetate to either groundwater promoted the growth and activity of dissimilatory Fe(III)-reducing microorganisms and removing soluble U(VI) from contaminated groundwater [174].

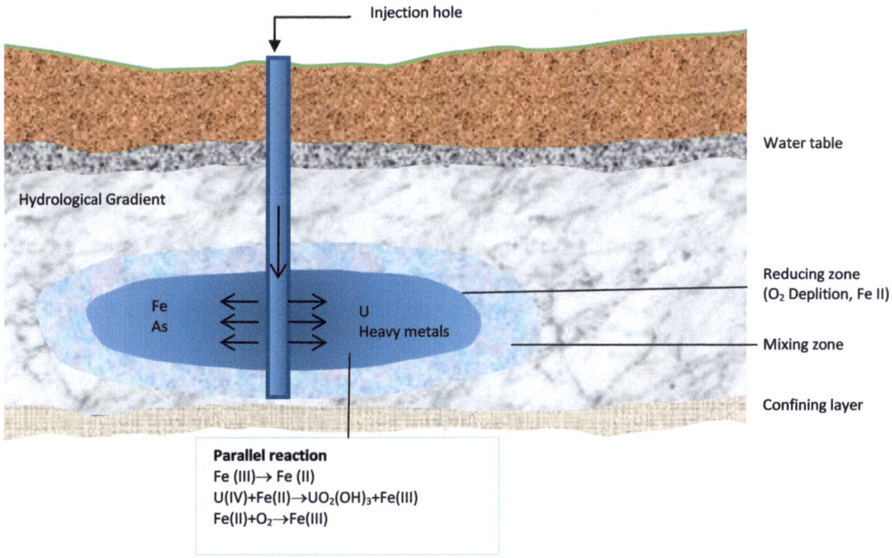

Fig. 10.5 Schematic illustration of a bioremediation well injection strategy, where biostimulants such as ethanol are injected underground to stimulate bacterial reduction of uranium. The well can also be used for bioaugmentation, in which uranium-reducing bacteria can be injected into the well [174]

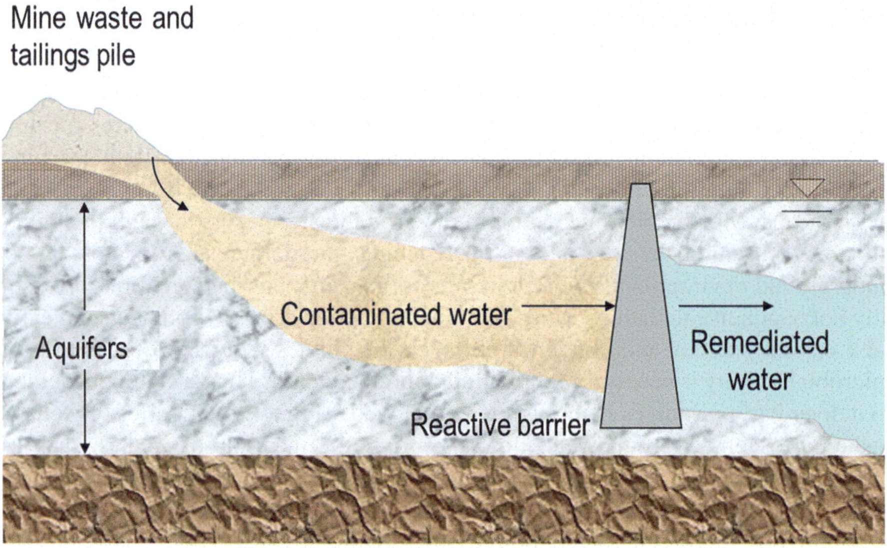

$$2CH_2O + SO_4^{2-} \rightarrow 2HCO_3^{-} + H_2S$$

$$Me^{2+} + S^{2-} \rightarrow MeS$$

Fig. 10.6 Schematic illustration of passive treatment using a permeable reactive barrier functioning in treating tailing heap leachates, especially AMD. The reactive barrier can be chemical or biological. For example, the incorporation of microbes can treat uranium through bioreduction processes, such as using dissimilatory Fe-bacteria or sulfur-reducing bacteria. In addition, a source of sulfur-reducing bacteria, such as cow manure or organic matter, is added to passive treatment systems to stimulate bioremediation

10.5.3 Bioreactive Barrier

A bioreactive barrier is a bioremediation strategy targeting leaching and groundwater. The application of bioreactor barriers in groundwater systems is schematically illustrated in Fig. 10.6. The strategies capitalize on the process of bioaugmentation, in which microorganisms are introduced to serve as a barrier to clean leaching water [175, 176].

10.6 Biogeochemical Challenges in Remediation Application in AUM

The challenge of applying bioremediation effectively in abandoned uranium mines lies in overcoming several limitations, fears, and failures.

10.6.1 Stoichiometry and Homeostasis Limits

There are several limitations to applying bioremediation in contaminated waters of abandoned uranium mining sites. One major limitation is the organisms' nature, in which uranium immobilization is not a benevolent gesture but a survival strategy. Most bioremediation agents do their job under environmental conditions that suit the organisms' needs. Thus, unless conducive bioremediation conditions are set, organisms use alternative substrates as energy sources other than uranium.

Consequently, environmental modification is needed to encourage organisms to take up or bioreduce uranium at an acceptable rate. Further, the organism must be exposed to low pollutant levels over a period to induce the production of the metabolic pathways needed to reduce uranium. As a result, bioremediation can also become very costly due to modifications, be labour-intensive, and take several months to achieve acceptable levels.

10.6.2 Environmental Condition Limits

Some environmental factors may also limit the applicability and effectiveness of the uranium bioremediation process. There are occasions when sediment or soil matrices prevent contaminant microorganisms from coming into contact with each other. The matrices may form a barrier that prevents the contaminated water from reaching the organisms. Worse is when microbes have preferential colonization, which may clog nutrient and water flow in the sediments' pores. Furthermore, preferential flow paths may severely reduce contact between the biostimulants and uranium, as well as between the bioremediation agents and uranium. Similarly, bioremediation in clay-rich, highly layered, or heterogeneous subsurface environments is limited in application because these environments interfere with oxygen- or other electron-acceptor-transfer processes [45, 121]. Therefore, the rate of uranium immobilization from the aqueous phase depends on the mass transfer rates of contaminants from the low-permeability matrix blocks into higher-permeability regions within fractures.

10.6.3 Discharge Pattern Limits

The application of bioremediation is limited to the type of contaminant discharge in abandoned uranium mining sites because the procedure is not responsive to sudden changes. High concentrations of heavy metals are likely to be toxic to microorganisms. Low contaminant concentrations may be below the threshold for organism survival. In particular, air bioremediation is inefficient, so radon emanated from abandoned uranium mines is hardly tackled. Consequently, bioremediation attempts fail and become costly because the microorganisms are sometimes sited specifically. Each new site presents a different set of conditions, and the microorganisms used for bioaugmentation are usually lab stock cultures whose conditions are far from those in the field.

10.6.4 Ecological Implication Fears

Another problem is that the applicable bioremediation technologies can cause environmental disruption beyond pollution damage. In most cases, biostimulation and bioaugmentation are unavoidable. These environments can disrupt other organisms and upset the ecological balance. For instance, during biostimulation procedures, the continued addition of acetate may result in a shift in the dominant terminal electron-accepting process from Fe(III) reduction to sulfate reduction, as well as complete degradation of the acetate under sulfate-reducing conditions and an apparent decrease in the rate of removal of soluble U(VI) from groundwater [174]. Therefore, maintaining Fe(III)-reducing conditions is critical for sustaining reductive precipitation of U(IV) and underscores the need to correlate U(VI) removal with alterations in microbial community composition in the subsurface.

Further, modifying the environment can be expensive, as with classical treatments, such as acetate or ethanol in biostimulation. Furthermore, in bioaugmentation, the long-term effects of introducing naturally occurring non-native bioremediation organisms into an area can alter the ecological balance when they become invasive. Worse, most high- and fast-acting bioremediation organisms are genetically altered, and their other side is less understood. As a result, there are always fears of horizontal transfer of the modified gene, the creation of new pathogens, and the possibility of mutations that could make the organism invasive.

10.6.5 Physicochemical Characteristics Limits

In many abandoned uranium mining and processing sites, ambient groundwater has a very low pH and high levels of aluminium, calcium, and nitrate. pH values below 3.4 are not favourable for most organisms' growth, requiring subsurface pH manipulation to a value more suited to microbial activity. However, increasing the pH in situ is undesirable because aluminium hydroxide precipitates form at approximate pH values above 4.5, and calcium solids precipitate at near-neutral pH. In addition, calcium is present in groundwater at levels that inhibit U(VI) reduction, especially at neutral pH. The formation of U-carbonate-calcium complexes also retard bioreduction [130]. The aqueous species, $CaUO_2(CO_3)_3^{2-}$ and $Ca_2UO_2(CO_3)_3$, also have an important role in the inhibition of microbial reduction and aqueous speciation of uranium(VI) in contaminated groundwater. Ca^{2+} significantly impacts the aqueous speciation of U(VI) and, consequently, the sorption and mobility of U(VI) in aquifers [177].

Nitrate also inhibits U(VI) reduction and is co-contaminated with uranium in groundwater [178]. Before reducing uranium, persistent nitrate must first and continuously be removed or transformed, as nitrate is a preferred electron acceptor over uranium and stimulation of metal-reducing bacteria is hindered by low pH [128]. In situ denitrification of bulk nitrate is likely to be incomplete because nitrate levels are usually very high on abandoned uranium mining sites. Any nitrite formed can oxidize U(IV) back to the more soluble U(VI). In addition, complete denitrification of nitrate generates N_2 in excess of its solubility limit. The presence of a gas phase can alter flow direction and decrease the solid surface area available for sorption, desorption, and redox reactions.

10.6.6 Processes and Counteraction Limits

The performance of the wetland treatment system depends on the input water quality and the specific mechanisms of metal removal. The wetland plants translocate oxygen from the shoots to the root rhizomes through their internal gas spaces and parenchyma. The roots and rhizomes, in turn, leak oxygen to the reduced environment. Much as this is an advantage, it also limits uranium bioremediation. If uranium immobilization is aerobic, these oxidizing conditions promote the precipitation of Fe^{3+} and Mn^{2+} oxyhydroxides, which may adsorb or co-precipitate with uranium ions [179]. However, the volume available for collecting and storing chemical precipitates, or the number of available removal sites, governs the aerobic systems' removal rate. The primary removal processes, including sedimentation, adsorption, ion exchange, and complexation, are finite. Thus, uranium removal or immobilization ceases unless new removal sites are generated. Hence, periodic removal of precipitated uranium from wetlands is necessary. Only well-vegetated systems can generate sufficient new removal sites to offer the possibility of long-term, low-maintenance treatment. Therefore, only site conditions may determine whether it is more economical to allow the wetland system to develop into a new ecosystem or to excavate the accumulated solids for proper disposal. Since the generation of new removal sites is a slow process, long-term treatment requires a large wetland with a relatively small input load, and finding such land is difficult. Further, the wetland concentrates uranium, increasing radiation risks and radon emissions.

If the primary uranium immobilization mechanism is anaerobic, oxygenating macrophytes affect the performance of bioremediation agents in sediments and subsurface environments. The system also works efficiently as long as sulfate-reducing bacteria reduce sulfate to sulfide, thus precipitating uranium sulfide complexes and generating alkalinity. The mechanism is limited by the amount of available carbon for sulfate-reducing bacteria, but the oxygen availability provides a more preferable electron acceptor than uranium. Therefore, uranium bioreduction ceases.

Furthermore, in the absence of organic-rich sediments, uranium-bearing particles are released back into the water column when macrophytes decompose. In reality, both aerobic and anaerobic processes occur simultaneously in the wetlands. This is a big challenge in designing constructed wetlands, where these processes do not counteract each other.

10.6.7 A Paradigm of Eco-Remediation

In most cases, successful laboratory stories have failed to scale up in real situations [180]. The main reason is that laboratory procedures do not compress the multiple interactions in nature, especially the roles of different ecological interactions and hierarchies. A natural system is self-organized within an ecological hierarchy, where interactions determine the roles of other organisms within it. Hence, bioremediation procedures should be used at higher ecological interaction, where multiple interactions are part of the processes [80]. Thus, uranium remediation in abandoned uranium mines can be effective when ecological engineering is employed to enhance natural ecological processes:

(i) Association of the uranium with organic particles in the water column
(ii) Provision of conditions whereby organic particles sink onto and into organic sediments
(iii) Provision of low E_H conditions in the sediments for metal-reducing microbial populations

Consequently, the oxidized uranium associated with particulate matter settles and is biomineralized in the sediments—effectively creating biogenic ores [37, 80, 181]. The advantage of using living material in the initial step is that it continues to grow, providing new organic material for sequestration. This is especially useful in flow-through systems, where the supply of U to be removed is continuous. Using the ecological approach, the removal of uranium follows a three-step process: (i) first is the sequestration of U on, in, and around aquatic plants and periphyton biofilms. Periphyton is a mixed community composed of algae, fungi, and bacteria that grow on the surfaces of macrophytic vascular plants and sediment [80]. Biogenic ligands produced by plants and microorganisms facilitate U(VI) removal from the mine water. The biofilms continuously renew the cellular surface area; (ii) the next step is removing the biological sequestrated uranium units from the water column.

The dead macrophytes, phytoplankton, and bacteria can aggregate into larger particles, sinking faster. Invertebrate shredders (e.g., zooplankton, diatoms, etc.) and small aquatic vertebrates (e.g., herbivorous fish) play an essential role in forming uranium-rich organic particulates. The faecal pellets are part of biogenic organic particles that may bind uranium and transfer it into the sediments, as the pellets are heavy and sink rapidly to the bottom [80]. However, the production of biogenic organic particles is attributed to the higher primary productivity and the generation of organic particulate matter, which removes the uranium and other metals from the water column; and finally, (iii) the reduction of U(VI) to U(IV) and transforming the ions into stable precipitates in the sediments. Once the biomass of U-biofilms reaches the sediment interface, it continues to decompose, releasing nutrients and colloidal-bound uranium. The dead organic material provides organic carbon and other nutrients to heterotrophic microbial consortia to maintain the low E_H, within which the U is transformed and, consequently, biomineralized [80]. The nutrients that sustain microbial populations in the anaerobic sediments come from natural organic matter (NOM) or algal particles. Plant and biofilm biomass encounter sediment primarily as decomposing material. The decomposition of the biomass in the sediments provides nutrients to sustain heterotrophic microbial communities, which further degrade organic matter and provide dissolved organic material for anaerobic metal-reducing microbial consortia [80, 182].

Dissolved oxygen is reduced first, followed by denitrification, UO_2CO_3 reduction to uraninite, and sulfate reduction to sulfide. Different microbial enzymes may catalyse reactions at each step in the redox ladder. There is a succession of microbial activity from denitrifiers, metal reducers, and sulfate reducers. The denitrification process is complete before the sulfate reducers are activated. Thus, the U(VI) reduction does not occur until the nitrate has been utilized by nitrate-reducing microorganisms [150]. When nitrate enters a reduced zone where U(IV) is present, the subsequent reduction of nitrate may remobilize the uranium through the facilitation of oxidation of U(IV) to U(VI).

Therefore, the anaerobic pond should be planted with high-nitrogen-demanding macrophytes, such as *Lemna* sp. and *Eichhornia* sp., and with denitrifying bacteria. Further, the intermediates of dissimilatory nitrate reduction (nitrite, nitrous oxide, and nitric oxide) can also oxidize and mobilize U(IV). Therefore, the accumulation of carbonate from microbial respiration may promote the formation of highly stable carbonato-U(VI) complexes, thereby increasing the thermodynamic favourability of U(IV) oxidation [47].

10.7 Concluding Remarks

Bioremediation of uranium shows high potential as an alternative to classic remediation under laboratory conditions. However, the transition from pure academic research to marketable technology applicable to abandoned uranium mines is slow to materialize. Like many technologies with good scientific foundations, its merits are marred by over-optimistic speculations and fraudulent claims. Further, just like this chapter, uranium bioremediation is usually considered without other influential ecological interactions under trial. This undermines its potential application in situ because natural aquatic environments, including abandoned uranium mining sites, are heterogeneous, with complex interactions among the water column, bed sediments, air, and the biotic compartment occurring continuously. It is characterized by multiple stressors, ligand heterogeneity and polydispersity, and non-equilibrium conditions.

Consequently, metals and metalloids undergo dynamic physical, chemical, and biological processes. For instance, sulfide minerals, acid-buffering capacity, organic matter, colloids, iron, and manganese oxide minerals affect the solubility, mobility, and bioavailability of other metal ions. The toxicity of uranium depends on its ability to cross biological barriers (e.g., the plasma membrane), which is partially regulated by the concentration or flux of internalized metals and by the medium's physicochemical conditions. When chemical interactions disrupt homeostasis, the biotic compartment's homeostatic regulation reverses the resulting stoichiometric imbalance. Thus, the homeostatic regulation of resource stoichiometry in an aquatic system is one of the primary regulatory processes that can be leveraged to enhance bioremediation technology and alleviate toxicity. Therefore, stimulating these biological processes to stabilize uranium in the subsurface is promising as a basis for U remediation technologies.

Thus, realistic functioning is only feasible if bioremediation procedures are embedded in ecological engineering processes that manipulate multiple interactions while emulating nature's rules. For instance, the cascade wetland procedure would be more effective than just monoculture-constructed wetlands in treating uranium. The long-term stability of biologically reduced uranium will be determined by the complex interplay of sediment mineralogy, aqueous geochemistry, microbial and invertebrate activity, and the potential of aquatic plants and U(IV) oxidants to stabilize it. Many of these factors have been studied under laboratory conditions; however, their impact on uranium cycling in natural subsurface environments remains poorly understood.

References

1. Hoffmann, D. (2000). In the laboratory of the planned economy - From labor allocation to labor recruitment in the Russian Zone of Occupation/German Democratic Republic (1945–1961). *Vierteljahr Zeitges, 48*, 531–666.
2. Mkandawire, M. (2005). *Fate and effect of uranium and arsenic in surface waters of abandoned uranium mines: Studies with Lemna gibba L.* (Shaker.
3. Mkandawire, M., Dudel, E. G., & Taubert, B. (2004). In A. P. Jarvis, B. A. Dudgeon, & P. L. Younger (Eds.), *Mine water—Process, policy and progress* (pp. 9–18). 2, University of Newcastle upon Tyne Press.
4. Diehl, P. *Uranium mining in Eastern Germany: The WISMUT legacy*, 2003).
5. Diehl, P. (1995). *Uranabbau in Europa—Die Folgen für Mensch und Umwelt. Report No. 1/95, 58 S.* BBU.
6. Abdelouas, A. (2006). Uranium mill tailings: Geochemistry, mineralogy, and environmental impact. *Elements, 2*, 335–341. https://doi.org/10.2113/gselements.2.6.335
7. Moyes, L. N., et al. (2000). Uranium uptake from aqueous solution by intrasection with goethite, lepidocrocite, muscovite and mackinawite: An X-ray Absorption spectroscopy study. *Environmental Science & Technology, 34.*
8. Meinrath, A., Schneider, P., & Meinrath, G. (2003). Uranium ore and depleted uranium in the environment, with reference to uranium in the biosphere from Erzgebirge/Sachsen, Germany. *Journal of Environmental Radioactivity, 64*, 175–193.

9. Morgenstein, M. E. (2003). Uranium-series and luminescence dating of volcanic lithic artefacts. *Archaeometry, 45*, 503–518.
10. Tomasek, L., Darby, S. C., Swerdlow, A. J., Placek, V., & Kunz, E. (1993). Radon exposure and cancers other than lung cancer among uranium miners in West Bohemia [see comments]. *Lancet, 341*, 919–923.
11. Enderle, G. J., & Friedrich, K. (1995). East German uranium miners (Wismut): Exposure conditions and health consequences. *Stem Cells, 13*, 78–89.
12. Al Hashimi, A., Evans, G. J., & Cox, B. (1996). Aspects of the permanent storage of uranium tailings. *Water, Air, and Soil Pollution, 88*, 83–92.
13. Pressyanov, D. S., Guelev, M. G., & Sharkov, B. G. (1995). Radon and radon progeny outdoors in a valley of enhanced natural radioactivity. *Atmospheric Environment, 29*, 3433–3439.
14. Clara, M., & Magalhães, F. A. (2002). An environmental problem limited by solubility. *Pure and Applied Chemistry, 74*, 1843–1850.
15. Fritzsche, A., Dienemann, H., & Dudel, E. (2006). Arsenic fixation on iron-hydroxide-rich and plant litter-containing sediments in natural environments. *Environmental Geology, 51*, 133–142.
16. Munoz, J. A., Gonzalo, A., & Valiente, M. (2002). Arsenic adsorption by Fe(III)-loaded open-celled cellulose sponge. Thermodynamic and selectivity aspects. *Environmental Science & Technology, 36*, 3405–3411.
17. Carvalho, I. G., et al. (2005). Environmental impact of uranium mining and Ore processing in the Lagoa Real District, Bahia, Brazil. *Environmental Science & Technology, 39*, 8646–8652.
18. Belzile, N., & Tessier, A. (1990). Interactions between arsenic and iron oxyhydroxides in lacustrine sediments. *Geochimica et Cosmochimica Acta, 54*, 103–109.
19. IAEA. (2001). *Manual of acid in situ leach uranium mining technology. Report No. IAEA-TECDOC-1239, 294*. International Atomic Energy Agency.
20. IAEA. (1998). *Classification of uranium reserves/resources*. International Atomic Energy Agency.
21. Beddow, H., Black, S., & Read, D. (2006). Naturally occurring radioactive material (NORM) from a former phosphoric acid processing plant. *Journal of Environmental Radioactivity, 86*, 289–312.
22. Whitehead, N. E., Brooks, R. R., & Peterson, P. J. (1971). The nature of uranium occurrence in the leaves of *Coprosma australis* (A. Rich) Robinson. *Australian Journal of Biological Sciences, 24*, 67–73.
23. Lovley, D. R., Widman, P. K., Woodward, J. C., & Phillips, E. J. P. (1993). Reduction of uranium by Cytochrome C3 of *Desulfovibrio vulgaris*. *Applied and Environmental Microbiology, 59*, 3572–3576.
24. Benders, J., Duff, M. C., Phillips, P., & Hill, M. (2000). Bioremediation and bioreduction of dissolved U(VI) by microbial mat consortium supported on silica gel particles. *Environmental Science and Technology, 34*, 3235–3241.
25. Markich, S. J. (2002). Uranium speciation and bioavailability in aquatic systems: an overview. *Scientific World Journal*.
26. Jerden, J. J. L., & Sinha, A. K. (2006). Geochemical coupling of uranium and phosphorous in soils overlying an unmined uranium deposit: Coles Hill, Virginia. *Journal of Geochemical Exploration, 91*, 56–70.
27. Cherdyntsev, V. V., Ivanovich, M., & Harmon, R. S. (1982). *Uranium series disequilibrium : Applications to environmental problems*. Clarendon Press.
28. Ivanovich, M., & Harmon, R. S. (1992). *Uranium-series disequilibrium: Applications to earth, marine, and environmental science* (2nd ed.). Clarendon Press/Oxford University Press.
29. ATSDR. (1999). Agency for Toxic Substances and Disease Registry, (Ed.) 462. U.S. Department of Health and Human Services
30. WHO. (2001). *Depleted uranium Sources, Exposure and Health Effects*. (Department of Protection of the Human Environment,. World Health Organization.

31. WHO. (2004). *Uranium in drinking-water: Background document for development of WHO Guidelines for drinking-water quality. 23*. World Health Organization.
32. Peterson, J., et al. (2007). In Argonne National Laboratory (Ed.), *Human health fact sheet. 4*. Argonne National Laboratory, EVS in collaboration with U.S. Department of Energy.
33. Meinrath, G. (1996). Coordination of uranyl (VI) carbonate complexation in aqueous solution. *Journal of Radioanalytical and Nuclear Chemistry, 211*, 349–362.
34. Meinrath, G. (1999). Aquatic chemistry of uranium: A review focusing on aspects of environmental chemistry. *Freiberge On-line Geoscience, 1*, 1–112.
35. Bernhard, G., Geipel, G., Brendler, V., & Nitsche, H. (1998). Uranium speciation in waters of different uranium mining areas. *Journal of Alloys and Compounds, 271–273*, 201–205.
36. Duff, M. C., Coughlin, J. U., & Hunter, D. B. (2002). Uranium co-precipitation with iron oxide minerals. *Geochimica et Cosmochimica Acta, 66*, 3533–3547. https://doi.org/10.1016/S0016-7037(02)00953-5
37. Martinez Aguirre, A., Garcia Leon, M., & Ivanovich, M. (1995). U and Th speciation in river sediments. *Science of the Total Environment, 173–174*, 203–209.
38. Helen, S., & J, T. R. (2007). A theoretical study of the inner-sphere disproportionation reaction mechanism of the pentavalent actinyl ions. *Inorganic Chemistry, 46*, 6311–6318.
39. Shilov, V., Yusov, A., & Fedoseev, A. (2007). Disproportionation of nonoxygenated U(V) bound in the complex with heteropoly anions P2W17O 6110−. *Radiochemistry, 49*, 361–366.
40. Brendler, V., Geipel, G., & Nitsche, H. (1998). *Possible impact of phosphate influx on the uranium speciation and migration in seepage waters*. Forschungszentrum Rossendorf e.V, Institute of Radiochemistry.
41. Meinrath, G. (1997). Uranium speciation by spectroscopy. *Journal of Radioanalytical and Nuclear Chemistry, 224*, 119.
42. Ulrich, K.-U., Rossberg, A., Foerstendorf, H., Zanker, H., & Scheinost, A. C. (2006). Molecular characterization of uranium(VI) sorption complexes on iron(III)-rich acid mine water colloids. *Geochimica et Cosmochimica Acta, 70*, 5469–5487.
43. Zanonato, P., et al. (2004). Hydrolysis of uranium(VI) at variable temperatures (10–85 °C). *Journal of the American Chemical Society, 126*, 5515–5522.
44. Meinrath, G. *Freiberg On-line Geosciences* , 1998).
45. Luo, J., et al. (2007). Modeling in-situ uranium(VI) bioreduction by sulfate-reducing bacteria. *Journal of Contaminant Hydrology, 92*, 129–148.
46. Liu, Y., & Xu, H. (2007). Equilibrium, thermodynamics and mechanisms of Ni2+ biosorption by aerobic granules. *Biochemical Engineering Journal, 35*, 174–182.
47. Gavshin, V. M., et al. (2005). Disequilibrium between uranium and its progeny in the Lake Issyk-Kul system (Kyrgyzstan) under a combined effect of natural and manmade processes. *Journal of Environmental Radioactivity, 83*, 61–74.
48. Milodowski, A. E., et al. (1990). Uranium-mineralized micro-organisms associated with uraniferous hydrocarbons in southwest Scotland. *Nature, 347*, 465–467.
49. Petrescu, L., & Bilal, E. (2006). Speciation of natural uranium and thorium in surface waters around a uranium mine (Bistrita Mts., Romania). *Geochimica et Cosmochimica Acta, 70*, A486.
50. Murakami, T., Ohnuki, T., Isobe, H., & Sato, T. (1997). Mobility of uranium during weathering. *American Mineralogist, 82*, 888–899.
51. Davis, J. A., Meece, D. E., Kohler, M., & Curtis, G. P. (2004). Approaches to surface complexation modeling of Uranium(VI) adsorption on aquifer sediments. *Geochimica et Cosmochimica Acta, 68*, 3621–3641.
52. Ohnuki, T., et al. (2004). The formation of autunite (Ca(UO2)2(PO4)2[square root]nH2O) within the leached layer of dissolving apatite: incorporation mechanism of uranium by apatite. *Chemical Geology, 211*, 1–14.
53. Allard, T., Ildefonse, P., Beaucaire, C., & Calas, G. (1999). Structural chemistry of uranium associated with Si, Al, Fe gels in a granitic uranium mine. *Chemical Geology, 158*, 81–103.

54. Hiess, J., Condon, D. J., McLean, N., & Noble, S. R. (2012). https://doi.org:335/6076/1610 [pii]). $^{238}U/^{235}U$ systematics in terrestrial uranium-bearing minerals. *Science, 335*, 1610–1614. https://doi.org/10.1126/science.1215507
55. Luo, W., & Gu, B. (2011). Dissolution of uranium-bearing minerals and mobilization of uranium by organic ligands in a biologically reduced sediment. *Environmental Science & Technology, 45*, 2994–2999. https://doi.org/10.1021/es103073u
56. Stefaniak, E. A., et al. (2009). https://doi.org:S0304-3894(09)00246-5 [pii]). Combined SEM/EDX and micro-Raman spectroscopy analysis of uranium minerals from a former uranium mine. *Journal of Hazardous Materials, 168*, 416–423. https://doi.org/10.1016/j.jhazmat.2009.02.057
57. Cai, C., et al. (2007). Mineralogical and geochemical evidence for coupled bacterial uranium mineralization and hydrocarbon oxidation in the Shashagetai deposit, NW China. *Chemical Geology, 236*, 167–179.
58. Gomez, P., et al. (2006). Modeling of geochemical processes related to uranium mobilization in the groundwater of a uranium mine. *Science of the Total Environment, 366*, 295–309.
59. Duff, M. C., Hunter, D. B., Bertsch, P. M., & Amrhein, C. (1999). Factors influencing uranium reduction and solubility in evaporation pond sediments. *Biogeochemistry, 45*, 95–114.
60. Bots, P., & Behrends, T. (2008). Uranium mobility in subsurface aqueous systems: the influence of redox conditions. *Mineralogical Magazine, 72*, 381–384. https://doi.org/10.1180/minmag.2008.072.1.381
61. Gadelle, F., Wan, J., & Tokunaga, T. K. (2001). Removal of Uranium(VI) from contaminated sediments by surfactants. *Journal of Environmental Quality, 30*, 470–478.
62. Vinay Kumar Reddy, K., Gopal Reddy, C., Vidya Sagar, D., Yadagiri Reddy, P., & Rama Reddy, K. (2012). https://doi.org:ncs005 [pii]). Environmental radioactivity studies in the proposed Lambapur and Peddagattu uranium mining areas of Andhra Pradesh, India. *Radiation Protection Dosimetry, 151*, 290–298. https://doi.org/10.1093/rpd/ncs005
63. Rodgher, S., et al. (2012). Evaluation of surface water quality in aquatic bodies under the influence of uranium mining (MG, Brazil). *Environmental Monitoring and Assessment*. https://doi.org/10.1007/s10661-012-2719-5
64. Shvareva, T. Y., & Albrecht-Schmitt, T. E. (2006). General route to three-dimensional framework uranyl transition metal phosphates with atypical structural motifs: The case examples of $Cs_2\{(UO_2)_4[Co(H_2O)_2]_2(HPO_4)(PO_4)_4\}$ and $Cs^{3+x}[(UO_2)_3CuH^{4-x}(PO_4)_5]\cdot H2O$. *Inorganic Chemistry, 45*, 1900–1902.
65. Smeaton, C. M., Weisener, C. G., Burns, P. C., Fryer, B. J., & Fowle, D. A. (2008). Bacterially enhanced dissolution of meta-autunite. *American Mineralogist, 93*, 1858–1864. https://doi.org/10.2138/am.2008.2836
66. Allan, V. J. M., Callow, M. E., Macaskie, L. E., & Paterson-Beedle, M. (2002). Effect of nutrient limitation on biofilm formation and phosphatase activity of a Citrobacter sp. *Microbiol-Sgm, 148*, 277–288.
67. Kalin, M. (2004). Passive mine water treatment: the correct approach? *Ecological Engineering, 22*, 299–304.
68. Hsieh, K. M., Murgel, G., Lion, L. W., & Shuler, M. (1994). Interactions of microbial biofilms with toxic trace metals: 1. Observation and modeling of cell growth, attachment, and production of extracellular polymer. *Biotechnology and Bioengineering, 44*, 219–231.
69. Lloyd, J. R., & Renshaw, J. C. (2005). Bioremediation of radioactive waste: radionuclide-microbe interactions in laboratory and field-scale studies. *Current Opinion in Biotechnology, 16*, 254–260.
70. Akhtar, K., Waheed Akhtar, M., & Khalid, A. M. (2007). Removal and recovery of uranium from aqueous solutions by Trichoderma harzianum. *Water Research, 41*, 1366–1378.
71. Bhainsa, K. C., & D'Souza, S. F. (1999). Biosorption of uranium(VI) by *Aspergillus fumigatus*. *Biotechnology Techniques, 13*, 695–699.

72. Bustard, M., & McHale, A. P. (1997). Biosorption of uranium by cross-linked and alginate immobilized residual biomass from distillery spent wash. *Bioprocess Engineering, 17*, 127–130.
73. Chmielowski, J., Woznica, A., Klapcinska, B., Pytlak, M., & Piech, W. (1994). Binding of uranium by yeast cell wall polysaccharides. *Bulletin of the Polish Academy of Sciences: Biological Sciences, 42*, 147–149.
74. Gonzalez-Munoz, M. T., Merroun, M. L., BenOmar, N., & Arias, J. M. (1997). Biosorption of uranium by *Myxococcus xanthus*. *International Biodeterioration and Biodegradation, 40*, 107–114.
75. Malekzadeh, F., Latifi, A. M., Shahamat, M., Levin, M., & Colwell, R. R. (2002). Effects of selected physical and chemical parameters on uranium uptake by the bacterium *Chryseomonas* MGF-48. *World Journal of Microbiology and Biotechnology, 18*, 599–602.
76. O'Flaherty, E. J. (1995). PBK modeling for metals: Examples with lead, uranium, and chromium. *Toxicology Letters, 82–83*, 367–372.
77. Sar, P., & D'Souza, S. F. (2001). Biosorptive uranium uptake by a *Pseudomonas* strain: Characterization and equilibrium studies. *Journal of Chemical Technology & Biotechnology, 76*, 1286–1294.
78. Ariff, A. B., Mel, M., Hasan, M. A., & Karim, M. I. A. (1999). The kinetics and mechanism of lead (II) biosorption by powderized Rhizopus oligosporus. *World Journal of Microbiology and Biotechnology, 15*, 291–298.
79. Prikryl, J. D., Jain, A., Turner, D. R., & Pabalan, R. T. (2001). UraniumVI sorption behavior on silicate mineral mixtures. *Journal of Contaminant Hydrology, 47*, 241–253.
80. Kalin, M., Wheeler, W. N., & Meinrath, G. (2005). The removal of uranium from mining waste water using algal/microbial biomass. *Journal of Environmental Radioactivity, 78*, 151–177.
81. Daković, M., Kovačević, M., Andjus, P., & Bačić, G. (2008). On the mechanism of uranium binding to cell wall of *Chara fragilis*. *European Biophysics Journal.*
82. Yang, J. & Volesky, B. (1999). Biosorption and elution of uranium with seaweed biomass. Conference IBS'99-Spain.
83. Bonfada, É., de Oliveira, G. M., Back, D. F., & Lang, E. S. (2005). Metallation of ligands with biological activity: Synthesis and X-Ray characterization of $[UO_2(PN)_2(H_2O)]Cl_2$\{PN = vitamin B6 pyridoxine[2-methyl-3-hydroxy-4,5-bis(hydroxymethyl) pyridine]\}. *Zeitschrift für Anorganische und Allgemeine Chemie, 631*, 878–881.
84. Haas, J. R., Bailey, E. H., & Purvis, O. W. (1998). Bioaccumulation of metals by lichens: Uptake of aqueous uranium by *Peltigera membranacea* as a function of time and pH. *American Mineralogist, 83*, 1494–1502.
85. Kalin, M., Wheeler, W. N., & Meinrath, G. (2004). The removal of uranium from mining waste water using algal/microbial biomass. *Journal of Environmental Radioactivity, 78*, 151–177.
86. Worms, I., Simon, D. F., Hassler, C. S., & Wilkinson, K. J. (2006). Bioavailability of trace metals to aquatic microorganisms: importance of chemical, biological and physical processes on biouptake. *Biochimie, 88*, 1721–1731.
87. Yang, J., & Volesky, B. (1999). Modelling uranium-proton ion exchange in biosorption. *Environmental Science & Technology, 33*, 4079–4085.
88. Donnellan, N., Rollan, A., Mchale, L., & Mchale, A. P. (1995). The effect of electric field stimulation on the biosorption of uranium by non-living biomass derived from *Kluyveromyces marxianus* IMB3. *Biotechnology Letters, 17*, 439–442.
89. Mkandawire, M., Taubert, B., & Dudel, E. G.N. Kalogerakis (Ed.). *Second European bioremediation conference* (pp. 357–360). Technical University of Crete, and European Community (High-level Scientific Conference Programme).
90. Volesky, B. A., & May Phillips, H. A. (1995). Biosorption of heavy metals by *Saccharomyces cerevisiae*. *Applied Microbiology and Biotechnology, 42*, 797–806.

91. Dodge, C. J., & Franscis, A. J. (1997). Biotransformation of binary and ternary citric acid complexes of iron and uranium. *Environmental Science & Technology, 31*, 3062–3067.
92. Merroun, M. L., et al. (2005). Complexation of uranium by cells and S-layer sheets of *Bacillus sphaericus* JG-A12. *Applied and Environmental Microbiology, 71*, 5532–5543. https://doi.org/10.1128/aem.71.9.5532-5543.2005
93. Shanbhag, P. M., & Chopping, G. R. (1981). Binding of uranyl by humic acid. *Journal of Inorganic and Nuclear Chemistry, 43*, 3369–3372.
94. Francis, A. J., & Dodge, C. J. (1998). Remediation of soils and wastes contaminated with uranium and toxic metals. *Environmental Science and Technology, 32*, 3993–3998.
95. Macaskie, L. E., & Basnakova, G. (1998). Microbially enhanced chemisorption of heavy metals: A method for the bioremediation of solutions containing long lived isotopes of neptunium and plutonium. *Environmental Science and Technology, 32*, 184–187.
96. Macaskie, L. E., Bonthrone, K. M., Yong, P., & Goddard, D. T. (2000). Enzymically mediated bioprecipitation of uranium by a *Citrobacter* sp.: A concerted role for exocellular lipopolysaccharide and associated phosphatase in biomineral formation. *Microbiology, 146*, 1855–1867.
97. Schmeide, K., et al. (2003). Interaction of uranium(VI) with various modified and unmodified natural and synthetic humic substances studied by EXAFS and FTIR spectroscopy. *Inorganica Chimica Acta, 351*, 133–140.
98. Mazen, A. M. A., Zhang, D., & Franceschi, V. R. (2003). Calcium oxalate formation in *Lemna minor*: Physiological and ultrastructural aspects of high capacity calcium sequestration. *The New Phytologist, 161*, 435–448.
99. Ma, J. F. (2000). Mini review role of organic acids in detoxification of aluminum in higher plants. *Plant & Cell Physiology, 41*, 383–390.
100. Mkandawire, M., Dudel, E. G., & Müller, C. (2005). In B. J. Merkel & A. Hasche-Berger (Eds.), *Uranium in the environment* (pp. 495–505). (Springer.
101. Branquinho, C., Brown, D. H., & Catarino, F. (1997). The cellular location of Cu in lichens and its effects on membrane integrity and chlorophyll fluorescence. *Environmental and Experimental Botany, 38*, 165–179.
102. Premuzic, E. T., Francis, A. J., Lin, M., & Schubert, J. (1985). Induced formation of chelating agents by *Pseudomonas aeruginosa* grown in presence of thorium and uranium. *Archives of Environmental Contamination and Toxicology, 14*, 759–768.
103. Chiang, P.-N., Wang, M. K., Chiu, C. Y., & Chou, S.-Y. (2006). Effects of cadmium amendments on low-molecular-weight organic acid exudates in rhizosphere soils of tobacco and sunflower. *Environmental Toxicology, 21*, 479–488.
104. Jin, C. W., He, Y. F., Tang, C. X., Wu, P., & Zheng, S. J. (2006). Mechanisms of microbially enhanced Fe acquisition in red clover (*Trifolium pratense* L.). *Plant, Cell & Environment, 29*, 888–897. https://doi.org/10.1111/j.1365-3040.2005.01468.x
105. Wasaki, J., et al. (2005). Root exudation, phosphorus acquisition, and microbial diversity in the rhizosphere of White Lupine as affected by phosphorus supply and atmospheric carbon dioxide concentration. *Journal of Environmental Quality, 34*, 2157–2166. https://doi.org/10.2134/jeq2004.0423
106. Meinrath, G., Volke, P., Helling, C., Dudel, E.-G., & Merkel, B. J. (1999). Determination and interpretation of environmental water samples contaminated by uranium mining activities. *Fresenius' Journal of Analytical Chemistry, 364*, 191–202.
107. Ebbs, S. D., Brady, D. J., Norvell, W. A., & Kochian, L. V. *Membrane transport*. American Society of Plant Biologists.
108. Kaplan, D., Prakash, N. T., & Abeliovich, A. (1998). Glutathione-induced recovery in *Chlorella* cells from metal toxicity. *Fresenius Environmental Bulletin, 7*, 153–159.
109. Maksymiec, W. (1997). Effect of copper on cellular processes in higher plants. *Photosynthetica, 34*, 321–342.
110. Klapheck, S., Schlunz, S., & Bergmann, L. (1995). Synthesis of phytochelatins and homophytochelatins in *Pisum sativum* L. *Plant Physiology, 107*, 515–521.

111. Shah, K., & Dubey, R. S. (1998). A 18 kDa cadmium inducible protein complex: Its isolation and characterisation from rice (*Oryza sativa* L.) seedlings. *Journal of Plant Physiology, 152*, 448–454.
112. Curie, C., & Briat, J.-F. (2003). Iron transport and signaling in plants. *Annual Review of Plant Biology, 54*, 183–206.
113. White, P. J., & Broadley, M. R. (2003). Calcium in plants. *Annals of Botany, 92*, 487–511. https://doi.org/10.1093/aob/mcg164
114. Van Assche, F., & Clijsters, H. (1990). Effects of metals on enzyme activity in plants. *Plant, Cell & Environment, 13*, 195–206.
115. Martinoia, E., Maeshima, M., & Neuhaus, H. E. (2007). Vacuolar transporters and their essential role in plant metabolism. *Journal of Experimental Botany, 58*, 83–102. https://doi.org/10.1093/jxb/erl183
116. Samardakiewicz, S., & Wozny, A. (2000). The distribution of lead in duckweed (*Lemna minor* L.) root tip. *Plant and Soil, 226*, 107–111.
117. Neumann, D., Zurnieden, U., Lichtenberger, O., & Leopold, I. (1995). How does *Armeria maritima* tolerate high heavy metal concentrations? *Journal of Plant Physiology, 146*, 704–717.
118. Vandenhove, H., Van Hees, M., Wannijn, J., Wouters, K., & Wang, L. (2007). Can we predict uranium bioavailability based on soil parameters? Part 2: Soil solution uranium concentration is not a good bioavailability index. *Environmental Pollution, 145*, 577–586.
119. Min, M., Xu, H., Chen, J., & Fayek, M. (2005). Evidence of uranium biomineralization in sandstone-hosted roll-front uranium deposits, northwestern China. *Ore Geology Reviews, 26*, 198–206.
120. Virk, H. S., Sharma, N., & Bajwa, B. S. (1999). Environmental radioactivity: a case study in Himachal Pradesh, India. *Journal of Environmental Radioactivity, 45*, 119–127.
121. Ginder-Vogel, M. et al. (2006). In situ biological uranium remediation within a highly contaminated aquifer. *SSRL Science Highlights*. http://www-ssrl.slac.stanford.edu/research/highlights_archive/u_remed.html.
122. Behrends, T., & Van Cappellen, P. (2005). Competition between enzymatic and abiotic reduction of uranium(VI) under iron reducing conditions. *Chemical Geology, 220*, 315–327.
123. Liger, E., Charlet, L., & Van Cappellen, P. (1999). Surface catalysis of uranium(VI) reduction by iron(II). *Geochimica et Cosmochimica Acta, 63*, 2939–2955.
124. DiSpirito, A. A., Talnagi, J. W., & Tuovinen, O. H. (1983). Accumulation and cellular distribution of uranium in *Thiobacillus ferrooxidans*. *Archives of Microbiology, 135*, 250–253.
125. Wall, J. D., & Krumholz, L. R. (2006). Uranium reduction. *Annual Review of Microbiology, 60*.
126. Ortiz-Bernad, I., Anderson, R. T., Vrionis, H. A., & Lovley, D. R. (2004). Resistance of solid-phase U(VI) to microbial reduction during in situ bioremediation of uranium-contaminated groundwater. *Applied and Environmental Microbiology, 70*, 7558–7560.
127. Rawlings, D. E., & Silver, S. (1995). Mining with microbes. *Biotechnology, 13*, 773–778.
128. Madden, A. S., Smith, A. C., Balkwill, D. L., Fagan, L. A., & Phelps, T. J. (2007). Microbial uranium immobilization independent of nitrate reduction. *Environmental Microbiology, 9*, 2321–2330. https://doi.org/10.1111/j.1462-2920.2007.01347.x
129. Fredrickson, J. K., et al. (2000). Reduction of U(VI) in goethite ([alpha]-FeOOH) suspensions by a dissimilatory metal-reducing bacterium. *Geochimica et Cosmochimica Acta, 64*, 3085–3098.
130. Wilkins, M., Livens, F., Vaughan, D., & Lloyd, J. (2006). The impact of Fe(III)-reducing bacteria on uranium mobility. *Biogeochemistry, 78*, 125–150.
131. Choppin, G. R. (2006). Actinide speciation in aquatic systems. *Marine Chemistry, 99*, 83–92.
132. Takashima, T., Fukunishi, N., Nishiki, T., & Konishi, Y. (2002). Anaerobic oxidation of dissolved hydrogen sulfide in continuous culture of the chemoautotrophic bacterium *Thiobacillus denitrificans*. *Kagaku Kogaku Ronbunshu, 28*, 25–30.

133. Anderson, R. T., et al. (2003). Stimulating the in situ activity of geobacter species to remove uranium from the groundwater of a uranium-contaminated aquifer. *Applied and Environmental Microbiology, 69*, 5884–5891. https://doi.org/10.1128/aem.69.10.5884-5891.2003
134. Neal, A. L., Clough, L. K., Perkins, T. D., Little, B. J., & Magnuson, T. S. (2004). In situ measurement of Fe(III) reduction activity of *Geobacter pelophilus* by simultaneous in situ RT-PCR and XPS analysis. *FEMS Microbiology Ecology, 49*, 163–169.
135. Lovley, D. R. (1995). Bioremediation of organic and metal contaminants with dissimilatory metal reduction. *Journal of Industrial Microbiology, 14*, 85–93.
136. Lovley, D. R. (2003). Cleaning up with genomics: Appling molecular biology to bioremediation. *Nature Reviews, 1*, 35–44.
137. Creanga, D. E., Poiata, A., Morariu, V. V., & Tupu, P. (2004). Zero-magnetic field effect in pathogen bacteria. *Journal of Magnetism and Magnetic Materials, 272–276*, 2442–2444.
138. Lovley, D. R. (2008). Extracellular electron transfer Extracellular electron transfer: Wires, capacitors, iron lungs, and more. *Geobiology, 5*, 225–231.
139. Anderson, R. T. & Lovley, D. R. in *Interactions of microorganisms with radionuclides* (eds Miranda J. Keith-Roach & Francis R. Livens) 205–223 Elsevier Science Ltd, 2002).
140. Vauclare, P., et al. (2002). Flux control of sulphate assimilation in *Arabidopsis thaliana*: adenosine 5′-phosphosulphate reductase is more susceptible than ATP sulphurylase to negative control by thiols. *The Plant Journal, 31*, 729–740.
141. Cohen, C. K., Fox, T. C., Garvin, D. F., & Kochian, L. V. (1998). The role of iron-deficiency stress responses in stimulating heavy-metal transport in plants. *Plant Physiology, 116*, 1063–1072.
142. Lloyd, J. R., & Lovley, D. R. (2001). Microbial detoxification of metals and radionuclides. *Current Opinion in Biotechnology, 12*, 248–253.
143. Yang, D. H., Andersson, B., Aro, E. M., & Ohad, I. (2001). The redox state of the plastoquinone pool controls the level of the light-harvesting chlorophyll a/b binding protein complex II (LHC II) during photoacclimation—Cytochrome b(6)f deficient *Lemna perpusilla* plants are locked in a state of high-light acclimation. *Photosynthesis Research, 68*, 163–174.
144. Fredrickson, J. K., Kostandarithes, H. M., Li, S. W., Plymale, A. E., & Daly, M. J. (2000). Reduction of Fe(III), Cr(VI), U(VI), and Tc(VII) by *Deinococcus radiodurans* R1. *Applied and Environmental Microbiology, 66*, 2006–2011.
145. Koike, H., Yoneyama, K., Kashino, Y., & Satoh, K. (1996). Mechanism of electron flow through the QB Site in photosystem II. 4. Reaction mechanism of plastoquinone derivatives at the QB Site in spinach photosystem II membrane fragments. *Plant and Cell Physiology, 37*, 983–988.
146. Doshi, S. M. (2006). *Bioremediation of acid mine drainage using sulfate-reducing bacteria. 72*. U.S. Environmental Protection Agency.
147. Spear, J. R., Figuaroa, L. A., & Honeyman, B. (1999). Modeling the removal of uranium U(VI) from aqueous solutions in the presence of sulfate reducing bacteria. *Environmental Science & Technology, 33*, 2667–2675.
148. Basnakova, G., Spencer, A. J., Palsgard, E., Grime, G. W., & Macaskie, L. E. (1998). Identification of the nickel uranyl phosphate deposits on Citrobacter sp. cells by electron microscopy with electron probe X-ray microanalysis and by proton induced X-ray emission analysis. *Environmental Science & Technology, 32*, 760–765.
149. Seidel, H., Mattusch, J., Wennrich, R., Morgenstern, P. & Ondruschka, J. Mobilization of arsenic and heavy metals from contaminated sediments by changing the environmental conditions. (2002).
150. Finneran, K. T., Housewright, M. E., & Lovley, D. R. (2002). Multiple influences of nitrate on uranium solubility during bioremediation of uranium-contaminated subsurface sediments. *Environmental Microbiology, 4*, 510–516.
151. Rittle, K. A., Drever, J. I., & Colberg, P. J. S. (1995). Precipitation of arsenic during bacterial sulfate reduction. *Geomicrobiology Journal, 13*, 1–11.

152. Mishra, S. R., Bharati, K., Sethunathan, N., & Adhya, T. K. (1999). Effects of heavy metals on methane production in tropical rice soils. *Ecotoxicology and Environmental Safety, 44*, 129–136.
153. Zayed, G., & Winter, J. (2000). Inhibition of methane production from whey by heavy metals—Protective effect of sulfide. *Applied Microbiology and Biotechnology, 53*, 726–731.
154. Pilon-Smits, E. (2005). Phytoremediation. *Annual Review of Plant Biology, 56*, 15–39.
155. Slaveykova, V., & Wilkinson, K. (2003). Effect of pH on Pb biouptake by the freshwater alga *Chlorella kessleri. Environmental Chemistry Letters, 1*, 185–189.
156. Slaveykova, V., & Wilkinson, K. J. (2002). Physicochemical aspects of lead bioaccumulation by *Chlorella vulgaris. Environmental Science & Technology, 36*, 969–975.
157. Overall, R. A., & Parry, D. L. (2004). The uptake of uranium by *Eleocharis dulcis* (Chinese water chestnut) in the Ranger Uranium Mine constructed wetland filter. *Environmental Pollution, 132*, 307–320.
158. Ebbs, S. D., Brady, D. J., & Kochian, L. V. (1998). Role of uranium speciation in the uptake and translocation of uranium by plants. *Journal of Experimental Botany, 49*, 1183–1190.
159. Lack, J. G., et al. (2002). Immobilization of radionuclides and heavy metals through anaerobic bio-oxidation of Fe(II). *Applied and Environmental Microbiology, 68*, 2704–2710.
160. Antunes, S. C., Pereira, R., & Gonçalves, F. (2007). Acute and chronic toxicity of effluent water from an abandoned uranium mine. *Archives of Environmental Contamination and Toxicology, 53*, 207–213.
161. Zänker, H., Ulrich, K.-U., Opel, K., & Brendler, V. (2007). In R. Cidu & F. Frau (Eds.), *IMWA Symposium 2007: Water in Mining Environments*.
162. Vymazal, J. (2005). Constructed wetlands for wastewater treatment. Backhuys Publishers.
163. Batty, L. C. & Younger, P. L. Growth of *Phragmites australis* (Cav.) Trin ex. Steudel in mine water treatment wetlands: effects of metal and nutrient uptake. *Environmental Pollution* 132, 85–93 (2004).
164. Lüderitz, V., & Gerlach, F. (2002). Phosphorus removal in different constructed wetlands. *Acta Biotechnologica, 22*, 91–99.
165. Zayed, A., Gowthaman, S., & Terry, N. (1998). Phytoaccumulation of trace elements by wetland plants: I. Duckweed. *Journal of Environmental Quality, 27*, 715–721.
166. Hallberg, K. B., & Johnson, D. B. (2003). Passive mine water treatment at the former Wheal Jane tin mine, Cornwall: Important biogeochemical and microbiological lessons. *Land Contamination & Reclamation, 11*, 213–220.
167. Müller, B., Granina, L., Schaller, T., Ulrich, A., & Wehrli, B. (2002). P, As, Sb, Mo, and other elements in sedimentary Fe/Mn layers of Lake Baikal. *Environmental Science & Technology, 36*, 411–420.
168. Sheoran, A. S., & Sheoran, V. (2006). Heavy metal removal mechanism of acid mine drainage in wetlands: A critical review. *Minerals Engineering, 19*, 105–116.
169. Schaller, J., et al. (2008). *Enrichment of uranium in particulate matter during Litter decomposition affected by Gammarus pulex L. Environ. Sci.*
170. Selig, U., Hübener, T., & Michalik, M. (2002). Dissolved and particulate phosphorus forms in a eutrophic shallow lake. *Aquatic Sciences, 64*, 97–105.
171. Dudel, G. E., Brackhage, C., Dienemann, H., Mkandawire, M., & Weiske, A. (2004). *Mine water* (Vol. in press). Springer.
172. Dienemann, C., Dienemann, H., Dudel, G. E., Brackhage, C., & Stolz, L.E. Coring & J. Bäthe (Eds.). *der Deutschen Gesellschaft für Limnologie (DGL) und der deuschen und österreichen Section der Societes Internationalis Limnologie (SIL)* (pp. 779–784) Eingenverlag der DGL.
173. Wu, W. M., et al. (2006). Pilot-scale in situ bioremediation of uranium in a highly contaminated aquifer. 1. Conditioning of a treatment zone. *Environmental Science & Technology, 40*, 3978–3985.
174. Vrionis, H. A., et al. (2005). Microbiological and geochemical heterogeneity in an in situ uranium bioremediation field site. *Applied and Environmental Microbiology, 71*, 6308–6318. https://doi.org/10.1128/aem.71.10.6308-6318.2005

175. Ellen, C. E. (2006). Treatment of uranium-contaminated waters using organic-based permeable reactive barriers. *Federal Facilities Environmental Journal, 17*, 19–35.
176. Fuller, C. C., Barger, J. R., Davis, J. A., & Piana, M. J. (2002). Mechanisms of uranium interactions with hydroxyapatite: Implications for groundwater remediation. *Environmental Science and Technology, 36*, 158–165.
177. Fox, P. M., Davis, J. A., & Zachara, J. M. (2006). The effect of calcium on aqueous uranium(VI) speciation and adsorption to ferrihydrite and quartz. *Geochimica et Cosmochimica Acta, 70*, 1379–1387.
178. Franklin, N. M., Stauber, J. L., Markich, S. J., & Lim, R. P. (2000). pH-dependent toxicity of copper and uranium to a tropical freshwater alga (Chlorella sp.). *Aquatic Toxicology, 48*, 275–289.
179. Davis, J. A., et al. (2006). Processes affecting transport of uranium in a suboxic aquifer. *Physics and Chemistry of the Earth, Parts A/B/C, 31*, 548–555.
180. Bennett, G. F. (1995). In situ bioremediation: When does it work? *Journal of Hazardous Materials, 42*, 109–110.
181. Ketterer, M. E., Wetzel, W. C., Layman, R. R., Matisoff, G., & Bonniwell, E. C. (2000). Isotopic studies of source of uranium in sediments of Ashtabula River, Ohio, USA. *Environmental Science and Technology, 34*.
182. Wielinga, B., Bostick, B., Hansel, C. M., Rosenzweig, R. F., & Fendorf, S. (2000). Inhibition of bacterial promoted Uranium reduction: Ferric (Hydr)oxide as competitive electron accepter. *Environmental Science & Technology, 34*, 2190–2195.
183. Raicevic, S., Wright, J. V., Veljkovic, V., & Conca, J. L. (2006). Theoretical stability assessment of uranyl phosphates and apatites: Selection of amendments for in situ remediation of uranium. *Science of the Total Environment, 355*, 13–24.

Chapter 11
Natural Attenuation of Uranium in Abandoned Mining Sites

Abstract Natural attenuation of uranium in abandoned mining sites is examined in this chapter as both a promising and inherently constrained remediation pathway. The discussion evaluates the geochemical and hydrological conditions that enable natural processes to immobilize or transform uranium while acknowledging the uncertainties and site-specific limitations that may restrict their effectiveness. Opportunities for cost-effective, minimally invasive management are weighed against challenges, such as variable groundwater flow, fluctuating redox conditions, and long-term monitoring needs. This chapter also outlines circumstances in which enhanced natural attenuation becomes necessary, detailing strategies, such as biostimulation, sorbent amendments, and hydrological controls. It emphasizes how geological setting, contaminant speciation, and risk thresholds guide the selection of ENA approaches and shape long-term stewardship decisions.

Keywords Biotransformation · Speciation · Sorption · Redox conditions · Bioaccumulation · Hydrogeology · Colloids · Biostimulation · Sedimentation · Phosphates

11.1 Introduction

One question commonly asked is: What would happen if abandoned uranium mines were left unattended for years, decades, or centuries? Will the associated issues and hazards naturally fade away? Yes, the hazards would reduce, but it may take ages. It is "yes" because nature takes control of itself through a series of processes that keep the system in balance at equilibrium. For instance, the issues that have become a legacy of abandoned uranium mines come to prominence only after mining, because mining operations disturb the ecosystem balance. When natural systems are disturbed, a homeostasis-like process occurs to counteract the induced imbalance. This ecosystem healing process is called natural attenuation.

In its strict sense, attenuation means "reduction". Thus, natural attenuation, also known as intrinsic remediation, reduces pollutants in the environment through

M. Mkandawire, *Fate and Effects of Uranium in Abandoned Uranium Mines*,
https://doi.org/10.1007/978-3-032-10447-2_11

natural processes. Processes such as dilution, dispersion, precipitation, sorption, biodegradation, bioaccumulation, volatilization, and chemical and biochemical stabilization of contaminants in an environment effectively reduce mobility, bioavailability, toxicity, or concentrations of contaminants to levels that are not overly harmful to human health and ecosystems [1]. Therefore, natural attenuation at an abandoned uranium mine represents a form of temporal dilution of the uranium and radionuclide concentration and reduction in radioactivity, which should not be confused with the physical dilution that takes place, for example, when contaminants or activity is discharged into watercourses and dispersed. In recent years, natural attenuation at abandoned mining sites has gained prominence as the least invasive environmental remediation approach. It forms an integral component of the design criteria for geological repositories that depend on geochemical processes to retard radionuclide migration to the biosphere.

The issues and concerns in abandoned mines are site-specific. However, the processes involved in the natural attenuation of abandoned uranium mines are similar, if not identical, regardless of mining technique (i.e., open pit, underground, or ISL) or site. Understanding the physical, chemical, and biological processes underlying natural attenuation in abandoned uranium mining sites is crucial for designing site remediation and restoration strategies. Thus, this chapter discusses the natural processes leading to contaminant migration and retention, specifically focusing the discussion on: (a) the mechanisms affecting natural attenuation as a factor of the geosphere media and radionuclides and (b) the applicability of natural attenuation strategies and identification for the need of enhanced natural attenuation at an abandoned uranium mining and processing site.

11.2 Natural Attenuation Mechanisms

11.2.1 General Mechanism

Natural attenuation of contaminants at a site can occur through either destructive or non-destructive mechanisms. The destructive mechanism involves an irreversible transformation from a toxic to non-toxic chemical species or, in the case of organics, degrading to new compounds or their precursor elements. For organic pollutants, degradation occurs primarily through biodegradation and hydrolysis, with biodegradation being by far the most prevalent mechanism [2]. However, metal and metalloid contaminants, mostly associated with abandoned mines, cannot be degraded. Consequently, biodegradation cannot happen. Instead, the pollutants are transformed from one species to another through changes in their oxidation state, which eventually render them non-toxic, insoluble, or immobile, removing them from the exposure pathway. A general term for the mechanism is biotransformation.

Biotransformation involves naturally occurring microorganisms, such as yeast, fungi, and bacteria, which can either reduce or oxidize metallic compounds into less toxic or non-toxic species, change their solubility, bio-volatilize them, or form biominerals. For instance, certain anaerobic microorganisms (e.g., iron- or sulfur-reducing bacteria) do not require oxygen as an electron acceptor. Thus, when such microorganisms each have an electron donor (i.e., mainly hydrocarbons), they use metals as electron acceptors to complete respiration. Further, many microorganisms use enzymatic processes to induce biomineralization for various reasons, including defence mechanisms. Others are purely geochemical processes, discussed in several chapters already and revisited in the following sections. Non-destructive attenuation mechanisms include sorption, dispersion, dilution, and volatilization.

11.2.2 Adoptable Natural Attenuation Strategies

The growing interest in using natural attenuation in abandoned uranium mining sites does not mean adopting a "walk-away" strategy. However, understanding the processes at work in natural attenuation at abandoned uranium mining and processing sites would help design remediation and restoration strategies that are low-cost, low-intensity "passive", or low-maintenance solutions rather than intrusive, and usually expensive, engineering techniques. Currently, two emerging strategies involving natural attenuation are being adopted for the remediation of abandoned mines, including uranium mining sites, namely (a) monitored natural attenuation (MNA) and (b) enhanced natural attenuation (ENA).

11.2.2.1 Monitored Natural Attenuation

An abandoned uranium mine is a highly dynamic system, regulated by natural processes over time. Thus, applying natural attenuation at the sites requires a detailed understanding of the processes operating there, gained through characterization and adequate monitoring. This is known as monitored natural attenuation (MNA), not a "technology" per se. Instead, scientists and engineers gather information on how a range of physical, chemical, and biological processes, unaided by deliberate human intervention, reduce the concentration, toxicity, or mobility of chemical or radioactive contaminants [3–6].

Monitored natural attenuation (MNA) relies on natural processes (e.g., biodegradation, dilution, and sorption) within soil and groundwater to reduce contaminant mass, toxicity, and mobility, while scientists monitor these processes to ensure effectiveness and site safety over time. Figure 11.1 shows how MNA is designed as a network of wells and sampling points to track contaminant plume stability and decreasing concentrations/mass. The effectiveness of natural attenuation processes is evaluated using background, source, plume centreline, and sentinel wells to

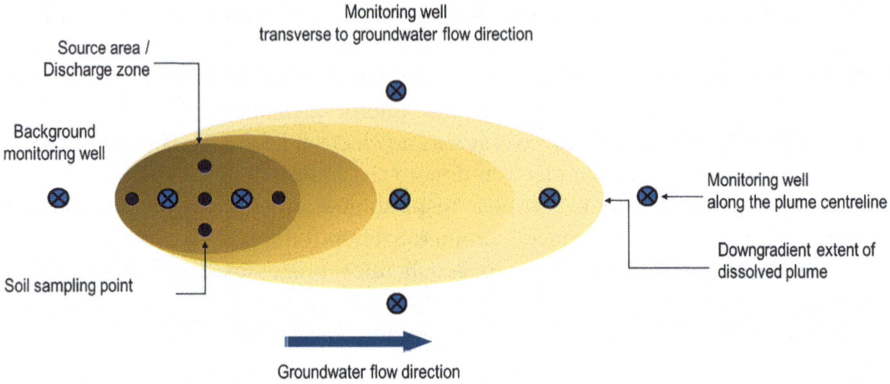

Fig. 11.1 Generalized application principle of MNA in abandoned uranium mines. (Modified from: https://frtr.gov/matrix/Monitored-Natural-Attenuation/)

ensure that cleanup objectives are achieved and that no unacceptable risks remain [2, 5, 7].

The adoption of MNA for uranium and radionuclides in abandoned mines has limitations and concerns. Many geochemical and biological processes contributed immensely to the retardation of uranium and radionuclides in the abandoned mines. However, these can also lead to a reconcentration of radiologically relevant elements, which can then form potential secondary sources of contamination. Further, mine waste and tailings heaps require long-term, effective containment before being considered for MNA, as they are prone to wind and erosion. Generally, MNA is not appropriate in abandoned mines where imminent site risks are present. MNA works best where the source of contamination has been removed. When the source is no longer present, natural processes can remove the remaining, smaller amount of contaminants in the sites. The sites are monitored regularly to ensure that pollutants attenuate quickly enough to meet site cleanup objectives and that contaminants are not spreading.

11.2.2.2 Enhanced Natural Attenuation (ENA)

When the natural processes that control and regulate contaminants' fate are well understood, they can be engineered to increase the system's attenuation capacity. This is known as enhanced natural attenuation (ENA). The ENA strategies, also known as enhancements, are adopted when the natural processes cannot reduce flux to the low-risk pollution levels within a reasonable timeframe. Generally, the upgrades must be sustainable, with continuing interventions until the enhancement is no longer required to reduce contaminant concentrations or fluxes. The decision process for selecting enhancements for application at an abandoned mining and processing site depends on site conditions (e.g., geology, hydrology, and depth of contaminants), potential deployment locations (source, plume, and receptor), and cost factors.

The ENA processes in abandoned mining and processing sites fall into two classes: (a) processes that reduce mass loading from the source to the plume and (b) processes that augment existing attenuation processes occurring within the plume and at the compliance point. Consequently, potential enhancements that meet sustainability requirements and reduce uranium and radionuclide mass flux are categorized into these two classes [8–12]. For instance, the enhancement for:

(a) Reduce mass loading from an abandoned mining site, including:

- Engineered structures to intercept and divert surface runoff, stormflow, and groundwater flow (also including diversion of competing electron acceptors).
- Methods for reducing infiltration in source areas (e.g., engineered covers and plant-based methods).
- Waste encapsulation (e.g., grouting; diffusion barriers, engineered cover over WRP, etc.)
- Techniques for lowering the hydraulic gradient in source areas (e.g., French drains and plant-based methods).
- Passive reduction of source mass (e.g., soil vapour extraction by barometric pumping).

(b) Reduce mass flux within a plume, including:

- Biological processes:

(i) Biostimulation: A process involving the modification of the environment to stimulate indigenous microorganisms' growth and activity, like the addition of various forms of rate-limiting nutrients and electron acceptors (e.g., phosphorus, nitrogen, oxygen, or carbon, etc.) or electron donors (organic substrates).
(ii) Bioaugmentation: The addition of archaea or bacterial cultures required to speed up the rate of immobilization of a contaminant.
(iii) Wetlands (natural and constructed).
(iv) Phyto-based methods.

- Abiotic processes:

(i) Contaminant degradation by biologically enhanced abiotic reactions.
(i) Permeable reactive barriers.

11.2.3 Specific Mechanisms in Abandoned Uranium Mines

Most processes do not reduce the total mass of uranium and radionuclides in abandoned mines. Instead, they affect their spatial and temporal distributions, except for radioactive decay. Mostly, the processes affect the migration of metals and related

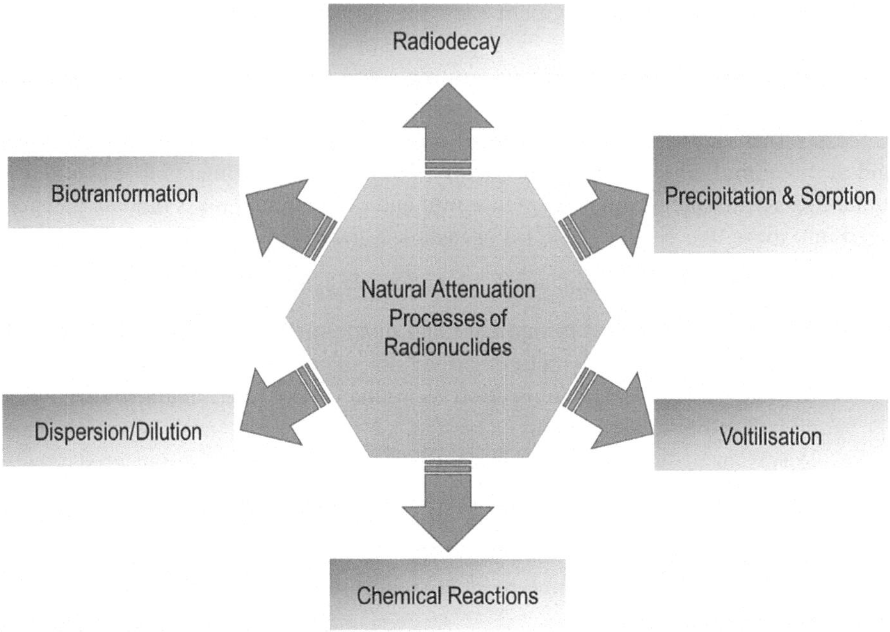

Fig. 11.2 Mechanism of natural attenuation specific to abandoned uranium mines with known uranium and radionuclide contamination

radionuclides. To be utilized for natural attenuation in abandoned mines, the processes must prevent or delay the arrival of uranium or a radionuclide at a receptor until they have decayed to levels of no concern. Generally, natural attenuation is applied at abandoned uranium mine and processing sites to immobilize the metals and radionuclides present and reduce radioactivity. The processes require a mechanism that involves a change in the valence state of the metal or radionuclide, resulting in immobilization (e.g., changing from U(VI) to U(IV)), and natural breakdown due to radioactive decay. Figure 11.2 shows the mechanism of natural attenuation in abandoned uranium mines and mine processing sites.

11.3 Factors Influencing Uranium Attenuation

All-natural attentional mechanisms of radionuclides, summarized in Fig. 11.2, work in natural attenuation processes to reduce uranium levels in abandoned mines. The potential and extent of natural attenuation of uranium in an abandoned mine depend on its chemical form. Its interaction with biota and the abiotic system, as influenced by the milieu conditions, should favour one or more of the attenuation processes. Consequently, it requires understanding uranium behaviour in aqueous and solid forms and its interactions with its milieu, especially the biosphere, hydrosphere, and geochemistry.

11.3.1 Aqueous Chemistry

The characteristics and behaviour of uranium in an aqueous environment are the most critical factors in uranium attenuation. Under oxidizing conditions, uranium exists as U(VI) as the uranyl cation (UO_2^{2+}). Under reducing conditions, uranium transforms into the insoluble U(IV) state. Reduction to U(IV) occurs readily, while further reduction to U(III) is less. U(V) has a high tendency to disproportionate, so the pentavalent ion UO^{2+} is usually considered unstable. However, the U(V) species can be stabilized at mineral surfaces.

Consequently, only U(VI) and U(IV) are important in the natural aquatic system. Oxidized uranium (U(VI)) is very mobile, controlled by the composition of the aqueous solution and complex reaction sequences. The relative stability, fixation, and migration depend on the U(VI) speciation in a specific condition of the site. System pH strongly influences the solubility, hydrolysis, sorption, complexation, and colloid formation of uranium.

Hydrolysis reactions are susceptible to the activity and concentration of hydrogen ions in solution. Figure 11.3 depicts uranium speciation at 25 °C in the absence of all ligands except hydroxide. Other than hydroxyl, the dominant ligands are carbonate and, in strongly acidic solutions, sulfate. Otherwise, U(VI) phosphate species are very stable in water. As already mentioned in the preceding chapter, sulfuric

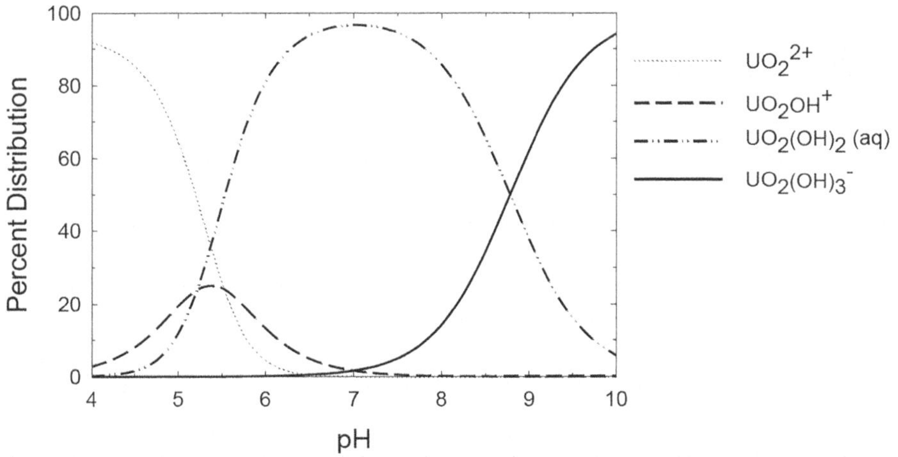

Fig. 11.3 Distribution of uranium species in a U(VI)-H_2O system at 25 °C, I = 0.1 M, PCO_2 = 0 bar, and U(VI) ions = 10^6 M in the absence of complexing ligands other than hydroxide. The plot shows that stepwise monomeric species dominate the distribution of uranium across the pH range. (Source: Golovich et al. [13])

acid and alkali carbonates are responsible for the chemical leaching of uranium in mine waste and tailings, leading to the dispersion of uranium from its source in abandoned uranium mines into receiving waters. Uranyl species are less prone to hydrolysis than U(IV). Consequently, colloids play a less significant role in the near-surface transport of uranium than thorium. However, this general rule does not hold in organic-rich environments, such as peat bogs, because high-molecular-weight organics (e.g., humic and fulvic acids) have a strong affinity for uranium, readily binding it in the U(VI) state.

11.3.2 Solid-State Geochemistry

At abandoned mining sites, uranium attenuation means immobilizing or fixing it to prevent it from entering our waterways or transforming into a less toxic form. The central issue in abandoned uranium mines is acid mine drainage, which can cause the release of uranium from its mineral form in waste rock. Thus, knowledge of uranium form in solid waste is very important. In most cases, abandoned mines retain rocks that still contain ore, except at processing sites. There are almost 200 discrete mineral phases of uranium occurring in the earth, as (a) primary minerals in the form of U(IV) oxides, silicates, phosphates, and carbonates and (b) a wide array of secondary U(VI) phases containing common and less abundant groundwater anions [14, 15].

In the near-surface zone, uranium is a relatively mobile element due to air exposure and exists mainly as stable U(VI) complexes. However, U(VI) may be reduced to U(IV), which forms precipitates or secondary uranyl minerals. Principally, the uranyl minerals and phases that precipitate from aqueous solutions include phosphates, silicates, arsenates, vanadates, and oxyhydroxides, several of which may occur simultaneously at an abandoned mining site. Table 11.1 lists selected mineral phases likely to be found at the abandoned sites. The formation of such phases is the dominant fixation mechanism for uranium in natural weathering environments and plays a determining role in the attenuation process in abandoned uranium mines. The amount of uranium released to surface or groundwaters from these secondary sources depends on the phases' solubility and dissolution rate as a function of pH and water composition.

Complexation of aqueous uranium increases the solubility of uranium minerals. Two known mechanisms can contribute to the increase in aqueous uranium:

(a) A Chelating Effect: This mechanism involves an oxygen-containing ligand (i.e., carbonate, phosphate, and hydroxide) binding to uranium in the mineral structure and subsequently releasing uranium complexes into solution.
(b) An Activity Equilibration Effect: This mechanism occurs when uranium is released from the mineral structure and then complexed by an aqueous ligand, reducing its activity in solution, which eventually allows further release of uranium from the mineral structure.

Table 11.1 Minerals of importance in solid geochemistry of uranium (VI)

Minerals		Chemical formula
Hydrates	Schoepite	$UO_3{\cdot}2H_2O$
	Becquerelite	$Ca(UO_2)_6O_4(OH)_6{\cdot}8H_2O$
	Compreignacite	$K_2(UO_2)_6O_4(OH)_6{\cdot}8H_2O$
	Calcium uranate	$CaUO_4$, CaU_2O_7
	Sodium uranate	Na_2UO_4, $Na_2U_2O_7$, $Na_6U_7O_{24}$, or $Na_2U_6O_{19}{\cdot}11H_2O$
Silicates	Soddyite	$(UO_2)_2(SiO_4){\cdot}2H_2O$
	Weeksite	$K_2(UO_2)(Si_2O_5)_3{\cdot}4H_2O$
	Boltwoodite	$K(H_3O)(UO_2)(SiO_4)$
	Uranophane	$Ca(UO_2)_2(SiO_3OH)_2{\cdot}5H_2O$
	Haiweeite	$CaO{\cdot}2UO_3{\cdot}6SiO_2{\cdot}5H_2O$
	Kasolite	$Pb(UO_2)(SiO_4){\cdot}H_2O$
	Sklodowskite	$Mg(H_3O)_2(UO_2)(SiO_4)_2{\cdot}4H_2O$
Phosphates	Autunite	$Ca(UO_2)_2(PO_4)_2$ Ca(UO2)2(PO4)2·8-12H2O
	Bassetite	Fe((UO2)(PO4))2(H2O)8
	Chernikovite	H3O(U2O)PO4(H2O)3
	Meta-ankoleite	KUO2PO4(H2O)
	Meta-autunite	Ca(UO2)2(PO4)2(H2O)6
	Metatorbernite	Cu(UO2)2(PO4)2·8H2O
	Phosphuranylite-groups	$KCa(H_3O)_3(UO_2)_7(PO_4)_4O_4{\cdot}8(H_2O)$
	Saleeite	$Mg(UO_2)_2(PO_4)_2{\cdot}10H_2O$
	Torbernite	$Cu[(UO_2)(PO_4)]_2{\cdot}10\text{–}12H_2O$
	Autunite	$Ca(UO_2)_2(PO_4)_2{\cdot}12H_2O$
Carbonates	Andersonite	$Na_2CaUO_2(CO_3)_3{\cdot}6H_2O$
	Grimselite	$K_3Na(UO_2)(CO_3){\cdot}H_2O$
	Andersonite	$Na2K_3-$ $UO_3(CO_3)3(H_2O)_6$
	Bayleyite	Mg_2UO_2(CO3)3(H2O)18
	Grimselite	NaK3UO2–(CO3)3H2O
	Liebigite	Ca2UO2(CO3)3(H2O)10
	Rutherfordine	UO2CO3
	Swartzite	CaMgUO2(CO3)3(H2O)12
	Schrockingerite	NaCa3UO2(CO3)3SO4F(H2O)
	Wyartite (U(V))	CaU5 + (UO2)2(CO3)O4(OH)(H2O)7
Arsenates	Carnotite	K2(UO2)2(VO4)2 K(UO2)2(VO4)2·13H2O K2(UO2)(V2O8)·3H2O
	Curienite	Pb2(UO2)(V2O8)·5H2O
	Francevillite	Ba, Pb(UO2)(V2O8)·5H2O
	Metatyuyamunite	Ca(UO2)(V2O8)·3H2O
	Strelkinite	Na2(UO2)(V2O8)·6H2O
Vanadates	Tyuyamunite	Ca(UO2)2(VO4)2·5-8H2O Ca(UO2)(V2O8)·9H2O
	Vanuralite	Al(OH)(UO2)2(V2O8)·3H2O

11.3.3 Milieu Settings and Condition

Aquatic and solid-state uranium chemistry is significantly influenced by the environmental conditions at an abandoned uranium mining site. Any changes over time of those properties that define the setting exert a significant influence over the site's natural attenuation processes. The relevant environmental factors that exert a substantial influence on uranium attenuation in abandoned mines include:

(a) Geological and Mineral Background: The site's geological and mineral background can influence its geochemical and physical conditions. For instance, an abandoned site's physicochemical properties, such as metal content, pH, and redox potential, depend on the site's background geochemistry. The site's background geochemistry determines the speciation and reactions that lead to the immobilization and fixation of uranium. Another key player in the immobilization of uranium is sorption/desorption. Adsorbing uranium species onto natural matrices keeps uranium from migrating deeper underground or from leaving the site with water flow. Generally, uranium attenuation is initially affected by its sorption/desorption reactions at the water/rock interface. Their sorption/desorption processes involve, among others, the mechanistic uptake of uranium by oxides, hydroxides and carbonates, clay minerals, silica, and organics. Further, the sorption is dependent on the speciation of uranium, like uranium hydroxide and carbonate species, onto goethite.

(b) Climate: The overall environmental conditions are determined by the region's climate. Climate parameters, such as temperature, humidity, precipitation, and sunlight intensity, to name a few, directly or indirectly control or influence the attenuation process, including reactions and interactions involving uranium and its milieu. Consequently, the speciation and sorption/desorption process of uranium are affected. In natural attenuation, the site ecology is fundamental. The ecological zones are climate-dependent, determining the occurrence and prevalence of microbial, plant, and animal species. For example, some floral or fauna species are better suited to humid sites with greater rainfall, relatively shallow depths to groundwater, and generally short flow paths to points of discharge, in comparison to what is typically found at arid sites.

(c) Hydrogeology of the Site: The flow rate and quality of groundwater in the abandoned mines profoundly influence natural attenuation by affecting uranium concentration, reactions, and transport. For instance, the high volume and flow rate of groundwater would dilute uranium concentrations as they move through and mix with clean groundwater. Aquifer heterogeneities lead to the potential for matrix diffusion, which affects the migration and lifetime of a plume.

(d) Changes in Environmental Factors That Can Occur Over Time: The following changes may influence natural attenuation in the environment.

- Short-term, natural weather phenomena such as flooding and off-normal precipitation events.

- Changes in the plume footprint and plume discharge locations due to seasonal variations in precipitation-infiltration or stream stage of a receptor.
- Gradual or episodic changes in the groundwater balance and gradient due to nearby development activities (i.e., future land use).
- Gradual changes in soil hydraulic properties and plant ecology may impact the performance of engineered covers.

11.4 Natural Attenuation of Uranium in Abandoned Mines

Nature may clean up uranium in the environment, and indeed in abandoned mines, through the mechanism presented earlier in Fig. 11.2, which will be discussed in detail in this chapter's preceding sections.

11.4.1 Radioactive Decay

Following logical thinking, uranium should naturally attenuate in abandoned mines over time through radioactive decay, as is the case with the predominant radioisotopes found in natural systems. All the isotopes of uranium are radioactive. Of the three naturally occurring isotopes, ^{238}U is the most abundant (99.275%; 4.468 × 10^5 year half-life), followed by ^{235}U (0.720%; 7.038 × 10^8year half-life) and ^{234}U (0.005%; 2.445 × 10^5 year half-life). Thus, logically again, it is evident from the decay half-lives of the isotopes that natural attenuation of uranium in abandoned mines through radioactive decay would be too negligible to observe within a century.

However, the natural attenuation of uranium through radioactive decay may be facilitated by alpha recoil processes. The alpha recoil process occurs when a radioactive daughter nucleus is displaced from its initial position by the energy of an alpha decay, in which an atomic nucleus emits an alpha particle. The released ionizing radiation has an energy content of 4–6 MeV. Observing the law of conservation of momentum, the emitted alpha particle and recoiling nucleus have well-defined energy after the decay, with most of the kinetic energy going to the alpha particle because of its smaller mass. Since ^{234}U is a decay product of ^{238}U, it generally occupies sites in minerals that have already been damaged by releasing an alpha particle during the ^{238}U decay process. The ^{238}U atoms transform into 234Th, which then decays to ^{234}U. Thus, the alpha recoil process may lead to two scenarios: (a) ejection of the ^{238}U daughter (234Th) directly into groundwater. ^{238}U atoms at the solid-liquid interface that emit decay alpha particles inward may experience a kinetic energy recoil enough to tear the 234Th progeny from the solid or (b) indirectly increase the potential for ^{234}U leaching from aquifer solids damaged by the recoil process. The fracture of the surficial solid layer is caused by kinetic recoil energy, making the 234Th more accessible for the enhanced dissolution that thorium typically experiences relative to uranium in mineral matrices. The occurrence of natural attenuation of uranium in abandoned mines through alpha recoil is ascertained by observing the ^{234}U:^{238}U activity ratio in groundwater. The activity ratio increases proportionately, where ^{234}U is released during dissolution or from the rapid decay of 234Th (and ^{234}Pa) following ejection from the solid surface.

11.4.2 Physical Processes

11.4.2.1 Advection, Diffusion, and Dispersion

The processes of advection, dispersion, and diffusion have a profound role in the subsurface transport of uranium. These processes affect the transported substance and the transport medium, resulting in complex distributions of uranium in natural groundwater systems. To understand these processes' role in uranium's natural attenuation in abandoned uranium mining sites, it is important to know how advection relates to diffusion and the dispersion of uranium from a contaminated site.

Advection is the transport of a solute by the bulk movement of groundwater and particles' movement within flowing water. Depending on the physicochemical milieu, geochemical reactions, and microbial activity, water ingress can cause uranium leaching from mine waste and tailings. Most mine waste piles and tailing heaps contain loose material, creating a highly porous medium that favours the flow of leachate to groundwater. If no new uranium is mobilized from their sink or solubilized from mineral material, groundwater advection reduces the uranium content of the abandoned mining site. The transport of uranium-contaminated leachate depends on groundwater bulk movement, average linear groundwater velocity, and advective velocity. Assuming that (a) the transport of mass does not affect the pattern of flow, (b) the densities of the leachate solute and groundwater do not differ significantly, causing the flow of mass and water to diverge, and (c) the linear average groundwater velocity is the same as the advective velocity of the mass, the one-dimensional flux through a porous medium can be expressed by:

$$J = V_x \times C n_e$$

where

J is the mass flux per unit area per unit time.
V_x is the average linear groundwater velocity in the direction of flow (i.e., advective velocity).
C is the concentration in mass per unit volume of solution.
n_e is the effective porosity of the geological medium, which is the porosity through which flow can occur.

Generally, most soluble uranium species have low advective velocities at low pH because they form positively charged ions. In higher pH, for example, above pH 8, $UO_2(OH)_2^-$ and $UO_2(CO_3)_2^{2-}$ dominate in the solution that the advective velocity significantly increases. The presence of clay minerals may lead to the accumulation of electrical charges, which force the negatively charged uranium species (e.g., $UO_2(OH)_2^-$ and $UO_2(CO_3)_2^{2-}$) to remain in the pores' centre, where the microscopic velocity is large, causing the solute to move at a maximum microscopic velocity. Conversely, solutes can move at a slower velocity than the average linear groundwater velocity if the geological medium exhibits semi-permeable membrane properties or sorption is involved.

The concentration of uranium in the plume gradually diminishes as advection and groundwater flow continue to spread it out. This dispersion effect arises from two phenomena: molecular diffusion and mechanical dispersion. When water flows through a porous medium, individual water particles have preferential pathways around the grains of a geological medium containing uranium. A particle of water may need a longer or a shorter time to move from one point to another, depending on the chosen pathway. For instance, water particles flow along a more direct path, through larger pores, or through the centre of pores, where there is less friction, rather than along grain boundaries, where there is more friction. Consequently, the different flow paths of water particles cause mechanical dispersion, mixing, and dilution of uranium leached from mine wastes or tailings within the bulk movement of groundwater.

The mechanical dispersion (*D*) is estimated as:

$$D = \sigma \times V_x$$

where dispersivity (σ) is the ability of a geological medium to disperse the solutes, which depends on the pore and grain size and distribution.

Dispersivity occurs in three dimensions, namely: (a) in the direction of flow, which is called longitudinal dispersivity (in the x-direction) and (b) dispersivity perpendicular to flow, which is transverse dispersivity, both in a horizontal plane to flow (in the y-direction) and (c) in a vertical plane to flow (in the z-direction, flow up or down in a groundwater medium). Generally, uranium's longitudinal dispersivity is estimated to be about 10 times larger than transverse dispersivity because the local variation in the velocity field is much more dominant in the direction of flow rather than perpendicular to it [16]. Transverse dispersivity is primarily caused by flow paths that branch out from the centreline of the solute movement due to the tortuosity of the geological medium [17, 18].

Molecular diffusion causes uranium to disperse in groundwater driven by a concentration gradient. The plume is surrounded by groundwater with low or zero uranium concentration, creating a concentration gradient that drives uranium to move stochastically from the plume into lower-concentration waters or zones. When the average pore flow velocity is zero, dispersion takes place through molecular diffusion only, which is described by Fick's first law [19]:

$$F = -D_d \cdot \delta_c / \delta_x$$

where

F is the diffusive flux (mol · cm^{-2} · d^{-1});
D_d is the diffusion coefficient (cm^2 · d^{-1});
δ_c/δ_x is the concentration gradient (mol · cm^{-3} · m^{-1}); and.
c is the concentration in the pore fluid (mol · cm^{-3}).

In diffusive transport in porous media, the pathway around grains must also be included to estimate the effective diffusion coefficient, (D^*):

$$D^* = \omega D_d$$

where ω is an empirically determined parameter.

The diffusive behaviour of uranium species depends on water mobility, the self-diffusion coefficient of the aqueous species, and electrostatic forces between the negatively charged surface and the charged uranium compounds. The mobility of uranyl species in the adsorbed layer decreased in the sequence: $UO_2^{2+} > UO_2CO_3 > UO_2\left(CO_3\right)_2^{2-}$. In the diffuse layer, $UO_2\left(CO_3\right)_2^{2-}$ has a higher diffusion coefficient and is faster than UO_2^{2+} [20]. In groundwater systems, aqueous alkaline-earth uranyl-carbonate species (i.e., $M_xUO_2\left(CO_3\right)_y^{2+2x-2y}$ with M = Mg, Ca, or Sr) often dominate uranyl speciation [21]. However, the effect of the alkaline-earth cation on the diffusion coefficient of uranyl-carbonate complexes remains unclear, and a single diffusion coefficient is used for convenience across different alkaline-earth uranyl-carbonate complexes.

11.4.2.2 Surface Runoff

Natural attenuation of uranium in abandoned mining sites can be enhanced by surface runoff. Surface runoff, also known as overland flow, is the water flow on the ground surface when excess water from precipitation cannot infiltrate into the ground sufficiently or rapidly. The attenuation of uranium through surface runoff can occur in two processes: (a) washout of soluble uranium species and (b) the washoff of particles containing uranium.

Washout Process

The process washout involves transporting soluble uranium species, mostly uranyl species, from the top layer of the mine and processing waste, tailing heaps, or soil covers at abandoned mining sites. The magnitude of washout is directly related to various factors and processes that affect runoff and the availability of easily soluble uranium in the top layer of the ground surface. Therefore, water infiltration from precipitation into the ground, which depends on the waste and tailing material's drainage properties, such as prior water saturation, grain-size distribution, and organic matter content, has significant implications for washout. Further, their magnitude of runoff depends on a range of additional variables, including (a) the duration and form of the precipitation events (i.e., rain and snow); (b) interception of the precipitation by the vegetation cover; (c) the sum of evapotranspiration processes; and (d) the climatic zone where the mine is located (e.g., in tropical regions with dry and wet seasons, there is a flush runoff at the beginning of rain season; while in temperate regions, the formation of snow layer and snowmelt processes result in the less overall runoff than that due to rainfall). The dissipation of raindrop kinetic energy dislodges mine waste and tail particles from their structure, thereby facilitating their washout.

A quantitative measure of uranium loss from an abandoned mining site in dissolved form is expressed as the washout coefficient (K_l):

$$K_l = \left(\frac{\int_0^T C_w(t) v(t) dt}{\int_s C_s ds} \right) = \frac{C_w^* V}{C_s^* S}$$

where.

$C_w(t)$, C_w^* are instantaneous and weighted average concentrations of uranium in the runoff, respectively;
$v(t)$ is the runoff intensity (L/s);
V is the total run-off volume (L);
C_s is the surface density of uranium (Bq/m^2);
C_s^* is the surface density of pollution averaged over a watershed (Bq/m^2);
S is the area of the site;
0, T are starting and termination times (s) of a run-off, respectively; and.
t is the time (s).

By dividing K_l by runoff depth, normalized washout coefficients (K_l^*) is obtained, which is equal to the ratio of a corresponding weighted average concentration and average density of pollution in an abandoned mining site:

$$K_l^* = \frac{K_l}{h} = \frac{K_l S}{V} = \frac{C_w^*}{C_s^*}$$

where h is the runoff depth.

The expected runoff depth multiplied by K_l^* can predict the input of soluble uranium washed away from the site with the runoff event.

Washoff Process

The washoff process involves the transport of loose particles containing uranium as water flows away from abandoned mines or processing sites. Uranium is either attached by sorption onto the surface or bound and locked in the particles. The particles can also be aggregates of uranium precipitates or co-precipitated uranium with other metals, in the form of nano- and microparticles, mostly of uraninite species. The washoff process of uranium from the site may be controlled by (a) physical structure and chemical properties of mine waste and tailing material; (b) properties of liquid precipitation (i.e., size and energy of droplets and distribution and intensity of rainfall); (c) hydrodynamic properties of surface and subsurface flows; (d) relief or topography of the abandoned mine sites especially the waste and tailing heaps; and (e) vegetation cover, including its seasonal characteristics in the site. Further, uranium's washoff behaviour also depends on sedimentology, specifically the grain-size distribution and mineralogy of the waste material and the tailings.

Similar to the washout process, the washoff coefficient (K_s) is used to quantify the fraction of total uranium lost through the washoff with solid particles as follows:

$$K_s = \left(\frac{\int_0^T C_a(t)m(t)dt}{\int_S C_s ds} \right) = \frac{C_a^* M}{C_s^* S}$$

where.

$C_a(t)$, C_a^* are instantaneous and weighted average specific activities of uranium on suspended particles;
$m(t)$ is the intensity of wash-off for suspended particles;
M is the total mass of washed-off particles.

Similarly, normalized coefficients of washoff (K_s^*) is obtained by dividing K_s by the amount of suspended matter washed off from a unit waste or tailing surface, which is equal to the ratio of a corresponding weighted average concentration and average density of pollution in a catchment area:

$$K_s^* = \frac{K_s}{q} = \frac{K_s S}{M} = \frac{C_a^*}{C_s^*}$$

where q is the suspension mass washed off from an abandoned mine site.

The expected removal of suspended material multiplied by K_s^* predicts the amount of uranium washed off by surface runoff.

11.4.2.3 Sedimentation

Natural attenuation of uranium in an abandoned mine site through sedimentation removes uranium from surface mine water. Surface mine water can be directly discharged from flooded mines or open-pit lakes, or acid mine drainage that affects receiving water reservoirs. Sedimentation is a process in which particles suspended in water settle out of the suspension under the effect of gravity. Thus, uranium's attenuation from the surface mine through the sedimentation process should be related to suspended particles, either being uranium locked in mineral particles, adsorbed on mine waste or tailing grains, or bound to particulate organic matter. The particles may also form as dissolved uranium precipitate, following biotic or abiotic processes that may affect its speciation.

The mine waste and tailing heaps can naturally undergo compaction, owing to the material's own weight. The mine waste and tailing material can also be consolidated and compacted into piles during mine site restoration. In both ways, the heaps and piles become increasingly compacted with depth, the excess porewater being expelled to the surface. The expulsion of porewaters may result in the release of dissolved or colloidal uranium from mine wastes and tailings into the overlying surface waters.

Sedimentation plays a special role in the attenuation of uranium in AMD discharging mine water, as well as mine water with a high content of high redox potential metals, and recalling that AMD is very rich in Fe^{2+}, which oxidizes rapidly to Fe^{3+} when exposed to air rich in O_2 as the AMD discharges to the surface. Fe^{3+} forms precipitates of iron oxide or oxo-hydroxide. During the process, uranium tends to co-precipitate with iron, forming aggregates that continue to agglomerate until their density exceeds that of water and, consequently, sediment to the bottom of the water reservoir or stream.

Colloidal uranium is subject to chemical interactions with the surrounding aqueous phase. Their transport behaviour and dispersal are controlled by the processes governing particles' movement. Among the most important determining factors is the water's hydrodynamic flow pattern, which is directly related to the size and orography of the abandoned mine site or the area where it is located, its geology and pedology, and the climatic conditions. The concentration of colloidal uranium in the water increases exponentially with water flow velocity. Then, the distance travelled by colloidal uranium in surface water is determined by grain-size distributions, grain shape, bottom surface roughness, and topography. Bottom surface roughness is particularly important in controlling the transport of larger particles, which are not typically suspended but roll or jump along the bottom of the surface water reservoir.

Suppose the receiving environment of the mine water rich in colloidal uranium is a stagnant water body; sedimentation would depend mainly on the water's viscosity, particle shape, and specific gravity. Thus, the uranium enrichment rate (R) into the sediment of such a water body can be written as:

$$R = K_d S_{\downarrow} \frac{\partial c}{\partial z}$$

where.

$S_{\downarrow}$ is the accumulation rate of particles with uranium and
K_d is the uranium distribution coefficient between sediment and water.

Accumulating sediments can result in areas with elevated uranium levels. Under certain hydrodynamic flow conditions, such water bodies can become a secondary source of uranium contamination, for instance, when the sediments are resuspended or disturbed. Further, geochemical processes under anoxic conditions in deeper sediment layers often immobilize uranium, making these layers a potential secondary source if disturbed mechanically or chemically.

11.4.2.4 Filtration

The filtration process is responsible for attenuating uranium bound to and transported on particulate matter or colloidal particles by flowing water. Globally, the process involves the physical, biological, or chemical separation of colloidal particles and mine water as the mine water flows through a porous medium with structures that allow only the fluid to pass while trapping the solids. In most cases, particle, solid, and mine water separation depends on pore size, filter media thickness, and biological activity. Thus, uranium attenuation in the waters of abandoned mines can occur through geological and biological processes. While both forms of filtration occur in surface and groundwater, geological filtration, a physical process, is more prominent in groundwater, whereas biological filtration is more prominent in surface water. The chemical process influences both biological and geological processes.

Uranium leachates from mine wastes and tailings migrate through the vadose zone, forming plumes in groundwater. Colloidal particles' transport depends on the geological media properties, such as permeability, hydraulic conductivity, pore-size distribution, heterogeneity, and fracture (size) distribution. The subsurface geological media may retard uranium-carrying colloidal particles when their sizes exceed the pore size—a screening effect that leads to particle exclusion. However, the particles' actual process is a complex combination influenced by surface interactions between the media and the particles. Thus, the long-lasting attenuation occurs through depth filtration, which results in effective retardation of particles, and through tortuous pathways, in which a particle follows a network of pores in the media. When colloidal particles are sufficiently small, they diffuse into dead-end pore spaces or intercrystalline pores in hard rocks and are physically retarded.

Uranium in surface-flowing mine water is usually found in wetlands. Various biotic processes and physical filtration processes remove uranium-carrying colloids from water. Further, uranium interactions with the biotic milieu can induce precipitation, aggregation, and biosorption, leading to removal through filtration. For instance, uranium can be removed from water in the aquatic system through phytofiltration in two ways, viz.:

(a) Physically retarding uranium-carrying colloidal particles directly from flowing with water, due to their root network forming pores, and reducing the flow rate of water, which facilitates sedimentation or.
(b) Taking up soluble uranium from the water pathway and sequestering or compartmentalizing it in different plant parts. Some plants exude low-weight organic compounds which bind and precipitate the uranium. If one of the above

processes occurs in the plant roots, phytofiltration, it is known as rhizofiltration. In some plants, rhizofiltration of uranium is facilitated by rhizobacterial activities. If plant seedlings are used, the process is called blastofiltration. Similarly, the use of excised plant shoots is caulofiltration.

11.4.2.5 Gaseous Transport

The natural attenuation of uranium through gaseous transport has not been documented anywhere. Uranium volatilization has been observed only in laboratory experiments, mostly involving uranium mononitride under vacuum at temperatures above 1600°C. However, it can be confidently stated that similar reaction processes cannot happen in the environment, primarily when driven by enzymatic and metabolic processes, such as those of bacteria. Nevertheless, uranium can be dispersed from the abandoned mining facility as particulate in a gaseous plume. Again, the extent to which dispersion through a gaseous plume contributes to natural attenuation from the abandoned mine site is not well studied.

Generally, uranium has been recorded at distances of up to 7.5 km from abandoned mines or mining sites, attributed to atmospheric dispersion resulting from uranium transported as dust particles from soil, organic matter, or mineral material [22–25]. Thus, desorption and sorption processes can contribute to uranium attenuation through gaseous transport. However, the uranium should be available as loose particulates in the environment. The rate of uranium suspension in the air depends on particle size and density, uranium particle content, wind turbulence, and chemical form. Further, the site's physical and meteorological properties, such as elevation, vegetation, wind direction and speed, and precipitation, influence the extent to which uranium's natural attenuation occurs through gaseous transport.

The best scenario for uranium to be transported through the atmosphere (unfortunately, the worst for the environment) is when the soils and mine material are loose and dusty at an abandoned mine. It is exacerbated when there is physical disturbance of the soil, enabling material released from an abandoned mine or processing site to rise to a few metres in height due to the plume's initial momentum (also see Fig. 11.4). The mine material particles, which are also, in most cases, radioactive, are then transported thousands of metres through air currents.

The transport range of particle-containing uranium depends on particle size, with larger particles settling more quickly than smaller ones when precipitation (i.e., rain or snow) or vertical wind velocity is not a factor. Such simplified meteorological conditions do not occur in nature but are assumed when estimating the transport of uranium-containing particles. For instance, considering the processes in a simplified meteorological condition illustrated in Fig. 11.4, the transport range of large particles can be calculated as follows:

If the dust particle can rise to a height of Z_0, it will take time (t) to fall back to the ground:

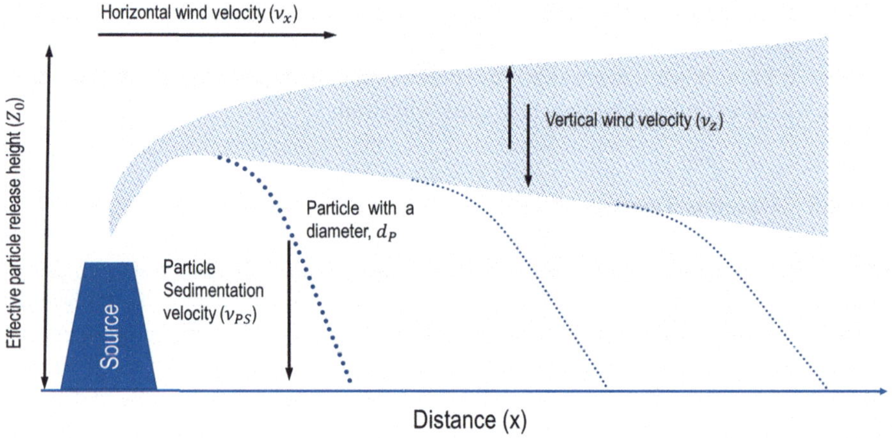

Fig. 11.4 Illustration of the transportation of particles of different sizes in a release plume. (Modified from Pöllänen et al. [26])

$$t = \frac{Z_0}{u_{PS}}$$

Assuming the wind is blowing at a constant direction and horizontal wind velocity (υ_x) during the transport, then the particle will travel a distance (x) as follows:

$$x = t\upsilon_x = \frac{z_0\upsilon_x}{\upsilon_{PS}}$$

Vertical velocity of air (υ_z) can essentially influence the x. When υ_z is constant during the transport, then x is given as:

$$x = \frac{z_0\upsilon_x}{\upsilon_{PS} + \upsilon_z}$$

The value of υ_z is positive for downward air flows and negative for upward air flows.

11.4.3 Chemical Processes

The formal oxidation state of uranium largely dictates its environmental chemistry and natural attenuation process in abandoned mines. Under ambient conditions, the thermodynamically stable uranium solid phases are either U(VI) or U(IV) compounds. In abandoned mines, uranium oxidation state U(0) hardly exists because it readily oxidizes to U(IV) and eventually to U(VI) if sufficiently oxidizing conditions pertain. Most abandoned mines have ambient oxidizing conditions, where the uranium oxidation state U(VI) predominates. In flooded mine and pit lakes, oxygen is usually limited, allowing U(IV) to dominate.

11.4.3.1 Aqueous Complexation

The intertwined controls of oxidation state, aqueous complexation, sorption to mineral and organic surfaces, and hydrologic transport govern natural attenuation of uranium in abandoned mine settings. In oxidizing waters, hexavalent uranium occurs as the uranyl ion, which forms highly stable hydroxo- and carbonato-complexes; in reducing waters, uranium is immobilized predominantly as tetravalent minerals (i.e., uraninite, UO_2(s), and coffinite, $USiO_4 \cdot nH_2O$) with very low solubility. These redox speciation transitions, together with complexation and surface interactions, dictate whether uranium partitions into the aqueous phase or is sequestered in solids during post-mining geochemical evolution [27–29].

Under circumneutral to alkaline, carbonate-bearing conditions typical of many pit lakes and mine aquifers, U(VI) speciation is dominated by uranyl-carbonate complexes. The stepwise formation reactions:

(Step 1)

$$UO_2^{+} + CO_3^{2-} \rightleftharpoons UO_2CO_3^{0}$$

(Step 2)

$$UO_2^{2+} + 2CO_3^{2-} \rightleftharpoons UO_2\left(CO_3\right)_2^{2-}$$

and (Step 3)

$$UO_2^{2+} + 3CO_3^{2-} \rightleftharpoons UO_2\left(CO_3\right)_3^{4-}$$

have large cumulative stability constants (e.g., $\log \beta_3^0 \approx 21.5$ log for $UO_2\left(CO_3\right)_3^{4-}$ at $I = 0.1$ M), enhancing solubility and mobility. In calcium-rich waters, ternary calcium-uranyl-carbonato complexes form:

$$Ca^{2+} + UO_2\left(CO_3\right)_3^{4-} \rightleftharpoons CaUO_2\left(CO_3\right)_3^{2-}$$

$$2Ca^{2+} + UO_2\left(CO_3\right)_3^{4-} \rightleftharpoons Ca_{2UO_2}\left(CO_3\right)_3^{0}$$

These species reduce electrostatic sorption to negatively charged surfaces and hinder bioreduction of U(VI) [30].

Complexation with dissolved organic matter (DOM), especially humic and fulvic substances, can further modulate U(VI) mobility. Field studies using asymmetrical flow field-flow fractionation show that at pH < 6, a significant fraction of aqueous uranium may be bound to 900–1400 Da organic complexes, elevating aqueous concentrations and altering bioavailability. Speciation models and measurements indicate that DOM-U(VI) complexes become less dominant as carbonate complexes take over at higher pH, but in acidic, organic-rich groundwaters, they can govern transport and attenuation behaviour [31–33].

Sorption and surface complexation provide critical attenuation pathways. Uranyl species adsorb to Fe(III) (hydr)oxides, aluminosilicate clay edge sites, and organic matter via inner-sphere carbonato ternary complexes and outer-sphere ion-pairing, processes that are strongly pH-dependent and competitive with carbonate complexation. The practical effect of sorption on transport is captured by the linear retardation factor (R),

$$R = 1 + \frac{\rho_b K_d}{\theta}$$

where ρ_b is bulk density, θ is porosity, and K_d is the distribution coefficient; site-specific variability in K_d (arising from speciation, ionic strength, and sorbent heterogeneity) translates directly into reductions in plume velocity and changes in spreading behaviour in reactive transport models [34, 35].

Inner-sphere complexation is the direct, coordinate-covalent bonding of a metal ion to a ligand (or surface functional group) through shared atoms, typically oxygen, nitrogen, or sulfur. In forming an inner-sphere complex, the metal replaces one or more of its hydration waters (or pre-adsorbed anions) and shares electrons with donor atoms on the ligand or mineral surface. Because the bond is specific and directional, inner-sphere complexes are stronger, less labile, and less sensitive to ionic strength than electrostatic attachments.

By contrast, outer-sphere association keeps the metal fully hydrated and relies on long-range electrostatics (e.g., Coulombic attraction to a charged surface). No direct bonds form; the hydration shell remains intact, making the interaction weaker, more reversible, and strongly influenced by ionic strength.

Phosphate provides a powerful geochemical sink for U(VI) through precipitation of sparingly soluble uranyl phosphates. In oxidizing, phosphate-bearing vadose and saturated zones, minerals of the autunite/meta-autunite group (e.g., Ca-autunite, Ba-meta-autunite, and metatorbernite) buffer dissolved uranium to well below drinking-water standards; long-term field observations demonstrate natural immobilization where uranyl phosphate assemblages persist for 10^5-year timescales under favourable pH and phosphate/carbonate activity ratios. Additionally, sequestration by uranyl silicates such as boltwoodite/uranophane, with dissolution strongly accelerated by bicarbonate, is central to attenuation near source zones [36–39].

Even for these low-solubility phases, complexation can destabilize solids under specific chemical conditions. Uranyl peroxide macroanions formed in alkaline, H_2O_2-rich systems (e.g., studtite and cage clusters) can dissolve meta-autunite group phosphates extensively, demonstrating that oxidants and strongly binding ligands may reverse attenuation. Similarly, bicarbonate promotes dissolution of Na-boltwoodite, increasing aqueous U(VI) via carbonate complexation; thermodynamic data quantify this behaviour and constrain models [28, 40].

The hydrologic expression of these reactions is captured by the 1-D advection-dispersion-reaction (ADR) equation:

$$R\frac{\partial C}{\partial t} = D_x\frac{\partial^2 C}{\partial x^2} - v\frac{\partial C}{\partial x} - \lambda C$$

where C is the dissolved U concentration; D_x, the longitudinal dispersion coefficient; v, the seepage velocity; R, the retardation factor; and λ, an effective first-order decay term that can include radioactive decay and reactive loss. Analytical solutions (e.g., Ogata-Banks; Bear) and split-operator numerical schemes (i.e., PHREEQC) are standard tools for predicting plume evolution when equilibrium complexation and kinetic reactions are coupled; they show that increases in carbonate or calcium (enhancing complexation) decrease the K_d and the R, steepening breakthrough, whereas phosphate additions or transitions to reducing conditions increase R via precipitation/sorption.

Reductive attenuation to U(IV) can be microbially mediated; dissimilatory Fe(III)-reducing bacteria can directly reduce U(VI) to UO_2(s), decreasing its solubility. However, the presence of Ca-uranyl-carbonato complexes can inhibit enzymatic reduction, and Fe(III) (hydr)oxides can both compete as electron acceptors and modulate calcium via adsorption, indirectly re-enabling reduction. Hence, complexation not only determines equilibrium solubility but also regulates the kinetics of bioreduction in sediments [29, 41].

These speciation-complexation controls also rationalize observations in pit lakes and mine aquifers: stratified, anoxic hypolimnia favour reduction and sedimentary sequestration, whereas oxic epilimnia and mixing events favour mobilization via carbonate complexation; elevated total dissolved solids and conductivity with depth reflect the accumulation of complexed and conservative ions in lower layers. The overall attenuation trajectory, therefore, depends on the coevolution of pH, Eh, carbonate alkalinity, calcium, phosphate supply, DOM, and hydrologic exchange with groundwater [27].

While autunite group phases and sorption can stabilize uranium for long periods, the system remains sensitive to perturbations in ligand availability, redox conditions, and hydrology. Case studies and thermodynamic analyses emphasize that equilibrium projections must be validated with solid-phase characterization because multiple uranyl precipitates and surface complexes can coexist, and apparent aqueous speciation may not uniquely predict the solid associations without spectroscopic or diffraction evidence [28].

11.4.3.2 Interaction with Natural Organic Matter

The role of uranium complexation by organic matter in the natural attenuation of uranium at abandoned mines and mining sites involves three primary mechanisms: (a) immobilization of uranium from its labile form through binding to organic materials; (b) organic matter serving as a mediator for uranium, thereby influencing reactions, speciation, and the binding of uranium to inorganic materials; and (c) the complexation of uranium with natural organic matter, which facilitates the dispersion and dilution of uranium from a site. These mechanisms may operate simultaneously or independently, depending on the specific environmental conditions present at an abandoned uranium mine.

Organic Matter as the Sink of Labile Uranium

Natural organic matter (NOM) includes humic/fulvic acids, biomacromolecules, and peat-derived organics. NOMs attenuate uranium mobility through (i) strong sorption/complexation of U(VI) (as uranyl, UO_2^{2+}), (ii) redox-driven transformation to poorly soluble U(IV), and (iii) physical sequestration within organic matrices. Uranyl binds inner-sphere to deprotonated carboxylate and phenolic groups of humic/fulvic substances, often as bidentate/multidentate complexes; computational and experimental studies consistently identify these motifs and their pH dependence. When NOM is associated with Fe(III) oxyhydroxides, uranyl partitioning reflects competition between mineral inner-sphere sorption and uranyl-NOM complexation, with NOM enhancing uptake at acidic pH and modulating speciation at circumneutral pH. Under carbonate-rich conditions, aqueous uranyl-carbonate species dominate and ternary Ca/Mg-uranyl-carbonate complexes form, which can inhibit sorption and favour mobility in oxic aquifers [42–45].

NOM also drives reductive immobilization. Dissimilatory Fe(III)-reducing bacteria conserve energy by coupling electron transfer to U(VI), producing U(IV) and thereby immobilizing uranium; humic substances act as electron shuttles that accelerate such microbial reductions. The resulting U(IV) occurs as nanocrystalline uraninite (UO_2) and, frequently, as noncrystalline/monomeric U(IV) bound to organic ligands or adsorbed to organic matter and organic-coated clays, forms that can dominate U(IV) speciation in anoxic sediments. These OM-associated U(IV) species are generally sparingly soluble, yet more labile than crystalline UO_2 in the presence of complexing ligands [29, 46, 47].

Photochemical pathways complement microbial reduction in illuminated waters. Chromophoric NOM undergoes photochemistry that generates reactive intermediates; recent work shows fulvic acids can serve as electron donors under light, precipitating U as UO_2/U(IV) phases even in oxic systems [48].

Co-precipitation and co-association further stabilize uranium. In sulfate-reducing environments, biogenic/abiotic FeS phases can abiotically reduce U(VI) to U(IV); however, FeS oxidation can subsequently accelerate the reoxidation of noncrystalline U(IV), highlighting the dynamic stability of U (IV) under fluctuating redox conditions. Phosphate availability within NOM-rich matrices promotes precipitation of low-solubility uranyl phosphates (autunite/meta-autunite group) and biologically induced U-phosphate biominerals, driving aqueous U towards sub-μM concentrations [49–52].

Physical sequestration lends a kinetic barrier: diffusion into nanoporous organic domains, association with flocs/biofilms, and incorporation into peat and detrital networks retain uranium and slow remobilization. Uranium commonly accumulates in peatlands through strong organic binding, with notable resistance to exchange and remobilization only by acid or carbonate leaching. Natural analogue studies of organic-rich bogs similarly show humic ligands can outcompete inorganic complexes for U, especially near pH ~6, reinforcing NOM's role in field settings [53, 54].

Overall efficacy depends on pH, alkalinity/carbonate, ionic strength, redox potential, ligand competition, and cation availability. At circumneutral pH and high carbonate, mobile uranyl-carbonate and Ca/Mg-bridged species predominate; conversely, lower carbonate activity, reactive NOM, and sustained anoxia favour NOM complexation and U(VI) → U(IV) conversion with long-term burial [55, 56].

Organic Mediation of Uranium Behaviour

In abandoned uranium mines, the speciation, transport, and long-term attenuation of U are tightly governed by interactions with natural organic matter (NOM) across dissolved, colloidal, and solid phases. At low dissolved organic matter (DOM) concentrations, uranyl ion forms stable aqueous complexes with humic/fulvic ligands that compete with mineral surfaces for U, suppressing adsorption and increasing breakthrough and lateral dispersion in aquifers and pit-lake margins; this effect is amplified where carbonate alkalinity and Ca/Mg are present, because mobile anionic U(VI)-carbonate and ternary M-U(VI)-CO_3 complexes dominate aqueous speciation and inhibit surface uptake. As a result, even small DOM inputs can measurably reduce retardation in sands and oxides adjacent to mine wastes, thereby slowing natural attenuation in oxic flow paths [43, 44, 56].

Where organic matter is present as particulate or solid phases, peat lenses, wood fragments, biofilms, or NOM coatings on clays and Fe oxides, U mobility can be reduced by specific inner-sphere complexation to deprotonated carboxylate/phenolic sites and by redox transformation. Under anoxic to suboxic microenvironments maintained by reactive carbon, indigenous metal-reducing bacteria couple electron transfer to U(VI) reduction, yielding sparingly soluble U(IV) sequestered as nanocrystalline uraninite and, frequently, as non-crystalline, monomeric U(IV) coordinated to organic ligands or adsorbed to organic-coated minerals. These OM-associated U(IV) pools are pervasive in organic-rich, anoxic mine-impacted sediments and materially contribute to attenuation over decades.

The stability of organically mediated U(IV) is dynamic under the redox and hydrologic fluctuations common to abandoned workings. Sulfidic niches (e.g., where sulfate reduction occurs) can generate FeS phases that abiotically reduce U(VI), enhancing immobilization; yet, upon oxygen ingress or transient oxidant pulses, FeS oxidation and reactive oxygen/sulfur species accelerate re-oxidation of labile, noncrystalline U(IV), remobilizing U to solution. Likewise, strong organic ligands (biogenic or introduced) can complex and mobilize OM-bound U(IV) far more readily than crystalline UO_2, creating a vulnerability of "attenuated" uranium to episodic leaching. These behaviours imply that organic sinks are effective but not invariably permanent; their persistence depends on sustained reducing conditions, limited oxidant delivery, and a ligand milieu that disfavours complex-promoted dissolution.

In surface-exposed mine waters, chromophoric NOM photochemistry adds nuance. Light absorption by fulvic/humic chromophores generates reactive intermediates that can reduce U(VI) and precipitate U(IV) phases within the photic zone, with subsequent burial enhancing long-term retention; simultaneously, photobleaching alters NOM composition and reactivity, potentially modifying downstream complexation and transport. The net effect of photo-processing on attenuation hinges on water clarity, irradiance, and residence time in pit lakes or flooded adits.

Collectively, organic mediation produces a "double-edged" control on attenuation: low-level dissolved NOM and carbonate promote U mobility along oxic pathways by sustaining dissolved complexes and suppressing sorption, whereas abundant solid OM in anoxic compartments fosters strong complexation and bioreduction that immobilize U as U(IV) or U-phosphate phases. Effective natural attenuation in abandoned mines therefore emerges where hydrologic compartmentalization maintains reducing, organic-rich zones with limited oxidant and ligand influx, while minimizing DOM-aided advective export in carbonate-bearing, oxic strata. Site evaluations should consequently resolve NOM abundance and phase (dissolved vs. solid), aromaticity and functionality, carbonate chemistry (including Ca/Mg), and redox dynamics to forecast whether organic processes will hinder or promote the long-term natural attenuation of uranium [43, 44, 56].

Organic Matter as Moieties of Uranium Dispersion

Organic carbon released from tailings and waste rock, principally as natural organic matter (NOM), humic and fulvic acids, and colloidal organic fragments, forms reactive moieties that complex, coat, and mobilize uranium. These organic-associated uranium phases can increase subsurface transport risk by producing small, high-surface-area particles and composite colloids that remain mobile across a wide range of hydrochemical conditions in abandoned-mine settings (i.e., tailing ponds, wetlands, and seepage zones) [57].

Natural organic matter in mine-impacted environments is a heterogeneous, polyfunctional assemblage of molecules spanning a continuum of molecular weight, polarity, and conformation; degradation of plant and microbial detritus yields organic colloids and supramolecular humic/fulvic fractions that present multiple ligand sites (carboxyl, phenolic, carbonyl, and aliphatic domains) capable of forming inner- and outer-sphere complexes with uranium in both the hexavalent and tetravalent states. In abandoned-mine aquatic systems, these organic phases act not only as dissolved complexing ligands but also as structural components of mobile colloids and composite particles (organic-coated mineral colloids and organo-metal oxide aggregates) that can transport uranium away from source zones more effectively than truly dissolved species [58, 59].

The occurrence and formation of U-bearing colloids within organic matrices have been observed across mining-impacted wetlands and tailings environments; nanoparticulate U(IV) and U(VI) phases can nucleate within or adsorb to humic aggregates, producing eigencolloids or pseudo-colloids whose small size (1–1000 nm), large specific surface area, and colloidal stability confer high mobility through porous media and surface waters [60]. Importantly, reductive transformation of U(VI) to U(IV) does not necessarily immobilize uranium: nanoparticulate U(IV) can remain colloidally dispersed and thus transportable, particularly when stabilized by organic coatings or when associated with other mobile colloids, such as clays or iron oxyhydroxides.

The influence of NOM on colloid stability and coagulation is strongly conditional on ionic strength, pH, and NOM composition. At low ionic strength, anionic polyelectrolytic character of humic substances alters net surface charge and inhibits coagulation, whereas at higher ionic strength the conformational extension of adsorbed macromolecules increases electrosteric repulsion and can further stabilize dispersions; these mechanisms explain why NOM-rich tailing ponds and wetlands are persistent sources of stable, transportable colloids [61, 62]. The hydrophobicity and alkyl carbon content of humic fractions correlate positively with their ability to cotransport U(VI) through water-saturated porous media, indicating that nonpolar domains in NOM facilitate strong association with uranium and enhance colloid-mediated mobility [63].

Interactions among different colloid types further modulate uranium fate: humic-metal oxide associations can either inhibit U sorption to mineral colloids or promote U desorption, depending on the oxide identity and binding mode, thereby producing complex, system-specific outcomes for U retention and release in heterogeneous suspensions in natural waters. Consequently, predicting natural attenuation in abandoned mines requires quantifying the concentrations of colloid-associated organic carbon, the fraction of OC on colloids, and the kinetics of U sorption/desorption under variable hydrological and geochemical perturbations, because these parameters control whether organic matter acts as a sink or a vector for uranium transport.

Nexus of Uranium, Organic Matter, and Inorganic Ions

A dynamic triad of processes governs the natural attenuation of uranium in abandoned mines. The processes include complexation with inorganic ligands (notably carbonate and phosphate), sorption to Fe-oxyhydroxides and clays, and association with natural organic matter (NOM), whose competitive and cooperative interactions control whether uranium is immobilized or exported as dissolved species or mobile colloids (i.e., processes and thresholds are strongly pH-, Eh-, and composition-dependent).

The aqueous speciation and solid-solution partitioning of uranium in mine-impacted systems are controlled first-order by pH and redox, which set the dominant oxidation state and ligand affinities: above approximately pH 6, uranyl-carbonate complexes become thermodynamically favoured and markedly increase U solubility, whereas reducing conditions that produce U(IV) generally lower solubility but do not guarantee permanent immobilization if U(IV) remains non-crystalline or colloid-associated. The pH dependence of U sorption to soils and minerals is therefore a primary control on mobility in abandoned-mine settings [64].

Carbonate chemistry exerts a powerful competitive influence on uranyl binding in multi-component waters. At circumneutral to alkaline pH, CO_3^{2-} forms highly stable uranyl-carbonate species that can displace organic ligands from pre-existing uranyl-humate associations and form binary or ternary carbonate-humate complexes; these ternary species (e.g., uranyl dicarbonato-humate) have been identified as potentially critical soluble carriers of U under carbonate-rich conditions and can dominate speciation when carbonate concentrations are sufficient [65]. The net effect is that carbonate availability can convert an organically complexed, potentially less mobile uranyl pool into highly soluble carbonate complexes, increasing downstream transport potential [65].

Interactions between iron minerals and Fe redox cycling add additional complexity, as NOM both competes with and facilitates U sorption to Fe-oxides. NOM can coat Fe-oxide surfaces, altering surface charge and sorption site availability, and can either enhance U uptake at low pH by providing additional binding sites or inhibit sorption at higher pH by stabilizing U in solution or on mobile organic colloids. Experimental and field studies show that the presence and concentration of humic/fulvic substances shift the pH window and kinetics of U sorption to haematite, goethite, and ferrihydrite, so that the same mineral assemblage may immobilize or release U depending on NOM composition and abundance [65, 66].

Reductive processes and mineral-organic aggregates further decouple oxidation state from immobility: biogenic or abiotic reduction often yields non-crystalline, monomeric U(IV), or U(IV) associated with amorphous Al-P-Fe-Si and Fe-OM aggregates, and these forms can be labile and transferred to mobile porewater colloids (Fe- and OM-rich), producing U(IV)-bearing colloids that are transportable into surface waters, thereby undermining the assumption that reduction necessarily sequesters uranium as stable uraninite [67].

Thus, multi-element competition and complexation (e.g., with V, P, As, Cu, Pb, and Zn) modulate site occupancy and sorption capacity: vanadium and other redox-active metals bind strongly to carboxyl and phosphoryl groups in NOM. They can be reduced and complexed in parallel with U, altering ligand availability and competitive sorption on mineral surfaces. These multi-component interactions mean that predicting natural attenuation requires coupled consideration of pH, Eh, carbonate and phosphate concentrations, NOM composition (hydrophobic vs hydrophilic fractions), Fe mineralogy, and competing metal loads, because small shifts in any of these variables can flip the system from net immobilization to enhanced colloid-facilitated transport [66–68] (Fig. 11.5).

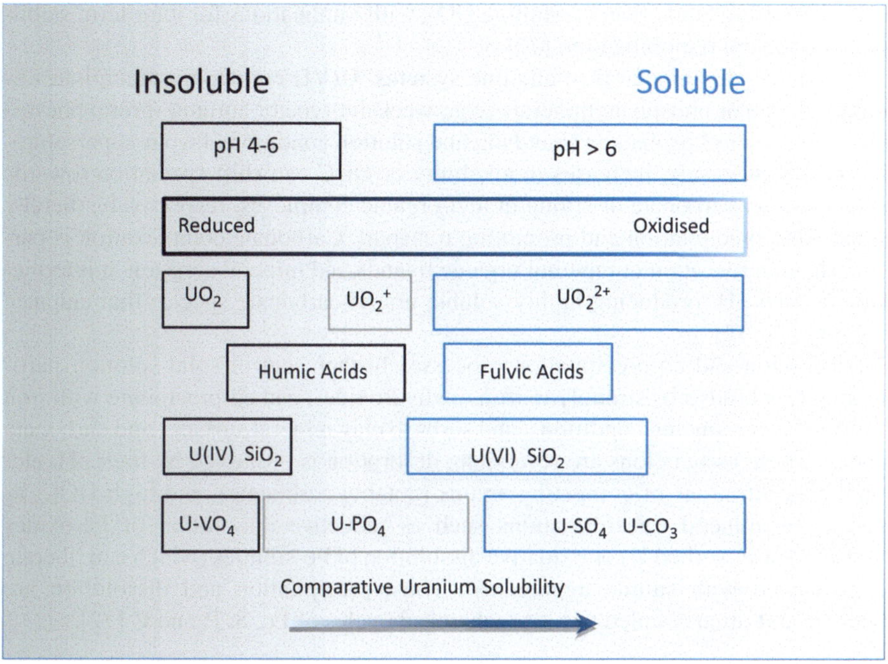

Fig. 11.5 The conceptualization of the factors affecting U solubility. (Source: Cumberland et al. [69])

11.4.3.3 Precipitation and Dissolution

Precipitation and dissolution of uranium in abandoned-mine settings are governed by coupled redox, pH, and ligand controls that determine whether uranium is sequestered as low-solubility solid phases (e.g., uraninite, U(IV)-silicates, U(VI) phosphates/ silicates, or co-precipitates with Fe-minerals) or remobilized by oxidation, carbonate complexation, high pH, or organic-ligand complexation; robust site characterization therefore requires paired aqueous speciation modelling, solid-phase oxidation-state determination, sequential extraction, and laboratory uptake tests to resolve the dominant attenuation versus mobilization pathways.

Precipitation of uranium in subsurface environments is a function of oxidation state, solution chemistry (pH, carbonate, phosphate, silicate, alkalinity, and dissolved organic carbon (DOC)), and the presence of reactive mineral surfaces (Fe-oxides, clays, carbonates, and sulfides). Under reducing conditions, uranium is commonly converted from soluble uranyl (U(VI), UO_2^{2+}) to tetravalent forms (U(IV)), and precipitation as uraninite (UO_2) or U(IV) silicate/organic-bound phases can remove U from solution. However, reduction does not universally produce crystalline uraninite; non-crystalline or monomeric U(IV) species and U(IV) associated with organic- or Fe-rich amorphous aggregates are frequently formed and can be more labile than crystalline UO_2, with implications for long-term stability and potential remobilization [70].

In oxic or circumneutral to alkaline systems, U(VI) commonly precipitates as uranyl silicate or phosphate minerals (e.g., weeksite-type or autunite-group phases) when silicate or phosphate is abundant, and solution conditions favour supersaturation [71]. Conversely, increases in alkalinity or DOC can shift speciation towards soluble uranyl-carbonate or strong uranyl-organic complexes, respectively, thereby suppressing precipitation and promoting transport. Carbonate complexation is particularly effective at outcompeting organic ligands and mineral sorption at intermediate to high pH, producing highly soluble uranyl-carbonate species that enhance mobility [71, 72].

Adsorption and co-precipitation processes further control solid-solution partitioning: U(VI) adsorbs strongly to iron oxyhydroxides, can co-precipitate with iron sulfides under reducing conditions, and sorbs to clay mineral surfaces and carbonate phases. These associations are reversible: desorption is promoted by high pH, elevated concentrations of competing anions (notably carbonate), and high DOC, as well as by mineral transformations such as reductive dissolution of Fe-oxides (which releases sorbed U) or oxidative dissolution of Fe-sulfides (which can liberate U associated with sulfide aggregates). Thus, precipitation and dissolution are dynamic and often coupled to biogeochemical cycles of Fe, S, P, and C [72].

11.4.3.4 Adsorption

Adsorption of uranium in near-surface and groundwater systems is dominated by inner-sphere complexation of the uranyl cation to oxygen-donor surface ligands. This process produces strong, often partially irreversible retention relative to simple ion exchange. Detailed laboratory studies on hydrous ferric oxides demonstrate that uranyl forms mononuclear, bidentate inner-sphere complexes at distinct strong and weak Fe-OH surface sites, and that a two-site surface-complexation model can reproduce adsorption over wide ranges of pH, carbonate partial pressure, and uranium loading. Adsorption densities and site occupancies control the transition from surface complexation to ternary carbonate surface species at high pH [73].

Mechanistic adsorption experiments on crystalline iron oxides and oxyhydroxides (goethite and haematite) show similar inner-sphere coordination geometries and reveal the importance of surface hydroxyl speciation and site density in controlling U(VI) uptake; these studies established the site-binding framework used to interpret uranyl sorption to Fe-oxides and to parameterize surface complexation models that predict pH- and ionic-strength-dependent behaviour in natural waters [74]. Layered aluminosilicates (e.g., smectites, vermiculite, and muscovite) provide both edge (i.e., aluminol/silanol) and basal interlayer sites where uranyl binds: edge sites typically support stronger, faster inner-sphere coordination, while basal/interlayer uptake is slower and diffusion-limited, producing the biphasic kinetic signatures commonly observed in clay sorption experiments and column tests [75].

Organic matter exerts multiple, composition-dependent controls on adsorption: humic and fulvic substances complex U(VI) via carboxyl and phenolic groups, coat mineral surfaces and alter surface charge, and can both enhance apparent uptake (by providing additional ligand sites and promoting surface nucleation) and inhibit mineral sorption (by stabilizing U as soluble organo-uranyl complexes or by forming mobile organo-mineral colloids). Quantitative treatments of U-humic interactions show variable coordination stoichiometries (1:1 to 1:2 U:HA) and competitive effects with carbonate and metal oxides that depend on pH and NOM composition [32, 34, 76].

Adsorption is frequently a precursor to longer-term sequestration by surface-mediated nucleation and incorporation: adsorbed uranyl can be structurally accommodated into iron oxide lattices or act as a nucleus for uranyl-phosphate or uranyl-silicate mineral growth (for example, metatorbernite), so that progressive incorporation, from reversible surface complexation to structural substitution and discrete uranyl mineral precipitation, links adsorption kinetics to secondary mineralization and long-term immobilization in natural systems [34, 77–79].

11.4.3.5 Mineralisation

Mineralization of uranium in abandoned sandstone systems proceeds by coupled redox and surface-complexation reactions in which soluble uranyl (UO_2^{2+}) is converted to sparingly soluble U(IV) phases (principally uraninite, UO_2), with Fe(II)-bearing minerals (including biogenic magnetite) acting as dominant abiotic reductants and sinks; carbonate speciation and surface ternary complexes strongly modulate rates and partitioning.

In oxidized porewaters, uranium is transported as uranyl, UO_2^{2+}, often complexed by carbonate to form species, such as $UO_2(CO_3)_2^{2-}$ and $UO_2(CO_3)_3^{4-}$. Under reducing microenvironments created by organic matter, sulfides, hydrocarbons, or Fe(III) reduction, U(VI) is reduced to U(IV) and precipitates as low-solubility phases (e.g., UO_2 or U(IV)-bearing mixed phases). Experimental and nanoscale spectroscopic work shows that magnetite (Fe_3O_4), including biogenic magnetite, can reduce surface-associated U(VI) to U(IV) and nucleate nanocrystalline UO_2, making magnetite a potent immobilizing phase in natural attenuation systems [80].

Reduction can be expressed as half-reactions and coupled redox balances. The uranyl reduction half-reaction (aqueous to solid UO_2) can be written conceptually as:

$$UO_2^{2+} + 2e^- \rightarrow UO_2(s)$$

Fe(II) oxidation supplying electrons is represented by:

$$2Fe^{2+} \rightarrow 2Fe^{3+} + 2e^-$$

Combining these gives a simplified abiotic Fe(II)-mediated reduction:

$$UO_2^{2+} + 2Fe^{2+} \rightarrow UO_2(s) + 2Fe^{3+}$$

In natural porewaters, the oxidized Fe(III) product commonly reprecipitates as Fe(III) oxides or is incorporated into magnetite; biogenic magnetite formation can itself be written in simplified form from Fe(II) and Fe(III) pools:

$$Fe^{2+} + 2Fe^{3+} + 4OH^{-} \rightarrow Fe_3O_4 + 2H_2O$$

Nanoscale studies demonstrate that surface-associated U(VI) is first converted to U(V) intermediates on magnetite surfaces and subsequently to U(IV) as UO_2 nanocrystals, consistent with stepwise surface reduction and internal nucleation of UO_2. These mechanistic observations are supported by experiments showing abiotic U(VI) removal by biogenic magnetite and vivianite under iron-reducing conditions [81].

Microbial cell walls and extracellular polymers provide high-affinity sorption sites (carboxylate and phosphoryl groups) that concentrate U(VI) and can facilitate enzymatic reduction when cells respire on uranium as an electron acceptor. In *Shewanella* and related bacteria, strong U(VI) association with cell surfaces in low-carbonate media enhances biotic reduction potential [82].

Carbonate ligands strongly stabilize aqueous uranyl complexes (e.g., $UO_2(CO_3)_3^{4-}$), decreasing sorption to mineral and cell surfaces and thereby reducing enzymatic and surface-catalysed reduction rates; however, ternary surface complexes (i.e., uranyl-carbonate-oxide) can retain a fraction of U(VI) on iron oxide surfaces and permit continued, slower surface-catalysed reduction [73, 81]. Surface-catalysed reduction models show that adsorption geometry and surface complexation control the kinetics of Fe(II)-mediated U(VI) reduction, so that mineralogy, Fe(II) adsorption behaviour, and carbonate speciation together determine whether enzymatic or abiotic pathways dominate [83].

In sandstone-hosted abandoned mines, the dominant long-term immobilization pathway is often abiotic reduction by Fe(II) and sequestration within secondary Fe(II/III) minerals (notably magnetite), with microbial activity driving Fe(III) reduction that produces the reactive Fe(II); carbonate chemistry and surface complexation modulate the rates and stability of the immobilized uranium, and co-precipitation or structural incorporation into magnetite and UO_2 nanocrystals yields the most persistent mineralized sinks.

11.4.4 Biological Processes

Biological processes underpin natural attenuation at abandoned uranium mines by transforming uranium speciation and transport through tightly coupled redox, sorption, and biomineralization pathways. Metal-reducing bacteria (e.g., *Geobacter* spp.) enzymatically reduce soluble U(VI) to sparingly soluble U(IV) (uraninite), often via outer-surface c-type cytochromes, thereby decreasing aqueous mobility under anoxic conditions. Phosphate-cycling microbes also induce U(VI)-phosphate biomineralization (meta-autunite and related phases) via phosphatase activity or phosphate solubilization, thereby immobilizing uranium even at low micromolar concentrations. Iron oxyhydroxides formed in post-mining environments act as major sinks via adsorption and mixed-valence incorporation; their reactivity and impurity associations modulate uranium retention and redox cycling. Sulfate-reducing consortia sustain reducing conditions and precipitate sulfides that indirectly stabilize U(IV), while exhibiting tolerance at tailings sites [84–87].

The effect of the biological process can be categorized into (a) biologically induced physical and chemical changes. The biologically induced physical category includes biofilms and extracellular polymers that drive bioclogging, reducing hydraulic conductivity, retarding advective flux, and enhancing residence times for reactive transport, thus facilitating contact between radionuclides, reactive minerals, and microbial cells. The biologically induced geochemical changes include microbial redox metabolism, phosphate release, and Fe cycling control uranium sorption, reduction, and mineral incorporation at solid-water interfaces, collectively attenuating dissolved uranium plumes [88].

11.4.4.1 Biologically Induced Physical Changes

Microbial growth in abandoned uranium mine aquifers and tailings imparts profound physical modifications to porous media, thereby enhancing natural attenuation by reducing advective transport, increasing residence time, and intensifying solute-surface interactions. The dominant mechanism is bioclogging: the accumulation of sessile biomass, extracellular polymeric substances (EPS), and biogenic precipitates at grain surfaces and pore throats, which progressively lowers effective porosity (ϕ) and hydraulic conductivity (K). At the continuum scale, flow reduction follows Darcy's law, $q = -K \nabla hq = -K \nabla hq = -K \nabla h$, so any biologically driven decrement in K directly attenuates volumetric flux q. Laboratory columns, microfluidic pore-scale experiments, and field-relevant models consistently show order-of-magnitude declines in K as biofilms mature, with the spatial pattern governed by inlet concentration, temperature, and flow regime [87, 89, 90].

The relationship between biomass accumulation and permeability can be formalized with Kozeny-Carman scaling. For a granular medium of specific surface area S_v and tortuosity τ, the intrinsic permeability k scales as

$$k \approx \frac{\phi^3}{S_v^2 \tau^2 (1-\phi)^2}$$

EPS production and cellular attachment effectively reduce the interconnected pore volume $\phi \rightarrow \phi - \Delta\phi_b$ and increase τ by creating labyrinthine flow paths; inserting these changes yields

$$\frac{k_b}{k_0} \approx \frac{\left(\phi - \Delta\phi_b\right)^3}{\phi^3} \frac{\tau_0^2}{\tau_b^2} \left(\frac{1-\phi}{1-\phi+\Delta\phi_b} \right)^2$$

where k_b and k_0 are permeabilities with and without biofilm, respectively. Even modest $\Delta\phi_b$ produces significant k reductions because of the cubic dependence on porosity. Pore-scale micromodels derived from μCT of reservoir sandstones confirm that biomass straining in narrow throats dominates the early decline, diverging flow and creating low-velocity eddies that expand the diffusion boundary layer and reduce local shear stresses, which in turn stabilize biofilm matrices against detachment [89, 91].

Bioclogging evolves in phases coupled to microbial growth kinetics. Percolation experiments demonstrate an inverse logistic attenuation of $K(t)$,

$$K(t) = K_\infty + \frac{K_0 - K_\infty}{1 + \exp\left[\alpha\left(t - t_{1/2}\right)\right]}$$

where K_0 is the initial conductivity, K_∞ the asymptote under mature biofilms, α a temperature-dependent rate constant, and $t_{1/2}$ the inflection time, which approximates the transition from exponential to stationary growth. Within 10–35 °C, the temperature dependence is non-monotonic: lower temperatures slow biomass production but favour denser EPS networks that can reduce K by >80–95%; at higher temperatures, rapid growth initially increases clogging, but shear-induced sloughing, channelization, and gas production can intermittently restore permeability. These dynamics are echoed in streambed hyporheic zones where layered biofilms impose vertical stratification, reducing hyporheic exchange flux by ~45%, elongating solute penetration times, and promoting anoxic microenvironments conducive to U(VI) reduction and sulfide precipitation [89, 91].

Physicochemical adhesion forces complement the mechanical consequences of EPS. Biofilm matrices, enriched in polysaccharides and proteins, modify zeta potential and surface roughness, amplifying deposition of suspended fines and colloids. This synergistic clogging by biofilms and particulates intensifies the decline in K relative to abiotic deposition alone, with persistent hydraulic barriers maintained by EPS bridging and multivalent cation cross-linking. Microbial gas production (CO_2 and H_2S) may further occlude pores via bubble entrapment, decreasing local saturation and augmenting relative permeability contrasts; microfluidic visualization under anaerobic conditions shows bubble-mediated diversion of flow that protects downstream biofilm clusters from shear, locking in low-permeability zones.

In fractured rock and coarse silts/fine sands typical of mine workings, bioclogging develops at characteristic radial distances from injection or recharge, governed by the interplay of flow velocity, pore-size distribution, and attachment kinetics. Under high near-well velocities, attachment is limited, and clogging is displaced outward, producing an annulus of reduced K that acts as a biogenic seal and reactive barrier. Macro-scale radial flow experiments and micromechanical analyses indicate two to three orders of magnitude K reduction in the clogging zone, which dramatically increases residence time and enhances sorption and redox transformations on mineral surfaces.

These biologically induced physical changes have direct implications for uranium fate. Lowered advective flux increases the Damköhler number $Da = k_{rxn}L/u$ by reducing the characteristic velocity uuu, shifting transport towards reaction-limited regimes where microbial reduction and phosphate biomineralization can outpace plume migration. The thickened diffusive boundary layer around biofilm-coated grains increases interfacial contact time and promotes solid-water interfacial processes, including U(VI) sorption to iron oxyhydroxides and mixed-valence incorporation. Concurrently, stratified bioclogging fosters redox heterogeneity: aerobic veneers at flow-exposed surfaces overlay anoxic interiors, establishing microzones that support sequential electron-acceptor use (Fe(III) → SO_4^{2-} → U(VI)) and stabilize U(IV) precipitates within protected niches. The emergent physical architecture, which narrowed channels, dead-end pores, and low-permeability layers, therefore acts as both a hydraulic attenuator and a biogeochemical reactor, coupling transport limitations to in situ uranium immobilization.

11.4.4.2 Biologically Induced Chemical Changes

Microbial activity in abandoned uranium mine settings alters aqueous speciation, interfacial chemistry, and solid-phase transformations, thereby regulating uranium mobility through coupled redox, sorption, and mineralization reactions. Under reducing conditions established by heterotrophic respiration, dissimilatory metal-reducing bacteria (notably *Geobacter* spp.) catalyse the reduction of hexavalent uranyl to sparingly soluble U(IV). At circumneutral pH, the net half-reaction is

$$UO_2^{2+} + 2e^- \rightarrow UO_2(s)$$

often mediated by outer-surface c-type cytochromes that deliver electrons to adsorbed U(VI) complexes at cell envelopes or mineral surfaces. Spectroscopic and genetic knock-out studies demonstrate that pili conductivity is not required. In contrast, deletion of key outer-surface cytochromes suppresses U(VI) reduction, confirming a direct enzymatic pathway to U(IV) nucleation at the cell-solution interface. When U(VI) is adsorbed to Fe(III) oxides, enzymatic electron transfer proceeds at the solid-water interface, and quinone shuttles can enhance the effective rate by bypassing kinetic barriers to direct contact. Field-mimicking incubations further indicate that Fe(III) availability controls *Geobacter* persistence during attenuation, underscoring the intimate link between iron and uranium redox cycling [84, 88, 92].

Iron cycling exerts a dual chemical control by both scavenging and redox buffering. Freshly precipitated iron oxyhydroxides (e.g., ferrihydrite, lepidocrocite, and goethite) provide high-affinity sorption sites for uranyl-carbonate and uranyl hydroxo species. At the same time, mixed-valence phases such as magnetite can incorporate reduced uranium in near-surface and bulk lattice sites. Co-precipitation experiments show partial reduction of U(VI) to U(V) and its stabilization via octahedral substitution during magnetite/green-rust formation, sometimes coexisting with surface-associated UO_2 when the Fe(II)/Fe(III) ratio is high. In post-mining technosols and drainage networks, these iron phases act as dynamic "uranium capacitors", repeatedly adsorbing U(VI) during oxic pulses and promoting reduction or incorporation under anoxia. Natural iron oxyhydroxides, however, can exhibit attenuated reactivity relative to synthetic analogues due to structural impurities (e.g., Si and organic coatings), limiting Fe(II)-promoted transformations and uranyl reduction; this compositional complexity must be accounted for in site-specific attenuation models [79, 93, 94].

Sulfate-reducing bacteria (SRB) reshape porewater chemistry by generating sulfide and alkalinity:

$$SO_4^{2-} + 2CH_2O \rightarrow H_2S + 2HCO_3^-$$

The produced HS^- reacts with Fe^{2+} to precipitate FeS and ultimately pyrite, consuming electron acceptors and maintaining a low redox potential; such conditions thermodynamically favour U(VI) reduction and the immobilization of trace co-contaminants. Community surveys at uranium tailings sites document SRB diversity and tolerance, with *Desulfotomaculum* dominating at high-U locales, which is consistent with a functional role in in situ attenuation. Mixed SRB-phosphate-solubilizing consortia can simultaneously drive reductive stabilization and enhance phosphate availability, shifting uranium partitioning towards low-solubility solids [85, 86, 95].

A complementary immobilization pathway is microbially induced phosphate biomineralization. Periplasmic and surface-bound alkaline phosphatases hydrolyse organophosphate substrates, elevating dissolved P_i locally and inducing precipitation of autunite-group U(VI)-phosphates at or near the cell surface:

$$Ca^{2+} + UO_2^{2+} + 2PO_4^{3-} + nH_2O \rightarrow Ca\left(UO_2\right)_2\left(PO_4\right)_2 \cdot nH_2O(s)$$

In Caulobacter spp., the PhoY phosphatase drives the formation of meta-autunite-like precipitates that confer U(VI) resistance and immobilization; crucially, bacteria create supersaturated microenvironments where abiotic precipitation would be thermodynamically inhibited at bulk low-μM U(VI) concentrations. Phosphate-solubilizing bacteria can also mobilize orthophosphate from mineral phases or organophosphate amendments, fostering rapid U-phosphate precipitation and EPS-assisted sequestration [85, 86, 93, 95, 96].

These biologically induced chemical changes operate within a framework of uranium aqueous speciation that strongly modulates kinetics. At circumneutral to alkaline pH with millimolar bicarbonate, U(VI) occurs predominantly as $UO_2\left(CO_3\right)_2^{2-}$ and $UO_2\left(CO_3\right)_3^{4-}$, which are less bioavailable for direct enzymatic reduction. The reduction potential for the U(VI)/U(IV) couple shifts with complexation; an Eh-pH representation using the Nernst relation:

$$E = E^{\circ} - \frac{RT}{nF}\ln\left(\frac{a_{\mathrm{UO}_2^{2+}}}{a_{\mathrm{UO}_2(\mathrm{s})}}\right)$$

indicates that increasing carbonate activity raises the potential required for reduction, slowing rates unless electron delivery is enhanced (e.g., via outer-membrane cytochromes or electron shuttles). Nonetheless, uranium bioreduction can proceed under elevated pH when Fe(III) oxides provide sorptive pre-concentration and reactive Fe(II) phases develop, yielding mixed U(IV)/non-uraninite products that are partially re-oxidizable under O_2 or nitrate influx [88, 97].

Interfacial sorption is a pervasive, rate-modulating chemical change. Surface complexation onto iron oxyhydroxides can be expressed through reactions such as

$$\equiv \mathrm{FeOH} + \mathrm{UO}_2^{2+} \rightleftharpoons \equiv \mathrm{FeO} - \mathrm{UO}_2^{+} + \mathrm{H}^{+}$$

$$\equiv \mathrm{FeOH} + \mathrm{UO}_2\left(\mathrm{CO}_3\right)_2^{2-} \rightleftharpoons \equiv \mathrm{FeO} - \mathrm{UO}_2\mathrm{CO}_3^{-} + \mathrm{HCO}_3^{-}$$

with conditional stability constants sensitive to ionic strength and competing ligands. In natural post-mining matrices, sorption capacity and reversibility depend on crystallinity and impurity load; site investigations show uranium capture and reconcentration on field-formed iron oxyhydroxides across hydrological cycles. Because sorption concentrates uranyl near electron-donor interfaces, it can act as a precursor to enzymatic or Fe(II)-mediated reduction, effectively coupling adsorption and redox on mineral surfaces [94, 97].

Microbial redox transformations are often described by Monod-type kinetics coupled to transport:

$$r_{\mathrm{U(VI)}} = V_{\max}\frac{C_{\mathrm{U(VI)}}}{K_s + C_{\mathrm{U(VI)}}}\frac{C_D}{K_D + C_D}f\left(E_h, pH, \text{speciation}\right)$$

where C_D is the electron donor concentration.

Thus, biologically induced chemical changes interact, inter alia, with the physical alterations of biofilms and bioclogging. Thicker diffusive boundary layers and micro-anoxic niches within biofilms elevate the Damköhler number for interfacial reactions, biasing the system towards reaction-dominated regimes where reduction, sulfide production, and phosphate precipitation outcompete advective export. The emergent result is a self-reinforcing chemical milieu that has low E_h, elevated P_i and Fe(II), and abundant reactive surfaces that sustain uranium immobilization even under fluctuating recharge conditions characteristic of abandoned workings and tailings repositories [79, 87].

11.5 Perspectives

11.5.1 Constituents of Natural Attenuation

Natural attenuation encompasses the suite of biogeochemical and physical processes that reduce contaminant concentrations, fluxes, or toxicity without the addition of engineered reagents, relying instead on mixing/dilution, sorption, redox transformations, and in situ biomineralization. At abandoned uranium mine sites, natural attenuation is governed by the coupling of microbial metabolism with mineral and aqueous chemistry, which control U speciation, mobility, and long-term stability. In groundwater and tailings porewaters, dissolved U is typically present as U(VI) carbonate or hydroxo complexes whose reduction to sparingly soluble U(IV) or precipitation as U(VI)-phosphate or U(VI)-carbonate minerals can immobilize plumes if appropriate redox and geochemical conditions persist. Natural attenuation is often a multiphase, mixed-control phenomenon in which biofilm development, iron and sulfur cycling, and dynamic sorption to iron oxyhydroxides jointly determine plume behaviour over seasonal to decadal timescales [1–6].

11.5.2 Core Mechanisms and Site Controls

Under anoxic conditions, dissimilatory metal-reducing bacteria (e.g., Geobacter spp.) catalyse the enzymatic reduction of adsorbed or aqueous U(VI) to U(IV), with outer-surface c-type cytochromes implicated as the terminal reductases; this pathway nucleates UO_2 at or near the cell surface and can proceed for U(VI) associated with Fe(III) oxides via interfacial electron transfer or redox shuttles [84, 98]. The availability of reducible Fe(III) phases modulates the persistence of these communities and the extent of U reduction, revealing a tight linkage between Fe and U cycles during attenuation. In parallel, sulfate-reducing bacteria (SRB) generate sulfide and alkalinity, maintaining low redox potential and precipitating FeS/pyrite, which indirectly stabilizes U(IV) and scavenges trace metals; SRB populations with notable U tolerance have been documented at mill tailings sites [99, 100].

Where phosphate is available or microbially released, phosphatase-mediated biomineralization forms autunite-group U(VI)-phosphates on cell surfaces, a process that can proceed in microenvironments even when the bulk solution is undersaturated, hence relevant at the low-μM U levels near drinking-water limits [95, 101]. Iron oxyhydroxides (ferrihydrite, lepidocrocite, goethite, and magnetite) function as "uranium capacitors", alternating between adsorption during oxic episodes and reduction/incorporation under anoxia. Field-formed, impurity-bearing Fe phases exhibit lower reactivity than synthetic analogues, which has implications for attenuation timescales and reversibility.[10215]– [17] In flow terms, bioclogging by biofilms and fines lowers hydraulic conductivity and increases residence time, elevating Damköhler numbers so that interfacial reactions outcompete advection; column, microfluidic, and hyporheic-zone studies document order-of-magnitude reductions in K and layered redox microenvironments that favour sustained attenuation [93].

11.5.3 Opportunities and Challenges

The principal opportunity of natural attenuation at abandoned mines is leveraging site geochemistry and native consortia to achieve cost-effective, minimally invasive risk reduction with lower secondary waste than active pump-and-treat or excavation. Where carbonate alkalinity is modest, organic electron donors are naturally available, and reactive Fe/S cycles are active, NA can drive U(VI) → U(IV) and/or U-phosphate precipitation at rates sufficient to match plume fluxes [102–104]. Conversely, major challenges arise from (i) carbonate complexation that stabilizes U(VI) and shifts the U(VI)/U(IV) redox potential, slowing bioreduction unless sorption concentrates uranyl at reactive interfaces; (ii) reoxidation risks: biogenic U(IV) phases can be partially remobilized upon oxic or nitrate incursions; and (iii) heterogeneity of natural Fe oxyhydroxides, whose coatings and substitutions dampen reactivity compared with lab surrogates [71, 105]. Long-term performance also depends on hydrologic stability; high-permeability pathways or episodic flushing can outpace reaction kinetics and export U before it is immobilized.

11.5.4 Favourable Conditions for Natural Attenuation

Natural attenuation is a favourable remedy component when multiple lines of evidence indicate that intrinsic processes already reduce concentrations or fluxes and will continue to do so within an acceptable timeframe. Indicators include sustained reducing conditions (E_h below the U(VI)/U(IV) boundary adjusted for carbonate), reactive mineral inventories (amorphous Fe(III) content, potential for magnetite/green-rust formation, and sulfide-generating capacity), evidence of U speciation change to U(IV) or U-phosphates by spectroscopy, and declining concentration trends along flow paths aligned with geochemical gradients rather than dilution alone [106, 107]. Sites with moderate flow velocities, biofilm-supporting media (fine sands/silts), and limited external oxidant loading are particularly amenable because the physical and chemical feedbacks, bioclogging, stratified redox niches, and sorption-reduction coupling reinforce attenuation.

11.5.5 Unfavourable Conditions of Natural Attenuation

Natural attenuation alone is unlikely to meet objectives where carbonate alkalinity and nitrate loads are high, hydraulic residence times are short, or oxidant pulses (seasonal water table rise, maintenance dewatering, or river stage reversal) periodically re-oxidize reduced phases. In sandy aquifers with high K and limited biofilm development, or in tailings with limited reactive Fe and P availability, attenuation may stall at reversible adsorption rather than durable sequestration. Empirical studies document partial reoxidation of biogenic U(IV) under air or in the presence of nitrate, and reduced reactivity of field iron phases, signalling a potential rebound if the redox buffer is not maintained [3, 107–109]. In such settings, relying on NA without augmentation risks exceeding the timeline and failing to comply.

11.5.6 When and how to Implement Enhanced Natural Attenuation

ENA integrates targeted, minimal interventions to strengthen intrinsic processes while preserving the low-impact ethos of natural attenuation. Favourable scenarios for enhanced natural attenuation include: (i) marginal E_h where slight electron-donor scarcity limits U(VI) reduction; (ii) insufficient phosphate to drive U-phosphate biomineralization at low U; (iii) heterogeneous flow where short-circuiting bypasses reactive zones; and (iv) impure Fe(III) phases with low redox reactivity. Evidence-based ENA strategies include:

11.5.6.1 Biostimulation

Biostimulation with low-dose electron donors (e.g., acetate or slow-release organics) to promote Geobacter-mediated U(VI) reduction and maintain Fe/S redox cycling, with careful control to avoid methanogenic overdrive and secondary water-quality impacts. Field and sediment-microcosm studies show that acetate additions shift communities towards Geobacter, until Fe(III) is depleted, after which SRB dominate; Fe(III) augmentation can revive *Geobacter* activity [92].

Phosphate amendments tailored to microbially mediated precipitation, leveraging phosphatase activity or phosphate-solubilizing bacteria to form autunite-group minerals near cells and within biofilms, are effective even at sub-μM U in microenvironments where bulk precipitation is thermodynamically disfavoured [52, 110, 111].

Reactive iron management, including introducing Fe(II/III) mixtures that co-precipitate magnetite/green-rust capable of incorporating U(V) and promoting surface UO_2 formation, thereby creating mixed-valence buffers against reoxidation;

site characterization must account for competitive impurities that deactivate Fe phases [110, 111].

Hydro-biological conditioning, deliberately creating low-flow, high-contact zones via permeable biobarriers or designed bioclogging to increase residence time and foster stable anoxia and sorption; pore- and reach-scale studies quantify K reductions and hyporheic-exchange damping under biofilm growth [52, 92, 111, 112].

Nitrate and oxygen load management, e.g., carbon-balanced dosing or in situ denitrification zones to prevent U(IV) reoxidation during recharge events; alkaline systems require special attention since U(VI) remains strongly complexed and less bioavailable [52, 111].

11.5.6.2 Bioaugmentation

Bioaugmentation involves introducing or enriching functional microbial consortia capable of accelerating uranium immobilization through reductive transformation and biomineralization pathways. For instance, augmenting groundwater with exogenous reducing and phosphate-solubilizing consortia enhances microbial U(VI) reduction and promotes U-phosphate precipitation, significantly increasing removal rates relative to natural communities alone. Bioaugmentation strengthens ENA by overcoming limitations, such as insufficient native microbial abundance, metabolic inhibition by high contaminant loads, and poor electron-donor utilization. Mechanistically, added consortia accelerate substrate consumption, broaden metabolic diversity, and reduce cell-surface precipitation that would otherwise inhibit microbial activity, thereby sustaining functional populations under challenging geochemical conditions. Broader bioremediation research confirms that bioaugmentation improves contaminant degradation or immobilization when endogenous microorganisms lack necessary catabolic or redox capabilities, offering a technically viable supplement to natural attenuation under site-specific constraints. In abandoned uranium mine settings, bioaugmentation therefore enhances the robustness and kinetics of NA, enabling more predictable plume stabilization and long-term contaminant sequestration through microbially driven reduction, sorption, and biomineralization [113].

11.5.7 Monitoring and Verification

Because NA/ENA performance is pathway-dependent and reversible, robust monitoring programs must track redox indicators (Fe(II)/Fe(III), sulfate/sulfide, nitrate, dissolved oxygen), U speciation and isotopes, iron mineralogy, and hydraulic conductivity trends associated with biofilm development. U(IV) or U-phosphate and oxidative stability under plausible perturbations are essential to demonstrate durability [114, 115]. Complementary microbial community analyses help diagnose functional capacity shifts and guide adaptive management (e.g., donor tapering and Fe or P re-balancing) [116–118].

11.6 Concluding Remarks

In summary, natural attenuation in abandoned uranium mines can be a scientifically defensible and economically attractive remedy where geochemistry, microbiology, and hydrology converge to immobilize U in forms with demonstrable stability under the site's disturbance regime. Where one or more controls are marginal, enhanced natural attenuation, implemented with precision and monitored against mechanistic endpoints, offers a pragmatic route to risk reduction while limiting intervention intensity, secondary waste, and the ecological footprint.

References

1. Favas, P. J. C., Pratas, J., Paul, M. S., & Prasad, M. N. V. (2019). In V. C. Pandey & K. Bauddh (Eds.), *Phytomanagement of polluted sites* (pp. 277–300). Elsevier.
2. Jusselme, M. D., Brondeau, F., & Balland-Bolou-Bi, C. (2023). In A. Mustafa & M. Naveed (Eds.), *Soil contamination—Recent advances and future perspectives*. IntechOpen.
3. Ford, R. G., & Wilkin, R. T. (2010). *Vol. 3 147.* U.S. Environmental Protection Agency, National Risk Management Research Laboratory.
4. Rittmann, B. E. (2003). In E. E. Moyer & P. T. Kostecki (Eds.), *Handbook for managing releases of gasoline containing MTBE* (Vol. 2, pp. 329–345). Association for Environmental Health and Sciences.
5. Beitinger, E., Jungbauer, H., & Rochmes, M. (1999). *TerraTech Heft* (Vol. 5).
6. O'Steen, B. (1999). EPA Region 4 perspective on the OSWER monitored natural attenuation policy. *Journal of Soil Contamination, 8*, 17–22.
7. Buchanan, R. J., & Sehayek, L. (1999). Science outpaces the policy: EPA-monitored natural attenuation (MNA). *Journal of Soil Contamination, 8*, 35–38.
8. Costello, C. (2005). *Acid mine drainage: Innovative treatment technologies. 52.* U.S. Environmental Protection Agency.
9. Abdel-Naby, H. H. & Dawood, Y. *Natural attenuation of uranium and formation of autunite at the expense of apatite within an oxidizing environment, south Eastern Desert of Egypt*. (2009).
10. Dudel, G. E., Brackhage, C., Dienemann, H., Mkandawire, M., & Weiske, A. (2004). In A. P. Jarvis, B. A. Dudgeon, & P. L. Younger (Eds.), *Mine water—Process, policy and progress* (Vol. 2). University of Newcastle upon Tyne Press.
11. Dudel, E. G., et al.A. Taboas, R. Vanbrabant, & G. Benda (Eds.). *8th international conference radioactive waste management and environmental remediation 8*. American Society of Mechanical Engineers and US Department of Energy.
12. Matson, J. V., & Schuhmann, R. J. (1999). Natural attenuation as a remedy not as an excuse. *Journal of Soil Contamination, 8*, 29–33.
13. Golovich, E., Wellman, D., Serne, R., & Bovaird, C. (2011). *Summary of uranium solubility studies in concrete waste forms and Vadose Zone environments*. U.S. Department of Energy.

14. Kelepertsis, A. E. (1981). The geochemistry of uranium and thorium in some Lower Carboniferous sedimentary rocks (Great Britain). *Chemical Geology, 34*, 275–288. https://doi.org/10.1016/0009-2541(81)90117-0
15. Abdelouas, A., Lutze, W., & Nuttall, H. E. (1999). Uranium: Mineralogy, geochemistry and the environment. *Mineralogical Society of America, 433–474.*
16. Cui, X., Zhu, C., Hu, M., Wang, X., & Liu, H. (2019). The hydrodynamic dispersion characteristics of coral sands. *Journal of Marine Science and Engineering, 7*, 291.
17. Zech, A., et al. (2019). A critical analysis of transverse dispersivity field data. *Groundwater, 57*, 632–639. https://doi.org/10.1111/gwat.12838
18. Huang, Y. H., Saiers, J. E., Harvey, J. W., Noe, G. B., & Mylon, S. (2008). Advection, dispersion, and filtration of fine particles within emergent vegetation of the Florida Everglades. *Water Resources Research, 44*. https://doi.org/10.1029/2007WR006290
19. Molz, F. (2015). Advection, dispersion, and confusion. *Groundwater, 53*.
20. Sun, H., Liu, S., Liu, S., & Wang, S. (2014). A comparative study of reduced graphene oxide modified TiO2, ZnO and Ta2O5 in visible light photocatalytic/photochemical oxidation of methylene blue. *Applied Catalysis B: Environmental, 146*, 162–168. https://doi.org/10.1016/j.apcatb.2013.03.027
21. Kerisit, S., & Liu, C. (2010). Molecular simulation of the diffusion of uranyl carbonate species in aqueous solution. *Geochimica et Cosmochimica Acta, 74*, 4937–4952. https://doi.org/10.1016/j.gca.2010.06.007
22. Mihalík, J., Tlustoš, P., & Száková, J. (2011). The impact of an abandoned uranium mining area on the contamination of agricultural land in its surroundings. *Water, Air, and Soil Pollution, 215*, 693–700. https://doi.org/10.1007/s11270-010-0518-6
23. Carvalho, F. P., et al. (2023). *The environmental behaviour of uranium. Report No. No. 488 378.* International Atomic Energy Agency.
24. Mkandawire, M. & Dudel, E. *Environmental behaviour of uranium in closed mining sites.* (2010).
25. Kalin, M. A. (1983). *Long-term ecological behaviour of abandoned uranium mill tailings: 1. Synoptic survey and identification of invading biota. Report No. EPS 4-ES-83-1.* Institute for Environmental Studies, University of Toronto.
26. Pollanen, R., & Toivonen, H. (1995). Skin dose calculations for uranium fuel particles below 500 mu-m in diameter. *Health Physics, 68*, 401–405.
27. Langmuir, D. (1997). *Aqueous environmental geochemistry*. Prentice Hall.
28. Shvareva, T. Y., et al. (2011). Thermodynamic characterization of boltwoodite and uranophane: Enthalpy of formation and aqueous solubility. *Geochimica et Cosmochimica Acta, 75*, 5269–5282. https://doi.org/10.1016/j.gca.2011.06.041
29. Lovley, D. R., Phillips, E. J. P., Gorby, Y. A., & Landa, E. R. (1991). Microbial reduction of uranium. *Nature, 350*, 413–416. https://doi.org/10.1038/350413a0
30. Bernhard, G., et al. (2001). Uranyl(VI) carbonate complex formation: Validation of the Ca2UO2(CO3)3(aq.) species. *Radiochimica Acta, 89*, 511–518. https://doi.org/10.1524/ract.2001.89.8.511
31. Ranville, J. F., Hendry, M. J., Reszat, T. N., Xie, Q., & Honeyman, B. D. (2007). Quantifying uranium complexation by groundwater dissolved organic carbon using asymmetrical flow field-flow fractionation. *Journal of Contaminant Hydrology, 91*, 233–246. https://doi.org/10.1016/j.jconhyd.2006.11.002
32. Reiller, P. E., Marang, L., Jouvin, D., & Benedetti, M. F. (2012). In B. Merkel & M. Schipek (Eds.), *The new uranium mining boom: Challenge and lessons Learned* (pp. 565–572). Springer.
33. Trenfield, M. A., et al. (2011). Dissolved organic carbon reduces uranium bioavailability and toxicity. 1. Characterization of an aquatic fulvic acid and its complexation with uranium[VI]. *Environmental Science & Technology, 45*, 3075–3081. https://doi.org/10.1021/es103330w
34. EPA. (1999). *The K_d model, methods of measurement, and application of chemical reaction codes. Report No. EPA 402-R-99-004A.* U.S. Environmental Protection Agency.

35. EPA. (1999). *Review of geochemistry and available K_d values for cadmium, cesium, chromium, lead, plutonium, radon, strontium, thorium, tritium (3H), and uranium*. U.S. Environmental Protection Agency.
36. Jerden, J. L., Sinha, A. K., & Zelazny, L. (2003). Natural immobilization of uranium by phosphate mineralization in an oxidizing saprolite–soil profile: Chemical weathering of the Coles Hill uranium deposit. *Virginia. Chemical Geology, 199*, 129–157. https://doi.org/10.1016/S0009-2541(03)00080-9
37. Jerden, J. L., & Sinha, A. K. (2003). Phosphate based immobilization of uranium in an oxidizing bedrock aquifer. *Applied Geochemistry, 18*, 823–843. https://doi.org/10.1016/S0883-2927(02)00179-8
38. Jerden, J. L., & Sinha, A. K. (2006). Geochemical coupling of uranium and phosphorous in soils overlying an unmined uranium deposit: Coles Hill, Virginia. *Journal of Geochemical Exploration, 91*, 56–70. https://doi.org/10.1016/j.gexplo.2005.12.003
39. Ilton, E. S., et al. (2006). The dissolution of synthetic Na-boltwoodite in sodium carbonate solutions. *Geochimica et Cosmochimica Acta, 70*, 4836–4849. https://doi.org/10.1016/j.gca.2006.06.1553
40. Lobeck, H. L., et al. (2020). Dissolution of poorly soluble uranyl phosphate phases in the Metaautunite subgroup under uranyl peroxide cage cluster forming conditions. *American Mineralogist, 105*, 182–193. https://doi.org/10.2138/am-2020-7106
41. Stewart, B. D., Amos, R. T., Nico, P. S., & Fendorf, S. (2011). Influence of uranyl speciation and iron oxides on uranium biogeochemical redox reactions. *Geomicrobiology Journal, 28*, 444–456. https://doi.org/10.1080/01490451.2010.507646
42. Sundararajan, M., Rajaraman, G., & Ghosh, S. K. (2011). Speciation of uranyl ions in fulvic acid and humic acid: A DFT exploration. *Physical Chemistry Chemical Physics, 13*, 18038–18046. https://doi.org/10.1039/C1CP21238A
43. Lenhart, J. J., Cabaniss, S. E., MacCarthy, P., & Honeyman, B. D. (2000). Uranium(VI) complexation with citric, humic and fulvic acids. *Radiochimica Acta, 88*, 345–353.
44. Dublet, G., et al. (2017). Partitioning of uranyl between ferrihydrite and humic substances at acidic and circum-neutral pH. *Geochimica et Cosmochimica Acta, 215*, 122–140. https://doi.org/10.1016/j.gca.2017.07.013
45. Djogic, R., & Branica, M. (1993). Uranyl mixed-ligand complex formation in artificial seawater. *Chemical Speciation & Bioavailability, 5*, 101–105. https://doi.org/10.1080/09542299.1993.11083209
46. Bone, S. E., Dynes, J. J., Cliff, J., & Bargar, J. R. (2017). Uranium(IV) adsorption by natural organic matter in anoxic sediments. *Proceedings of the National Academy of Sciences of the United States of America, 114*, 711–716.
47. Chardi, K. J., et al. (2022). Ligand-induced U mobilization from chemogenic uraninite and biogenic noncrystalline U(IV) under anoxic conditions. *Environmental Science & Technology, 56*, 6369–6379. https://doi.org/10.1021/acs.est.1c07919
48. Wang, X.-y. et al. Photochemical reactivity of humic substances in an aquatic system revealed by excitation-emission matrix fluorescence. Frontiers in Chemistry 9–2021 (2021). https://doi.org/10.3389/fchem.2021.679286
49. Lee, S. Y., et al. (2013). Abiotic reduction of uranium by mackinawite (FeS) biogenerated under sulfate-reducing condition. *Journal of Radioanalytical and Nuclear Chemistry, 296*, 1311–1319. https://doi.org/10.1007/s10967-013-2438-6
50. Loreggian, L., Sorwat, J., Byrne, J. M., Kappler, A., & Bernier-Latmani, R. (2020). Role of iron sulfide phases in the stability of noncrystalline tetravalent uranium in sediments. *Environmental Science & Technology, 54*, 4840–4846. https://doi.org/10.1021/acs.est.9b07186
51. Munasinghe, P. S., Elwood Madden, M. E., Brooks, S. C., & Elwood Madden, A. S. (2015). Dynamic interplay between uranyl phosphate precipitation, sorption, and phase evolution. *Applied Geochemistry, 58*, 147–160. https://doi.org/10.1016/j.apgeochem.2015.04.008

52. Newsome, L., Morris, K., & Lloyd, J. R. (2015). Uranium biominerals precipitated by an environmental isolate of Serratia under anaerobic conditions. *PLoS One, 10*, e0132392. https://doi.org/10.1371/journal.pone.0132392
53. Zielinski, R. A., & Meier, A. L. (1988). The association of uranium with organic matter in Holocene peat: An experimental leaching study. *Applied Geochemistry, 3*, 631–643. https://doi.org/10.1016/0883-2927(88)90095-9
54. Halbach, P., von Borstel, D., & Gundermann, K.-D. (1980). The uptake of uranium by organic substances in a peat bog environment on a granitic bedrock. *Chemical Geology, 29*, 117–138. https://doi.org/10.1016/0009-2541(80)90009-1
55. Götz, C., Geipel, G., & Bernhard, G. (2011). The influence of the temperature on the carbonate complexation of uranium(VI): A spectroscopic study. *Journal of Radioanalytical and Nuclear Chemistry, 287*, 961–969. https://doi.org/10.1007/s10967-010-0854-4
56. Dong, W., & Brooks, S. C. (2006). Determination of the formation constants of ternary complexes of uranyl and carbonate with alkaline earth metals (Mg2+, Ca2+, Sr2+, and Ba2+) using anion exchange method. *Environmental Science & Technology, 40*, 4689–4695. https://doi.org/10.1021/es0606327
57. Park et al. Fabrication and characterization of 3D-printed bone-like β-tricalcium phosphate/polycaprolactone scaffolds for dental tissue engineering. Journal of Industrial and Engineering Chemistry 46, 175–181 (2017). https://doi.org/10.1016/j.jiec.2016.10.028
58. Wang, D., Hu, L., Yang, K., Huang, S., & Tian, L. (2024). Enhancement of U(VI) sorption in natural soils at prospective high-level radioactive waste geological disposal sites with Fe(OH)3-humic acid composite colloids. *Journal of Radioanalytical and Nuclear Chemistry, 333*, 1203–1212. https://doi.org/10.1007/s10967-024-09369-x
59. Wang, Q., Cheng, T., & Wu, Y. (2014). Influence of mineral colloids and humic substances on uranium(VI) transport in water-saturated geologic porous media. *Journal of Contaminant Hydrology, 170*, 76–85. https://doi.org/10.1016/j.jconhyd.2014.10.007
60. Pan, D., et al. (2024). *6. International ATALANTE conference on nuclear chemistry for sustainable fuel cycles 177*. International Atomic Energy Agency (IAEA).
61. Hakim, A., & Kobayashi, M. (2021). Aggregation and aggregate strength of microscale plastic particles in the presence of natural organic matter: Effects of ionic valence. *Journal of Polymers and the Environment, 29*, 1921–1929. https://doi.org/10.1007/s10924-020-01985-4
62. Wang, C. I., et al. (2025). Comparison for colloidal stability and aggregation behavior of fulvic and humic acids: Effects of cations and pH. *Frontiers in Soil Science, 5–2025*. https://doi.org/10.3389/fsoil.2025.1452870
63. Bae, S., Lenhart, J. J., & Hwang, Y. S. (2022). The effect of ionic strength, pH and natural organic matter on heteroaggregation of CeO2 nanoparticles with montmorillonite clay minerals. *Environmental Engineering Research, 27*, 210470–210470. https://doi.org/10.4491/eer.2021.470
64. Echevarria, G., Sheppard, M. I., & Morel, J. (2001). Effect of pH on the sorption of uranium in soils. *Journal of Environmental Radioactivity, 53*, 257–264. https://doi.org/10.1016/S0265-931X(00)00116-8
65. Steudtner, R., Müller, K., Schmeide, K., Sachs, S., & Bernhard, G. (2011). Binary and ternary uranium(vi) humate complexes studied by attenuated total reflection Fourier-transform infrared spectroscopy. *Dalton Transactions, 40*, 11920–11925. https://doi.org/10.1039/C1DT11089F
66. Zhao, D., Wang, X., Yang, S., Guo, Z., & Sheng, G. (2012). Impact of water quality parameters on the sorption of U(VI) onto hematite. *Journal of Environmental Radioactivity, 103*, 20–29. https://doi.org/10.1016/j.jenvrad.2011.08.010
67. Wang, Y., et al. (2013). Mobile uranium(IV)-bearing colloids in a mining-impacted wetland. *Nature Communications, 4*, 2942. https://doi.org/10.1038/ncomms3942
68. Coney, L., et al. (2007). Geochemical and mineralogical investigation of the Permian–Triassic boundary in the continental realm of the southern Karoo Basin, South Africa. *Palaeoworld, 16*, 67–104. https://doi.org/10.1016/j.palwor.2007.05.003

69. Cumberland, S. A., Douglas, G., Grice, K., & Moreau, J. W. (2016). Uranium mobility in organic matter-rich sediments: A review of geological and geochemical processes. *Earth-Science Reviews, 159*, 160–185. https://doi.org/10.1016/j.earscirev.2016.05.010
70. Foster, R. I., Kim, K.-W., & Lee, K. (2020). Uranyl phosphate (MUO2PO4, M = Na+, K+, NH4+) precipitation for uranium sequestering: Formation and physicochemical characterisation. *Journal of Radioanalytical and Nuclear Chemistry, 324*, 1265–1273. https://doi.org/10.1007/s10967-020-07154-0
71. Kalintsev, A., et al. (2021). Uranium carbonate complexes demonstrate drastic decrease in stability at elevated temperatures. *Communications Chemistry, 4*, 120. https://doi.org/10.1038/s42004-021-00558-3
72. Vandenhove, H., Vanhoudt, N., Duquène, L., Antunes, K., & Wannijn, J. (2014). Comparison of two sequential extraction procedures for uranium fractionation in contaminated soils. *Journal of Environmental Radioactivity, 137*, 1–9. https://doi.org/10.1016/j.jenvrad.2014.05.024
73. Waite, T. D., Davis, J. A., Payne, T. E., Waychunas, G. A., & Xu, N. (1994). Uranium(VI) adsorption to ferrihydrite: Application of a surface complexation model. *Geochimica et Cosmochimica Acta, 58*, 5465–5478. https://doi.org/10.1016/0016-7037(94)90243-7
74. Ching-kuo Daniel, H., & Langmuir, D. (1985). Adsorption of uranyl onto ferric oxyhydroxides: Application of the surface complexation site-binding model. *Geochimica et Cosmochimica Acta, 49*, 1931–1941. https://doi.org/10.1016/0016-7037(85)90088-2
75. Payne, T. E., & Airey, P. L. (2006). Radionuclide migration at the Koongarra uranium deposit, Northern Australia—Lessons from the Alligator Rivers analogue project. *Physics and Chemistry of the Earth, Parts A/B/C, 31*, 572–586. https://doi.org/10.1016/j.pce.2006.04.008
76. Suzuki, Y., Sato, T., Isobe, H., Kogure, T., & Murakami, T. (2005). Dehydration processes in the meta-autunite group minerals meta-autunite, metasaléeite, and metatorbernite. *American Mineralogist, 90*, 1308–1314. https://doi.org/10.2138/am.2005.1568
77. Bachmaf, S. & Merkel, B. J. in *Clays in natural and engineered barriers for radioactive waste confinement* - 4 *International meeting Book of abstracts.* 1011.
78. Giammar, D. *Geochemistry of uranium at mineral-water interfaces: Rates of sorption-desorption and dissolution-precipitation reactions*. (2019).
79. Roberts, H. E., et al. (2017). Uranium(V) incorporation mechanisms and stability in Fe(II)/Fe(III) (oxyhydr)Oxides. *Environmental Science & Technology Letters, 4*, 421–426. https://doi.org/10.1021/acs.estlett.7b00348
80. Pan, Z., et al. (2020). Nanoscale mechanism of UO2 formation through uranium reduction by magnetite. *Nature Communications, 11*, 4001. https://doi.org/10.1038/s41467-020-17795-0
81. Veeramani, H., et al. (2011). Products of abiotic U(VI) reduction by biogenic magnetite and vivianite. *Geochimica et Cosmochimica Acta, 75*, 2512–2528. https://doi.org/10.1016/j.gca.2011.02.024
82. Haas, J. R., Dichristina, T. J., & Wade, R. (2001). Thermodynamics of U(VI) sorption onto Shewanella putrefaciens. *Chemical Geology, 180*, 33–54. https://doi.org/10.1016/S0009-2541(01)00304-7
83. Liger, E., Charlet, L., & Van Cappellen, P. (1999). Surface catalysis of uranium(VI) reduction by iron(II). *Geochimica et Cosmochimica Acta, 63*, 2939–2955. https://doi.org/10.1016/S0016-7037(99)00265-3
84. Orellana, R., et al. (2013). U(VI) reduction by diverse outer surface c-Type cytochromes of Geobacter sulfurreducens. *Applied and Environmental Microbiology, 79*, 6369–6374. https://doi.org/10.1128/AEM.02551-13
85. Yung Mimi, C., & Jiao, Y. (2014). Biomineralization of uranium by PhoY phosphatase activity aids cell survival in Caulobacter crescentus. *Applied and Environmental Microbiology, 80*, 4795–4804. https://doi.org/10.1128/AEM.01050-14
86. Morrison, K. D., et al. (2021). Influence of uranium concentration and pH on U-Phosphate biomineralization by Caulobacter OR37. *Environmental Science & Technology, 55*, 1626–1636. https://doi.org/10.1021/acs.est.0c05437

87. Lahrouch, F., et al. (2021). Uranium retention on iron oxyhydroxides in post-mining environmental conditions. *Chemosphere, 264*, 128473. https://doi.org/10.1016/j.chemosphere.2020.128473
88. Jeon, B.-H., et al. (2004). Microbial reduction of U(VI) at the solid−water interface. *Environmental Science & Technology, 38*, 5649–5655. https://doi.org/10.1021/es0496120
89. Wang, S., Chen, L., Gong, Y., Ma, S., & Yang, Y. (2024). The effect of bioclogging on the hydraulic conductivity of saturated porous media at different recharge water temperatures. *Hydrogeology Journal, 32*, 1057–1070. https://doi.org/10.1007/s10040-024-02785-z
90. Gaol, C. L., Ganzer, L., Mukherjee, S., & Alkan, H. (2021). Investigation of clogging in porous media induced by microorganisms using a microfluidic application. *Environmental Science: Water Research & Technology, 7*, 441–454. https://doi.org/10.1039/D0EW00766H
91. Zhang, Z., et al. (2025). The impact of biofilm-induced dynamic layered clogging on hyporheic exchange in streambed. *Water, 17*, 2717.
92. Barlett, M., Zhuang, K., Mahadevan, R., & Lovley, D. (2012). Integrative analysis of Geobacter spp. and sulfate-reducing bacteria during uranium bioremediation. *Biogeosciences, 9*, 1033–1040. https://doi.org/10.5194/bg-9-1033-2012
93. Lv, Y., et al. (2021). Optimization of environmental conditions for microbial stabilization of uranium tailings, and the Microbial Community response. *Frontiers in Microbiology, 12–2021*. https://doi.org/10.3389/fmicb.2021.770206
94. Stewart, B. D., Cismasu, A. C., Williams, K. H., Peyton, B. M., & Nico, P. S. (2015). Reactivity of uranium and ferrous iron with natural iron oxyhydroxides. *Environmental Science & Technology, 49*, 10357–10365. https://doi.org/10.1021/acs.est.5b02645
95. Zhong, J., et al. (2021). Isolation and identification of uranium tolerant phosphate-solubilizing Bacillus spp. and their synergistic strategies to U(VI) Immobilization. *Frontiers in Microbiology, 12–2021*. https://doi.org/10.3389/fmicb.2021.676391
96. Franz Charles, M. A. P., et al. (2001). Incidence of virulence factors and antibiotic resistance among enterococci isolated from food. *Applied and Environmental Microbiology, 67*, 4385–4389. https://doi.org/10.1128/AEM.67.9.4385-4389.2001
97. Williamson, A. J., et al. (2014). Microbial reduction of U(VI) under alkaline conditions: Implications for radioactive waste geodisposal. *Environmental Science & Technology, 48*, 13549–13556. https://doi.org/10.1021/es5017125
98. Shelobolina, E. S., et al. (2007). Importance of c-Type cytochromes for U(VI) reduction by Geobacter sulfurreducens. *BMC Microbiology, 7*, 16. https://doi.org/10.1186/1471-2180-7-16
99. Chang, Y.-J., et al. (2001). Diversity and Characterization of Sulfate-Reducing Bacteria in Groundwater at a Uranium Mill Tailings Site. *Applied and Environmental Microbiology, 67*, 3149–3160. https://doi.org/10.1128/AEM.67.7.3149-3160.2001
100. Chang, Y.-J., et al. (2000). Diversity and characterization of sulfate-reducing bacteria in groundwater at a uranium mill tailings site. *Applied and Environmental Microbiology, 67*.
101. Chen, Y., et al. (2025). Microbial-induced mineralization of uranium(VI) for radioactive contamination remediation: Progress and challenges. *Journal of Radioanalytical and Nuclear Chemistry*. https://doi.org/10.1007/s10967-025-10661-7
102. Acharya, C. (2015). In L. B. Sukla, N. Pradhan, S. Panda, & B. K. Mishra (Eds.), *Environmental microbial biotechnology* (pp. 119–132). Springers.
103. Wan, J., et al. (2005). Reoxidation of bioreduced uranium under reducing conditions. *Environmental Science & Technology, 39*, 6162–6169. https://doi.org/10.1021/es048236g
104. Seder-Colomina, M., et al. (2018). Carbonate facilitated mobilization of uranium from Lacustrine sediments under anoxic conditions. *Environmental Science & Technology, 52*, 9615–9624. https://doi.org/10.1021/acs.est.8b01255
105. Bone, S. E., Dynes, J. J., Cliff, J., & Bargar, J. R. (2017). Uranium(IV) adsorption by natural organic matter in anoxic sediments. *Proceedings of the National Academy of Sciences, 114*, 711–716. https://doi.org/10.1073/pnas.1611918114

106. ITRC. *A decision framework for applying monitored natural attenuation processes to metals and radionuclides in groundwater. Report No. APMR-1*. Interstate Technology & Regulatory Council.
107. Li, C., et al. (2024). Simulation of the natural attenuation of groundwater contaminated by acid in situ leaching at a uranium mine. *Journal of Radioanalytical and Nuclear Chemistry, 333*, 4907–4917. https://doi.org/10.1007/s10967-024-09624-1
108. Westrop, J. P., et al. (2023). Nitrate-stimulated release of naturally occurring sedimentary uranium. *Environmental Science & Technology, 57*, 4354–4366. https://doi.org/10.1021/acs.est.2c07683
109. Nolan, J., & Weber, K. A. (2015). Natural uranium contamination in major U.S. Aquifers linked to nitrate. *Environmental Science & Technology Letters, 2*, 215–220. https://doi.org/10.1021/acs.estlett.5b00174
110. Robinson, C., Shaw, S., Lloyd, J. R., Graham, J., & Morris, K. (2025). Bioremediation of uranium contaminated sites through the formation of U(vi) phosphate (bio)minerals. *Environmental Science: Water Research & Technology, 11*, 725–736. https://doi.org/10.1039/D4EW00846D
111. Wang, G., et al. (2023). The remediation of uranium-contaminated groundwater via bioreduction coupled to biomineralization with different pH and electron donors. *Environmental Science and Pollution Research, 30*, 23096–23109. https://doi.org/10.1007/s11356-022-23902-z
112. Anderson Robert, T., et al. (2003). Stimulating the in situ activity of Geobacter species to remove uranium from the groundwater of a uranium-contaminated aquifer. *Applied and Environmental Microbiology, 69*, 5884–5891. https://doi.org/10.1128/AEM.69.10.5884-5891.2003
113. Wang, G.-H., et al. (2024). Enhanced indigenous consortia for the remediation of uranium-contaminated groundwater by bioaugmentation: Reducing and phosphate-solubilizing consortia. *The Science of the Total Environment, 912*, 168954. https://doi.org/10.1016/j.scitotenv.2023.168954
114. Long, P. E., & Yabusaki, S. B. (2007). *Evaluating the Efficacy of Uranium Bioremediation in the Subsurface: Technical Bases and Performance Indicators*. Pacific Northwest National Laboratory.
115. Long, P. E., Yabusaki, S. B., Meyer, P. D., Murray, C. J., & N'Guessan, A. (2008). *L*. Technical Basis for Assessing Uranium Bioremediation Performance.
116. Van Nostrand Joy, D., et al. (2011). Dynamics of Microbial Community Composition and Function during In Situ Bioremediation of a Uranium-Contaminated Aquifer. *Applied and Environmental Microbiology, 77*, 3860–3869. https://doi.org/10.1128/AEM.01981-10
117. Holmes, D. E., et al. (2013). Enrichment of specific protozoan populations during in situ bioremediation of uranium-contaminated groundwater. *The ISME Journal, 7*, 1286–1298. https://doi.org/10.1038/ismej.2013.20
118. N'Guessan, A. L., Vrionis, H. A., Resch, C. T., Long, P. E., & Lovley, D. R. (2008). Sustained removal of uranium from contaminated groundwater following stimulation of dissimilatory metal reduction. *Environmental Science & Technology, 42*, 2999–3004. https://doi.org/10.1021/es071960p

Chapter 12
Concluding Perspectives

Abstract This concluding chapter synthesizes the key ideas developed across the book's chapters and provides a concise synopsis of the work as a whole. It revisits uranium occurrence, radioactivity, uses, market dynamics, and major global deposits while integrating discussions of mine abandonment, extraction methods, ore processing, and uranium behaviour across environments shaped by the three principal mining techniques. The chapter also unifies insights on dispersion pathways, ecotoxicology, radioecology, biogeochemistry, and natural attenuation. By weaving these strands into a coherent narrative, it highlights the scientific advances developed across this book and distils them into the final thematic reflections that frame the broader implications of uranium legacy sites.

Keywords Uranium occurrence · Radioactivity · Mine abandonment · Extraction techniques · Ore processing · Mining waste · Dispersion processes · Ecotoxicology · Radioecology · Natural attenuation

12.1 Conclusion

This book offers a general overview of the fate and effects of uranium in abandoned mines, presenting core principles that can be adapted to the climatic, geological, hydrological, and geochemical conditions of any particular site. It has examined most aspects of what an abandoned uranium mine can pose and, in doing so, has mainly assumed that mines were left without proper closure procedures. That assumption reflects the historical reality at many legacy sites. It frames the practical problems addressed here: dispersed wastes, unmarked transfer stations, exposed tailings and waste rock, and the absence of long-term stewardship. While the specific behaviour of uranium at any one location will always be site-specific, modulated by local climate, geology, hydrogeology, and geochemistry, the generalized concepts developed in this book provide the fundamentals needed to build site-representative assessments and management plans.

M. Mkandawire, *Fate and Effects of Uranium in Abandoned Uranium Mines*,
https://doi.org/10.1007/978-3-032-10447-2_12

12.2 Scope Synopsis

The focus of this book has been on the fate and effects of uranium in abandoned mines. By treating fate as a life cycle concept, the text traces uranium from its liberation during mining through various processes, such as weathering, transport, transformation, and ultimately to its sequestration or exposure in soils, sediments, surface water, groundwater, and biota. It describes the physical legacy of mining, including borehole drill cuttings, stripped topsoil, overburden, weakly enriched waste rock, subgrade ores, tailings, and, occasionally, ore that remains in place when market conditions change. This legacy creates a mosaic of potential sources of uranium contamination.

This book also examines the various mining methods that produce these wastes, namely open pit, underground, and in situ leaching operations, highlighting how each method leaves a distinct pattern of materials, geometries, and pathways that influence the subsequent behaviour of uranium. Furthermore, it explores uranium behaviour in waste and tailings heaps, contrasting consolidated point sources, such as engineered tailings impoundments, with unconsolidated, dispersed wastes that are more challenging to characterize and manage.

It has been demonstrated that predicting uranium dispersal is theoretically simpler when waste is consolidated into identifiable point sources, since source inventories, geometries, and boundary conditions can be more easily defined. In contrast, unconsolidated wastes, such as scattered rock piles, unmarked transfer stations, and residual ore left in trenches, create diffuse sources that complicate monitoring, modelling, and remediation efforts.

12.3 Key Controls on Uranium Fate

The environmental behaviour of uranium is governed by a set of interlinked physical, chemical, and biological controls. Redox conditions and oxidation states play a crucial role in determining whether uranium exists primarily as reduced U(IV) phases, which are generally relatively immobile, or as oxidized U(VI) species that can form soluble complexes. Factors such as pH and the presence of complexing ligands, like carbonate, phosphate, and dissolved organic matter, significantly influence speciation and mobility. Sorption to mineral surfaces and incorporation into secondary minerals can hinder transport, while microbial activity and seasonal hydrological changes can induce redox oscillations that episodically mobilize or immobilize uranium. Furthermore, the mineralogy of the ore and the presence of co-occurring metals and metalloids further shape the chemical environment and the range of contaminants to be considered in any risk assessment.

12.4 Predicting Dispersal and Reactive Transport

The migration of uranium from a source to a receiving environment is a complex process. It is influenced by reactive transport mechanisms, during which uranium undergoes various transformations as it travels with available hydrology and interacts with geochemical materials along its migration path. Reactive transport combines physical advection and dispersion with chemical reactions, including redox transformations, complexation, precipitation and dissolution, sorption and desorption, and biological uptake and decay. These interconnected processes lead to concentration patterns in groundwater and surface water, driven by factors such as source strength, waste geometry, flow pathways, residence times, and reaction kinetics. This book underscores the importance of integrating hydrological connectivity with geochemical reactivity in conceptual models: for instance, a high-flux pathway with minimal reactive surface area may quickly transport uranium to a receptor, while slow, diffuse flow through reactive sediments might attenuate uranium but create long-term secondary reservoirs of contamination.

Consolidated point sources are generally easier to inventory and model, allowing for more reliable predictions of dispersal and the design of targeted controls. In contrast, unconsolidated sources necessitate broader reconnaissance, more extensive monitoring, and adaptable modelling approaches to accommodate spatial heterogeneity and episodic releases. In all scenarios, predictive efforts are enhanced by iterative cycles of measurement, model refinement, and adaptive management.

12.5 Chemical Toxicity and Co-contaminants

A recurring conclusion of this work is that the chemical toxicity of uranium is frequently a more immediate and actionable concern at abandoned sites than its long-term radiotoxicity. The chemical effects of uranium, particularly on renal function, can pose health risks at concentrations independent of radiological dose. Equally important are the daughter radionuclides and co-occurring metals and metalloids that remain in the environment after ore processing. Daughter products that were not extracted during mining can be both chemically and radiologically potent, and metals such as arsenic, lead, selenium, and molybdenum are commonly associated with uranium ores. They can dominate the site's toxicological profile. Effective assessment and remediation, therefore, require a broad contaminant scope that includes parent uranium, its decay products, and associated trace elements.

12.6 Best Practices for Uranium Mine Closure

Proper closure of a uranium mine should be conceived as an integrated, long-term program that begins well before final extraction and continues indefinitely through post-closure stewardship. Best practice closure is adaptive, evidence-based, and community-centred. It combines rigorous source control with hydrological and geochemical management, robust monitoring, and legal and institutional arrangements that ensure accountability over time.

12.6.1 Preclosure Planning

Closure planning must start early and be informed by a comprehensive inventory of wastes, tailings, and residual ore, together with detailed mapping of waste geometries and potential exposure points. Baseline environmental characterization of soils, sediments, surface water, and groundwater is essential to establish pre-closure conditions and to design monitoring networks. Closure plans should be developed in consultation with local communities, regulators, and independent experts and should include clear criteria for successful closure and mechanisms for long-term funding and oversight.

12.6.2 Source Control

Where feasible, removal of the most mobile and toxic materials is the preferred option. When removal is impractical, engineered containment, capping, consolidation, and stabilization should be used to limit water infiltration, reduce erosion, and minimize dust generation. Consolidation of dispersed wastes into managed, well-designed repositories simplifies monitoring and reduces the number of potential release points. Capping systems should be designed to withstand local climatic extremes and to limit oxygen ingress where reducing conditions are desirable to limit uranium mobility.

12.6.3 Hydrological Management

Hydrological controls are central to preventing contaminant migration. Surface water diversion, runoff control, and the design of drainage systems that minimize contact between water and reactive wastes reduce the potential for leaching. Groundwater management may require cutoffs, hydraulic containment, or pump-and-treat systems when plumes threaten receptors. Designs must account for extreme weather events and potential climate change impacts that could alter recharge patterns and flow paths.

12.6.4 Geochemical Stabilization

Geochemical approaches can complement physical controls. Where appropriate, in situ stabilization techniques, such as redox manipulation to promote U(IV) precipitation, sorbent addition, or reactive permeable barriers, can reduce uranium mobility. Stabilization strategies must be chosen with an eye to long-term stability; for example, inducing reducing conditions may immobilize uranium but could also mobilize other elements or be reversed by future changes in hydrology or biogeochemistry. Any geochemical intervention should be accompanied by monitoring that can detect unintended consequences.

12.6.5 Monitoring and Stewardship

Long-term monitoring is non-negotiable. Monitoring programs should measure chemical and radiological endpoints, include daughter radionuclides and co-contaminant metals, and be designed to detect both chronic trends and episodic releases. Monitoring data should feed into adaptive management frameworks that allow remedial actions to be adjusted as conditions change. Legal and institutional arrangements must secure funding and responsibility for monitoring and maintenance over the timescales required to manage legacy risks.

12.6.6 Community Engagement and Legal Framework

Closure and post-closure stewardship must be characterized by transparency and active community participation. It is essential that communities have access to pertinent information, a role in decision-making, and avenues to express their concerns. Regulatory frameworks should mandate detailed closure plans, financial assurances for long-term care, and enforceable standards for both chemical and radiological safety. Furthermore, whenever feasible, closure efforts should consider the beneficial reuse of reclaimed land in alignment with residual risk and community priorities.

12.7 Closing Thoughts

This book aimed to elucidate the potential risks associated with uranium migration by detailing the complex geochemical interactions between uranium and both biotic and abiotic systems. These interactions, influenced by specific physicochemical conditions, play a critical role in determining the risk level at abandoned mines. It provides a comprehensive overview of uranium behaviour, highlighting that the factors affecting its fate and effects are highly site-specific, depending on factors, such as hydrogeological context, mining methods, ore grade and type, and waste disposition.

The generalized framework comprises source characterization, geochemical controls, hydrological connectivity, reactive transport, and the importance of chemical and co-contaminant hazards. It serves as a practical foundation for site-level application. When these essential elements are integrated with thorough local data, transparent community engagement, and legally enforceable long-term stewardship, they establish a solid basis for the responsible closure and management of uranium mining legacies.

The manufacturer's authorised representative in the EU is Springer Nature Customer Service Centre GmbH, Europaplatz 3, 69115 Heidelberg, Germany. If you have any concerns regarding our products, please contact ProductSafety@springernature.com

Printed and bound by CPI Group (UK) Ltd, Croydon, CR0 4YY
07/10/2026
02229357-0013